Lecture Notes in Computer Science   8799

Commenced Publication in 1973
Founding and Former Series Editors:
Gerhard Goos, Juris Hartmanis, and Jan van Leeuwen

Editorial Board

David Hutchison
Lancaster University, UK

Takeo Kanade
Carnegie Mellon University, Pittsburgh, PA, USA

Josef Kittler
University of Surrey, Guildford, UK

Jon M. Kleinberg
Cornell University, Ithaca, NY, USA

Alfred Kobsa
University of California, Irvine, CA, USA

Friedemann Mattern
ETH Zurich, Switzerland

John C. Mitchell
Stanford University, CA, USA

Moni Naor
Weizmann Institute of Science, Rehovot, Israel

Oscar Nierstrasz
University of Bern, Switzerland

C. Pandu Rangan
Indian Institute of Technology, Madras, India

Bernhard Steffen
TU Dortmund University, Germany

Demetri Terzopoulos
University of California, Los Angeles, CA, USA

Doug Tygar
University of California, Berkeley, CA, USA

Gerhard Weikum
Max Planck Institute for Informatics, Saarbruecken, Germany

T0183184

Edleno Moura   Maxime Crochemore (Eds.)

String Processing and Information Retrieval

21st International Symposium, SPIRE 2014
Ouro Preto, Brazil, October 20-22, 2014
Proceedings

 Springer

Volume Editors

Edleno Moura
Universidade Federal do Amazonas
Instituto de Computação
Manaus, Brazil
E-mail: edleno@icomp.ufam.edu.br

Maxime Crochemore
King's College London
London, UK
E-mail: maxime.crochemore@kcl.ac.uk

ISSN 0302-9743　　　　　　　　　e-ISSN 1611-3349
ISBN 978-3-319-11917-5　　　　　　e-ISBN 978-3-319-11918-2
DOI 10.1007/978-3-319-11918-2
Springer Cham Heidelberg New York Dordrecht London

Library of Congress Control Number: 2014949633

LNCS Sublibrary: SL 1 – Theoretical Computer Science and General Issues

© Springer International Publishing Switzerland 2014
This work is subject to copyright. All rights are reserved by the Publisher, whether the whole or part of the material is concerned, specifically the rights of translation, reprinting, reuse of illustrations, recitation, broadcasting, reproduction on microfilms or in any other physical way, and transmission or information storage and retrieval, electronic adaptation, computer software, or by similar or dissimilar methodology now known or hereafter developed. Exempted from this legal reservation are brief excerpts in connection with reviews or scholarly analysis or material supplied specifically for the purpose of being entered and executed on a computer system, for exclusive use by the purchaser of the work. Duplication of this publication or parts thereof is permitted only under the provisions of the Copyright Law of the Publisher's location, in ist current version, and permission for use must always be obtained from Springer. Permissions for use may be obtained through RightsLink at the Copyright Clearance Center. Violations are liable to prosecution under the respective Copyright Law.
The use of general descriptive names, registered names, trademarks, service marks, etc. in this publication does not imply, even in the absence of a specific statement, that such names are exempt from the relevant protective laws and regulations and therefore free for general use.
While the advice and information in this book are believed to be true and accurate at the date of publication, neither the authors nor the editors nor the publisher can accept any legal responsibility for any errors or omissions that may be made. The publisher makes no warranty, express or implied, with respect to the material contained herein.

Typesetting: Camera-ready by author, data conversion by Scientific Publishing Services, Chennai, India

Printed on acid-free paper

Springer is part of Springer Science+Business Media (www.springer.com)

Preface

This volume contains the papers presented at SPIRE 2014, 21st International Symposium on String Processing and Information Retrieval, held on October 20–22, 2014 in Ouro Preto (Brazil). There were 45 submissions and 3 Program Committee members reviewed each submission. The committee decided to accept 20 full papers and 6 short papers. The program also includes 3 invited talks that do not appear in the proceedings, delivered by Professors Gonzalo Navarro (University of Chile), Paolo Boldi (University of Milano) and Berthier Ribeiro-Neto (Google Inc.).

Following the scope of SPIRE, the articles in this edition include not only fundamental algorithms in string processing and information retrieval, but also application areas such as computational biology, Web mining and recommender systems. Given its interdisciplinary nature, SPIRE offers a unique opportunity for researchers from these different areas to meet and to network. Further, the conference was held in conjunction with the 9th Latin American Web Congress (LA-WEB 2014), which addresses current research in topics of interest to the SPIRE audience.

On behalf of SPIRE's Steering Committee and as chairs we thank the Program Committee members for their valuable contribution in the selection of articles, the local organizing chairs Alvaro R. Pereira Jr. and Fabricio Benevenuto for the support they provided, and all the authors and attendees for this year symposium. We adopted EasyChair as our conference management system and we also thank the EasyChair team for the nice and extremely helpful system.

August 2014

Edleno Silva De Moura
Maxime Crochemore

Organization

Program Committee Chairs

Maxime Crochemore   King's College London, UK and Université
Paris-Est, France

Edleno Silva De Moura   Universidade Federal do Amazonas, Brazil

Steering Committee

Nivio Ziviani   Federal University of Minas Gerais, Brazil

Ricardo Baeza-Yates   Yahoo! Research, Spain

Berthier Ribeiro-Neto   Google Inc., and Federal University
of Minas Gerais, Brazil

Oren Kurland   Faculty of Industrial Engineering and
Management, Technion, Israel

Moshe Lewenstein   Bar-Ilan University, Israel

Ely Porat   Bar-Ilan University, Israel

Conference Local Chairs

Álvaro R. Pereira Jr.   Universidade Federal de Ouro Preto, Brazil

Fabricio Benevenuto   Universidade Federal de Minas Gerais, Brazil

Program Committee

Giambattista Amati   Fondazione Ugo Bordoni, Italy

Amihood Amir   Bar-Ilan University, Israel and Johns Hopkins
University, USA

Alberto Apostolico   University of Padoa, Italy and Georgia
Tech., USA

Ricardo Baeza-Yates   Yahoo! Research, Spain

Hideo Bannai   Kyushu University, Japan

Nieves R. Brisaboa   Universidade da Corua, Spain

Ayelet Butman   Holon Institute of Technology, Israel

Carlos Castillo   Qatar Computing Research Institute, Israel,
Qatar

Edgar Chavez   Universidad Michoacana, Mexico

Ferdinando Cicalese   University of Salerno, Italy

Raphael Clifford	University of Bristol, UK
Maxime Crochemore	King's College London, UK and Université Paris-Est, France
Edleno Silva De Moura	Universidade Federal do Amazonas, Brazil
Carsten Eickhoff	ETH Zurich, Switzerland
David Fernandes	Universidade Federal do Amazonas, Brazil
Johannes Fischer	TU Dortmund, Germany
Raffaele Giancarlo	Dipartimento di Matematica Università di Palermo, Italy
Marcos Goncalves	Federal University of Minas Gerais, Brazil
Roberto Grossi	Università di Pisa, Italy
Inge Li Gørtz	Technical University of Denmark, Denmark
Shunsuke Inenaga	Kyushu University, Japan
Markus Jalsenius	University of Bristol, UK
Gareth Jones	Dublin City University, Ireland
Jaap Kamps	University of Amsterdam, The Netherlands
Tsvi Kopelowitz	University of Michigan, USA
Gregory Kucherov	CNRS/LIGM, France
Juha Kärkkäinen	University of Helsinki, Finland
Gad M. Landau	University of Haifa, Israel and NYU-Polytechnic, USA
Thierry Lecroq	University of Rouen, France
Chia-Jung Lee	CIIR, University of Massachusetts Amherst, USA
Avivit Levy	Shenkar College, Israel
Mauricio Marin	Yahoo! Research, Chile
Andrew McGregor	University of Massachusetts Amherst, USA
Alistair Moffat	The University of Melbourne, Australia
Viviane P. Moreira	Instituto de Informatica - UFRGS, Brazil
Ian Munro	University of Waterloo, Canada
Gonzalo Navarro	University of Chile, Chile
Yakov Nekrich	University of Waterloo, Canada
Alexandros Ntoulas	Zynga Inc., USA
Ely Porat	Bar-Ilan University, Israel
Simon Puglisi	University of Helsinki, Finland
Berthier Ribeiro-Neto	Google Inc., USA and Federal University of Minas Gerais, Brazil
Benjamin Sach	University of Warwick, UK
Kunihiko Sadakane	National Institute of Informatics, Japan
Rodrygo L.T. Santos	University of Glasgow, UK
Srinivasa Rao Satti	Seoul National University, South Korea

Rahul Shah	Louisiana State Univeristy, USA
Fabrizio Silvestri	ISTI - CNR, Italy
Torsten Suel	Polytechnic Institute of NYU, USA
Chris Thachuk	University of Oxford, UK
Dekel Tsur	Ben Gurion University, Israel
Oren Weimann	University of Haifa, Israel
David Woodruff	IBM Almaden, USA
Nivio Ziviani	Federal University of Minas Gerais, Brazil
Guido Zuccon	Queensland University of Technology, Australia

Table of Contents

Compression

Indexing

Genome and Related Topics

Sequences and Strings

Search

Mining and Recommending

Strategic Pattern Search in Factor-Compressed Text

Simon Gog[1,2], Alistair Moffat[1], and Matthias Petri[1]

[1] Department of Computing and Information Systems,
The University of Melbourne, Australia 3010
[2] Institute of Theoretical Informatics,
Karlsruhe Institute of Technology, Germany

Abstract. We consider the problem of pattern-search in compressed text in a context in which: (a) the text is stored as a sequence of factors against a static phrase-book; (b) decoding of factors is from right-to-left; and (c) extraction of each symbol in each factor requires $\Theta(\log \sigma)$ time, where σ is the size of the original alphabet. To determine possible alignments given information about decoded characters we introduce two Boyer-Moore-like searching mechanisms, including one that makes use of a suffix array constructed over the pattern. The new mechanisms decode fewer than half the symbols that are required by a sequential left-to-right search such as the Knuth-Morris-Pratt approach, a saving that translates directly into improved execution time. Experiments with a two-level suffix array index structure for 4 GB of English text demonstrate the usefulness of the new techniques.

Keywords: string search, pattern matching, suffix array, Burrows-Wheeler transform, succinct data structure, disk-based algorithm, experimental evaluation.

1 Introduction and Background

String search, or pattern search, is a classic problem in computing. Given a sequence T of n symbols and a pattern P of m symbols, both over an alphabet Σ of size σ, the requirement is to identify all locations in T at which P appears as a substring. Two paradigms for tackling this problem have emerged – if both T and P vary with every problem instance, the best that can be hoped for is linear $\Theta(n+m)$ time processing. But if T is regarded as being fixed, and only P varies with each instance, then the cost of pre-processing T to build an index can be regarded as being amortized down to zero. Large numbers of algorithms and data structures have been developed for both types of pattern search, as well as for variants of the basic problem; see, for example, Navarro and Raffinot [11]. Our work in this paper fits in the second "static T" category, but also requires application of techniques suited to the first "dynamic T" paradigm.

Sequential Pattern Search. The Knuth-Morris-Pratt (KMP) [10] and Boyer-Moore (BM) [1] methods remain significant more than 35 years after they were first developed. The KMP approach scans T from left to right, extending a prefix of P known to be in alignment with T; if a non-matching symbol is encountered, the pattern is shifted right by the amount indicated in a pre-computed table that is based on P (and not on T). In the BM method, the checking is from right-to-left in P. Two shift tables are used, the "good

E. Moura and M. Crochemore (Eds.): SPIRE 2014, LNCS 8799, pp. 1–12, 2014.
© Springer International Publishing Switzerland 2014

suffix" table that has a similar function to the KMP table; and a "bad symbol" table, which, for each symbol in Σ, records the rightmost occurrence of it in P.

Horspool [9] noted that use of the bad symbol shift associated with the symbol in T tentatively matched against $P[m-1]$ was sufficient for fast execution on average, since it never results in negative shifts, and is likely to create long shifts on average. The combination of right-to-left processing and a bad-symbol shift array is referred to as the BMH mechanism. Sunday [16] noted that the symbol in T *after* the last one of the current alignment could also be used for the same purpose, since it too must be part of the next alignment after the shift of P has taken place. Together, these two approaches then lead to Smith's [15] proposal to make use of the larger of the Horspool bad-shift and the Sunday bad-shift. We make use of these ideas in the development below. A 1997 web site[1] developed by Christian Charras and Thierry Lecroq gives details and examples of all of these methods, plus many more, as does the book of Navarro and Raffinot [11] and a survey of recent results by Faro and Lecroq [3].

Index-Based Pattern Search. There is again a myriad of methods in this category. Best-known are the suffix array, the suffix tree, and compressed/succinct variants thereof. The FM-INDEX of Ferragina and Manzini [5] represents T in compressed form (that is, in fewer than $n \log \sigma$ bits) yet still provides pattern search in $O(m \log \sigma)$ time. The combination of compact space and fast access make the FM-INDEX highly applicable for in-memory searching applications. On the other hand, the access pattern within the FM-INDEX is highly non-sequential, and it is not suitable for use on secondary storage devices such as mechanical disk and SSD memory.

The RoSA. In recent work, Gog et al. [8] introduce an indexed data structure for pattern search called the RoSA, or *reduced space on-disk suffix array*, as a mechanism to support exact pattern search. As with the previous LOF-SA structure of Sinha et al. [14], the RoSA supports efficient pattern search over very large static sequences by constructing a suffix array, and partitioning it into on-disk blocks. Each suffix block contains at most b pointers, and is formed so that every string addressed by the block has a unique common prefix, known as the *block prefix string*. The value of b is fixed at the time the index is constructed; Gog et al. [8] make use of $b = 4,096$ in their experiments. The first innovative feature of the RoSA is that the in-memory index is structured as a *condensed BWT*, that contains all of the block prefix strings, so that the suffix block a given pattern falls in to (if it exists at all) can be efficiently determined. The use of the condensed BWT, and careful engineering in regard to storage of bitvectors and point-ers, mean that the in-memory index can be as small as just a few percent of the original string. In experiments using multi-gigabyte files of English text, Gog et al. [8] show that the RoSA's in-memory index requires as little as 2% of the original text, with *count* queries – in which the objective is to determine how many occurrences there are of the pattern P, but not their exact locations – requiring at most two accesses to secondary storage: one to fetch a suffix block of at most b pointers, and a second to fetch a section of the original text T. The second access is required to verify that the pattern does in-deed exist, and is not a false-positive arising from the use of a *bit-blind tree* [4] during the within-block search [8].

---

[1] http://www-igm.univ-mlv.fr/~lecroq/string/

Suffix blocks have common prefix strings allowing for *block reductions* to be identified which map blocks of the suffix array to subsections of other blocks, allowing disk space to also be reduced. For the same test file, the space required by the suffix array is less than $2n$ bytes. Including the text as well, the total storage cost of the RoSA structure is less than $3n$ bytes when representing English text [8], substantially better than the $5n$ or $9n$ required by uncompressed suffix array structures.

In a followup paper, Gog and Moffat [7] further reduce the RoSA's space cost in two key ways. First, they re-use the block prefix strings as a static phrase-book for greedy-dictionary compression, and show that the condensed BWT index can be used to decode factors with only a small amount of overhead space. Second, Gog and Moffat approximate each of the stored suffix pointers (which, because of the compression of T, is a factor address rather than a byte address), truncating it to a multiple of R, a second parameter selected at the time the index is constructed. The first change reduces both the space required for T and the space required by each suffix pointer, since there are fewer factors in T than there are bytes; the second change saves a further approximately $\log_2 R$ bits from each suffix pointer. Using both techniques, a complete two-level RoSA structure for the same 62.5 GB file of English text requires less than $2n$ bytes when $R = 64$, with searching time approximately doubled compared to the $R = 1$ situation [7].

Our Contribution. We further enhance the RoSA by exploring alternative sequential pattern matching options, improving querying costs dramatically for large values of R, exactly the ones that give rise to the most compact index. In particular, we introduce two Boyer-Moore-like searching mechanisms: one that makes use of a suffix array constructed over the pattern; and one that makes use of a shift matrix covering a total of $\sigma \times (m + 1)$ different position/symbol combinations. Both mechanisms decode fewer than half the symbols that are required by the previous KMP approach, a saving that translates directly into improved execution time. We give detailed experiments with a two-level suffix array index structure for 4 GB of English text demonstrate the usefulness of the new techniques.

2 Search in Factorized Text

Each suffix block in the RoSA is structured as a bit-blind tree (see Gog et al. [8] for a description and an example) so that it can be quickly queried after it has been read from disk. The drawback of the bit-blind tree is that once the potential location in T for the pattern P has been identified via the index, it must be checked to ensure that it is not a false match. In the original RoSA the suffix pointers indicate byte offsets in T, and checking is easy. For a pattern of m symbols, a second disk access fetches a block of T, and then m character comparisons are required.

But when the suffix pointers address "div R" approximated factor numbers, the checking process is more costly. Now a sequence of $R + m - 1$ factor identifiers is supplied, and P must be searched for in the variable-length string that those factors represent. Table 1 lists the low-level operations that apply to compressed factors, and the cost of each such operation when the factors are represented via the condensed BWT structure. To decode a single factor identified by the reference f, function *length_of_factor*(f) is

Table 1. Operations used during factor decoding, where f is a reference to a factor and is assumed to include the necessary state variables. See Figure 4 of Gog and Moffat [7] for details.

Operation	Returns...	Time
length_of_factor(f)	the length in symbols of the factor	$O(1)$
final_symbol(f)	the rightmost symbol of the factor	$O(1)$
next_symbol(f)	the next (from the right) symbol of the factor	$O(\log \sigma)$

called to initialize the state variables and determine the length $len(f)$ of that factor (in terms of decoded symbols); then the rightmost symbol is accessed using *final_symbol*(f); and finally a loop iterates $len(f) - 1$ times, calling *next_symbol*(f) to fill in the remaining symbols of that factor, from right to left. Each call to the latter function requires $O(1)$ rank and $O(\log \sigma)$ select operations, where σ is the cardinality of the alphabet – for example, $\sigma = 256$ for byte streams, and perhaps $\sigma \approx 10^6$ or more for streams of word tokens. The *final_symbol*(f) is a much faster operation than *next_symbol*(f); indeed, it is the final component of all calls to *next_symbol*(f). In our implementation *final_symbol*(f) (and the equivalent in each call to *next_symbol*(f)) is implemented as a local $O(\log \sigma)$ binary search, to determine the current symbol. We note that the binary search could be replaced with an $O(1)$-time bitvector operation [13] for a slight further speed advantage, and this is what is presumed in Table 1.

That is, we face the set of constraints outlined in the abstract: (a) the text is stored as a sequence of factors against a static phrase-book; (b) decoding of factors is from right-to-left; and (c) extraction of each symbol in each factor requires $H_0(\mathsf{T})$ time on average, where $H_0(\mathsf{T})$ is the zeroth order entropy of the original text.

In their presentation, Gog and Moffat [7] describe the use of the KMP pattern search algorithm, with factors expanded on-demand as required in a left-to-right manner, starting with the first one. If $f[i]$ is the i th factor in the fragment of T that is being searched, $0 \le i \le R + m - 1$, then $cum[i] = \sum_{j=0}^{i-1} len(f[j])$ is the relative starting point in T of the i th factor. With this arrangement, the last valid offset at which pattern alignment is possible is given by $F = cum[R+1] - 1$; that is, at the final symbol of the first R factors. The other $m - 1$ factors that are passed to the search function are a worst-case allowance to ensure that the entirety of the pattern is covered.

If it is assumed that P appears in the fragment of T at an alignment that is equally likely to be any value between 0 and F, then the expected number of symbols decoded by the KMP-based approach is given by

$$\frac{F}{2} + (m - 1) + \frac{F/R}{2},$$

where $F/2$ is the cost of reaching the starting point of the matching alignment; $m - 1$ is the number of further symbols that must be checked to confirm the alignment; and F/R is the average number of symbols per factor.

Gog and Moffat [7] explored a range of values of R in their experiments, working with English text and a RoSA stored on SSD secondary memory. They demonstrated that when $R = 16$, around half a millisecond is required by the KMP sequential search phase, with overall search times (for *count* queries) under two milliseconds; but that when R is increased to 256, more than 80% of query computation was spent on the pattern-search phase, more than 4 milliseconds out of a total query time a little over 5 milliseconds. Our goal is to reduce that ratio by replacing the KMP search module by methods specifically targeted for the constraints listed above.

3   Strategic Search

Searching in the factorized text representation follows a restricted access model, and it is no longer appropriate to assume a random-access machine model of computation. For example, if i corresponds to the first symbol of a new factor of length p, the cost of accessing $T[i]$ is $O(p \log \sigma)$. However, after accessing i, positions $T[i+1 \ldots i+p-1]$ can be referenced without incurring additional decoding costs.

On the other hand, the search pattern is still represented in plain text. Thus, compared to accessing T, operations on P are relatively inexpensive. This imbalance in costs opens up two aspects of the pattern matching process as being potential targets for investigation: (1) text access strategies during pattern alignment which are aware of the underlying factor representation and focus towards the end of factors whenever possible; and (2) more complex pattern pre-processing steps – including the construction of larger shift tables – which enable longer shifts during the matching process.

Maintaining a Rightward Focus. Both KMP and BMH align P at a certain position i in the text T, and compare symbols in $T[i..i+m-1]$ against their tentative equivalents in $P[0..m-1]$. The KMP approach starts with $T[i]$ and seeks to build matching prefixes, whereas BMH starts with $T[i+m-1]$ and seeks to builds matching suffixes. But starting at the ends of factorized pattern can be expensive. Instead, we suggest that the rightmost factor fully contained within the current alignment of P be identified, and then decoded from the right. As each symbol is extracted from the factor, the pattern is checked at the equivalent position; if a mismatch is identified, the pattern is shifted as far to the right as is consistent with the characters that have been decoded. Figure 1 shows an example of the rightward focused alignment process, searching for the string ACTTTGCCGTATAAGACG for which $m = 18$. Presuming that the shifts can be calculated as shown, only three different alignments are explored before the match is found, and just 23 symbols are decoded while that is taking place.

Determining Shifts Using A Suffix Array. Each time a mismatch occurs, the alignment position is shifted. Algorithms such as KMP or BMH analyze P at the beginning of the search process to pre-compute the possible shifts. The rightward focus requires that for a given pattern P and a short fragment F drawn from the text T, that the fragment be shifted as far leftward over P as is consistent with the currently-decoded symbols. That is, the problem is now flipped – the objective is to determine locations in P at which F occurs, in order to determine possible pattern alignments.

For example, consider Figure 1, in which the substring TG is decoded during the first alignment of P. Determining the rightmost occurrence in P to the left of the current alignment of the longest suffix of the substring TG gives rise to the second alignment. Because two further factors are now within the span of P in its proposed alignment, the

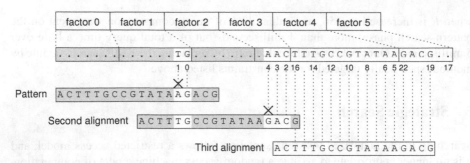

Fig. 1. Search process with partial factor decoding and rightward focus. Numbers show the order in which symbols are decoded. Grayed-out symbols in T are neither decoded nor accessed.

focus shifts to the right. Two symbols (labeled 2 and 3) are decoded from the fourth factor, and match P. But when a third symbol is decoded, the resulting fragment AAC is in conflict with the corresponding positions in P. Indeed, substring F = AAC does not occur in P at all. However, the suffix AC of F is a prefix of P, and so a complete shift is still not possible. In general, if F is not found in P, the pattern is re-aligned to match the longest suffix of F that matches a prefix of P, including the empty suffix if there are none longer. If the empty suffix is detected, the result is a complete shift of P past all currently-decoded characters.

To carry out the required search, we pre-process P to compute a suffix array SA for it – in effect, constructing a one-off index for the pattern that can be used to locate occurrences of text fragments F. Index construction might be prohibitively expensive for regular pattern search, but the high costs associated with accessing T mean that it can be considered in this context. Algorithm 1 describes how each shift is computed, given the inputs F, P, the current alignment *fpos* of F in P, and a suffix array SA over P.

There are two stages. In the first stage (steps 3–9), *backwards search* is carried out using SA. That is, F is processed from right to left (as symbols are decoded from the factor) to determine ranges (sp, ep) within SA that always match the currently-decoded fragment F. For each range (sp, ep) it is determined if the current suffix of F is a prefix of P (*head_overlap*) by determining if $0 \in SA[sp \ldots ep]$. The first stage of the algorithm terminates once F is completely processed, or if the range (sp, ep) becomes empty. The second stage (steps 10–18) determines the correct shift amount. If F occurs in P, the rightmost occurrence to the left of *fpos* corresponds to the next alignment. If no such occurrence exists, a check is made as to whether a suffix of F has matched a prefix of P (recorded by variable *head_overlap*) at any stage. If not, P is moved completely past the occurrence of F. Figure 1 does not illustrate an instance of this step.

Algorithm 1 is a high-level description, and several details have been omitted. The actual implementation uses the usual technique of building SA over the reverse P^r of P, similar to the suffix automaton of the reverse pattern of the BDM algorithm, allowing iterative determination of ranges for suffixes of F; and also makes use of an inverse suffix array for P^r, in order to expedite step 5.

Algorithm 1. Searching for a fragment F in a restricted section of string P.

0: Decide whether factor $F[0 \ldots p-1]$ has an alignment with $P[0 \ldots m-1]$ to the left of offset *fpos*. Array SA is a suffix array for P. Returns the shift such that P can be aligned to F, the rightmost matching prefix of F in P or past F if no matching prefix is found. Symbols in F are decoded on demand using the functions shown in Table 1.

```
 1: function fragment_shift(P[0...m − 1], F[0...p − 1], fpos)
 2:     (sp, ep) ← (0, m − 1), i ← p, head_overlap ← 0, shift ← 0
 3:     while |(sp, ep)| > 0 and i ≠ 0 do                          ▷ Search for F in P
 4:         (sp, ep) ← refine_interval(SA, P, (sp, ep), F[i − 1])
 5:         if 0 ∈ SA[sp...ep] then                        ▷ Suffix of F matches prefix of P
 6:             head_overlap ← head_overlap + 1
 7:         end if
 8:         i ← i − 1
 9:     end while
10:     if |(sp, ep)| > 0 then                                    ▷ Found F in P
11:         candidate ← max(SA[i] ∈ SA[sp...ep] | SA[i] < fpos)
12:         if candidate ≠ ∅ then                           ▷ Found to the right of fpos?
13:             shift ← fpos − candidate
14:         end if
15:     end if
16:     if shift = 0 then                              ▷ No full match to the left of fpos
17:         shift ← fpos + p − head_overlap                ▷ Compute prefix match, if any
18:     end if
19:     return shift
20: end function
```

Determining Shifts Using A BMH Matrix. The suffix-array based approach has two potential disadvantages: the time taken to build the suffix array for P^r may dominate the matching time; and it is unable to fully exploit non-contiguous fragments. To see the issue posed by the latter concern, consider Figure 1 again, and suppose that the "C" decoded at label 2 had in fact been an "A". Working solely with the right-focused factor, in this case the suffix array would generate a shift of one, ignoring the additional (and conflicting) information provided by the "TG" fragment that is also available in factor 1.

To address these concerns, we have explored a second mechanism, based even more closely on the Boyer-Moore-Horspool pattern search algorithm. The BMH mechanism uses the last position within the alignment to determine the shift, regardless of where in the pattern a mismatch occurs, based on a "bad symbol" table S. For each symbol $s \in \Sigma$, $S[s]$ stores the value $m - 1 - \ell_s$, where ℓ_s is the index of the last occurrence of s in $P[0 \ldots m-2]$, $\ell_s = \max\{0 \le k < m - 1 \mid P[k] = s\}$, or $S[s] = m$ if s does not appear in $P[0 \ldots m-2]$. When shifting on from an explored alignment i, BMH sets $i \leftarrow i + S[T[i+m-1]]$, using the symbol in T currently placed against $P[m-1]$ as a single reference point against which the proposed next alignment is located.

In the RoSA context $T[i+m-1]$ may not be known. On the other hand, the preprocessing on P is not required to be $O(m)$. The solution is to extend the shift table S, and make it two dimensional, setting

Table 2. Example of BMH shift matrix S for P = ACTTTGCCGTATAAGACG

	A	C	T	T	T	G	C	C	G	T	A	T	A	A	G	A	C	G	–
	0	1	2	3	4	5	6	7	8	9	10	11	12	13	14	15	16	17	18
A	1	1	2	3	4	5	6	7	8	9	10	1	2	1	1	2	1	2	3
C	1	2	1	2	3	4	5	1	1	2	3	4	5	6	7	8	9	1	2
G	1	2	3	4	5	6	1	2	3	1	2	3	4	5	6	1	2	3	1
T	1	2	3	1	1	1	2	3	4	5	1	2	1	2	3	4	5	6	7

$$S[s,p] \leftarrow p - \begin{cases} -1 & \text{if } s \notin P[0 \ldots p-1] \\ \max\{0 \leq k < p \mid P[k] = s\} & \text{otherwise}, \end{cases}$$

where $s \in \Sigma$ is a symbol in the alphabet; p is the length of a prefix of the pattern, $0 \leq p \leq m$; and $S[s,p]$ records the interval in P between offset p and the rightmost previous occurrence of symbol s. This shift table is similar to the δ_1 table of Colussi [2] which uses it as part of a more complex string matching algorithm. An example is shown in Table 2 for the same pattern as is used in Figure 1; clearly, the table requires $O(m\sigma)$ time to construct. Note also that the shift for the character aligned one past the end of the pattern can also be computed, as was first proposed by Sunday [16].

To apply the table S, every decoded character in T that overlaps the current alignment i (including $T[i+m]$, if it has been decoded) is used as an index into S, together with the corresponding offset in P. Each of those indicated values in S represents a minimum shift amount; hence, the largest of them also represents a minimum shift. Resuming the previous example, if the symbol labeled 2 in Figure 1 was an "A" rather than a "C", three elements of S would be considered: $S["A", 16]$, $S["G", 5]$, and $S["T", 4]$. The corresponding shifts (Table 2) are 1, 6, and 1; the maximum of these, 6, is used as the overall shift amount. More generally, if the current proposed alignment of P commencing at $T[i]$ is to be shifted, then the update performed is given by:

$$i \leftarrow i + \max\{S[T[i+j], j] \mid 0 \leq j \leq m \text{ and } T[i+j] \text{ has been decoded}\}.$$

Smith [15] also proposed the use of a "max" shift amount, based on the two shift vectors $S[*][m-1]$ and $S[*][m]$, shown as the last two columns in Table 2; and that Raita [12] explored the notion of checking non-sequential symbols from P, with his proposal to compare T against $P[m-1]$, $P[0]$, and $P[m/2]$ before looking at any other symbols.

4 Experiments

Methodology and Implementation. Our experimental study extends the original RoSA implementation, adding four further factor matching algorithms. Including the original KMP implementation of Gog and Moffat [7], we are able to explore: Exhaustive left-to-right matching (denoted EXH); KMP; Boyer-Moore-Horspool (BMH); the new suffix array based approach (SA); and the two-dimensional multiple-BMH technique

(MBMH). We measure the execution time cost of the factor matching process, together with the relative percentage of decoded symbols for different sample rates R, and different pattern lengths m, in all cases using the same block parameter $b = 4{,}096$ as was employed by Gog and Moffat [7]. We do not measure the other steps in the RoSA query process; they were explored in detail in the two previous studies [7,8], and those components of the implementation are reused here without alteration. All algorithms are implemented in C++11 and compiled using GCC 4.8.1 with optimizations. The suffix array for P^r was created using Yuta Mori's LIBDIVSUFSORT library[2] version 2.0.1.

Data Sets, Queries and Test Environment. We use the WEB-4G prefix of the data set used in the experimental evaluation of Gog and Moffat [7], and generate 1000 patterns for each length $m \in \{4, 10, 20, 40, 100\}$, with each pattern occurring 10–100 times in the collection. We built two RoSA indexes with $b = 4{,}096$, and factor approximation rates $R = 16$ and $R = 256$. Our machine was equipped with 148 GB RAM and we used one Intel Xeon core (E5640) running at 2.67 Ghz, approximately 1.6 times the clock speed of the 1.7 GHz Macbook Air used by Gog and Moffat. We report only the times required to perform the factor matching step, which does not include the use of the condensed BWT, does not include the access of a suffix block, does not include the use of the bit-blind tree within the block, and does not include fetching the set of factor identifiers. For the relative runtimes of these phases see Figure 5 of Gog and Moffat [7].

Symbols Decoded. In uncompressed text, the number of comparisons performed by an algorithm is generally a good indicator of run time performance. However, decoding symbols is the dominant cost when matching in factor compressed text. Figure 2 (top) shows the number of decoded symbol per query, for different sample rates and pattern lengths. When $R = 16$, the number of decoded symbols is roughly one order of magnitude smaller than for $R = 256$, as there are 16 times more factors which can contain the pattern. The relative performance of each algorithm remains similar. The two algorithms which employ left-to-right processing – EXH and KMP – decode the most symbols. The classic BMH approach is more efficient, as alignment is performed right-to-left, and so some factors are only partially decoded. As the pattern length increases, this effect is more visible, since the percentage of symbols that can be skipped per alignment increases. The two advanced methods – SA and MBMH – perform much better than the classical pattern matching algorithms. For $R = 16$ and $m = 4$, the new methods decode roughly half as many symbols as BMH, and decode a third of the symbols of EXH and KMP. For larger patterns and larger match regions, the difference is even more marked. For $R = 256$ and $m = 100$, both SA and MBMH on average decode 20% of the symbols of BMH, and one eighth of the symbols required by KMP and EXH.

The fraction of "possible" symbols decoded is shown in the bottom half of Figure 2, calculated as the percentage of symbols decoded compared to a "blind search" algorithm which aligns P to all positions amongst the first R factors. That is, the denominator is the count of symbols in all of the first R factors, plus the whole of the further factors required to span a further $m - 1$ symbols. If the occurrence of P is uniformly distributed over the positions within the first R factors, as was assumed earlier,

---

[2] Available at `https://code.google.com/p/libdivsufsort`

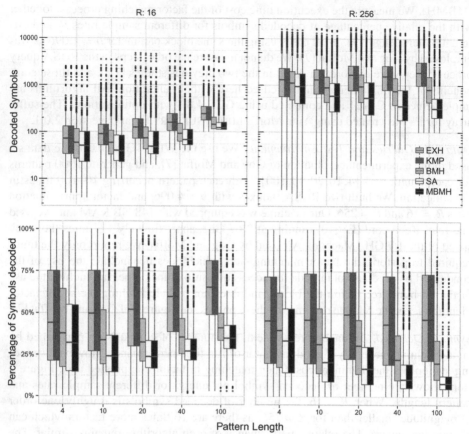

Fig. 2. Total symbols decoded (top) and percentage of decoded symbols (bottom) per query for queries of length $m \in \{4, 10, 20, 40, 100\}$, using $b = 4096$ and $R \in \{16, 256\}$ over WEB-4G. The boxes represent the median and quartiles of the measured distributions, and the whiskers depict elements within 1.5 times of the corresponding inter-quartile ranges.

then algorithms that access all symbols in each alignment should on average decode a little over 50% of that total number of symbols. For $R = 256$, both exhaustive decoding algorithms (EXH and KMP) do indeed decode close to 50% of the available symbols. For $R = 16$ the total number of positions covered by the first R factors is much smaller relative to the pattern; hence, for EXH and KMP, the percentage of decoded symbols is over 60% when $R = 16$.

Regular BMH decodes a smaller fraction of the symbols than the two exhaustive schemes. For $R = 256$ and large patterns, around 25% of all symbols are decoded. The SA and MBMH approaches significantly outperform the other three. The relative difference between the different methods increases as the pattern size increases, because longer patterns allow larger shifts to be performed. For $R = 256$ the percentages of decoded symbols for BMH, SA and MBMH decreases as the pattern length increases. For the small sample rate ($R = 16$), this is not the case as the number of decoded symbols

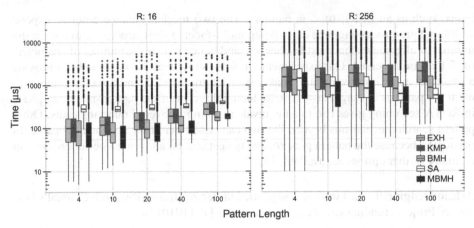

Fig. 3. Time in microseconds per query for queries of length $m \in \{4, 10, 20, 40, 100\}$, using $b = 4096$ and $R \in \{16, 256\}$ over WEB-4G

for large patterns (roughly 100, as shown in the top graph) is close to the length of the pattern. As a match always occurs in our experiments, at least m symbols have to be decoded by all methods. The two new approaches – SA and MBMH – require similar numbers of characters to be decoded, across all of the configurations tested.

Factor Matching Runtime Performance. As already noted, factor matching time can be a dominant component of overall ROSA search time, especially for large values of R. Replacing left-to-right search methods by right-to-left ones (BMH, SA, MBMH) substantially decreases the number of symbols decoded. Figure 3 shows how these savings translate into reduced execution times, showing the cost of the matching phase in microseconds for each of the different matching algorithms. When $R = 256$, the speed differentials match the arrangement shown in Figure 2. For shorter patterns the relativities are similar, but BMH outperforms both EXH and KMP for larger patterns. The original BMH approach is three times slower than SA, and five times slower than MBMH, validating the decision to search for more complex shift mechanisms. For $R = 16$, the SA method is slower than all other methods, whereas MBMH remains fast. This is caused by the additional time spent to construct the suffix array – small values of R don't allow the pre-processing investment to be sufficiently recouped. Suffix array construction takes between 150 to 350 microseconds in these experiments, and the SA method is only viable for large R and $m \geq 10$, whereas MBMH remains fast in all instances.

5   Conclusion

We have described enhanced string searching mechanisms that provide accelerated pattern matching when a particular combination of constraints applies, most notably, when access to elements of the text T is not $O(1)$ per operation. The particular application for the new methods is in the ROSA large-scale suffix array data structure, which provides indexed pattern search over large texts for which it is not possible to hold a compressed index, such as an FM-INDEX, in memory. The two mechanisms we describe expend additional pre-processing time on building multi-faceted shift structures, to reduce the number of alignments that must be checked, and hence reduce the number of characters of T that are accessed. Our experimental results show that the new methods provide a two-fold speed improvement for patterns of length $m = 20$, and a six-fold improvement for patterns of length $m = 100$. Reducing the cost of the pattern-matching phase in RoSA searching gives rise to an equivalent saving in overall querying costs. Other pattern search mechanisms beyond the two canvassed here may also be applicable. In particular, because the pattern pre-processing cost can be allowed to be super-linear in m, many further options are available [3].

Acknowledgment. This work was supported under Australian Research Council's Discovery Projects funding scheme (project number DP110101743).

Software. The RoSA software is available at https://github.com/mpetri/RoSA; it is based on the Succinct Data Structure Library (SDSL) [6].

References

1. Boyer, R.S., Moore, J.S.: A fast string searching algorithm. C. ACM 20, 1075–1091 (1977)
2. Colussi, L.: Fastest pattern matching in strings. J. Alg. 16, 163–189 (1994)
3. Faro, S., Lecroq, T.: The exact online string matching problem: A review of the most recent results. ACM Comput. Surv. 45(2), 13:1–13:42 (2013)
4. Ferragina, P., Grossi, R.: The string B-tree: A new data structure for search in external memory and its applications. J. ACM 46(2), 236–280 (1999)
5. Ferragina, P., Manzini, G.: Indexing compressed text. J. ACM 52(4), 552–581 (2005)
6. Gog, S., Beller, T., Moffat, A., Petri, M.: From theory to practice: Plug and play with succinct data structures. In: Proc. Symp. Experimental Algorithms, pp. 326–337 (2014)
7. Gog, S., Moffat, A.: Adding compression and blended search to a compact two-level suffix array. In: Proc. Symp. String Processing and Inf. Retrieval, pp. 141–152 (2013)
8. Gog, S., Moffat, A., Culpepper, J.S., Turpin, A., Wirth, A.: Large-scale pattern search using reduced-space on-disk suffix arrays. IEEE Trans. Knowledge and Data Engineering 26(8), 1 (2014)
9. Horspool, R.N.: Practical fast searching in strings. Soft. Prac. & Exp. 10(6), 501–506 (1980)
10. Knuth, D.E., Morris, J.H., Pratt, V.R.: Fast pattern matching in strings. SIAM J. Comp. 6(1), 323–350 (1977)
11. Navarro, G., Raffinot, M.: Flexible Pattern Matching in Strings: Practical On-Line Search Algorithms for Texts and Biological Sequences. Cambridge University Press (2002)
12. Raita, T.: Tuning the Boyer-Moore-Horspool string searching algorithms. Soft. Prac. & Exp. 22(10), 879–884 (1992)
13. Raman, R., Raman, V., Rao, S.S.: Succinct indexable dictionaries with applications to encoding k-ary trees and multisets. In: Proc. ACM-SIAM Symp. Discrete Algorithms, pp. 233–242 (2002)
14. Sinha, R., Puglisi, S.J., Moffat, A., Turpin, A.: Improving suffix array locality for fast pattern matching on disk. In: Proc. ACM SIGMOD Int. Conf. Management of Data, pp. 661–672 (2008)
15. Smith, P.D.: Experiments with a very fast substring search algorithm. Soft. Prac. & Exp. 21(10), 1065–1074 (1991)
16. Sunday, D.M.: A very fast substring search algorithm. C. ACM 33(8), 132–142 (1990)

Relative Lempel-Ziv
with Constant-Time Random Access

Héctor Ferrada[1,*], Travis Gagie[2,**],
Simon Gog[3,***], and Simon J. Puglisi[2,†]

[1] Department of Computer Science
University of Chile, Chile
[2] Department of Computer Science
University of Helsinki, Finland
[3] Institute of Theoretical Informatics
Karlsruhe Institute of Technology, Germany

Abstract. Relative Lempel-Ziv (RLZ) is a variant of LZ77 that can compress well collections of similar genomes while still allowing fast random access to them. In theory, at the cost of using sublinear extra space, accessing an arbitrary character takes constant time. We show that even in practice this works quite well: e.g., we can compress 36 *S. cerevisiae* genomes from a total of 464 MB to 11 MB and still support random access to them in under 50 nanoseconds per character, even when the accessed substrings are short. Our theoretical contribution is an optimized representation of RLZ's pointers.

1   Introduction

Advances in DNA sequencing have led to the creation of massive genomic databases. In many cases these databases hold collections of genomes from individuals of the same species or closely related species. Such genomes tend to be very similar, so referential compression schemes such as Ziv and Lempel's LZ77 [14] perform very well on them (see, e.g., [1,13]). Supporting fast random access to LZ77-compressed texts is problematic [12] but several authors have proposed variants of LZ77 on which random access is easier: e.g., Kreft and Navarro's LZ-End [7], Kuruppu, Puglisi and Zobel's Relative Lempel-Ziv (RLZ) [8,9] and Deorowicz and Grabowski's GDC [2].

In theory RLZ implemented with compressed bitvectors offers constant-time random access, which is faster than LZ-End, GDC or schemes such as block graphs [4] or FOLCA [10] that are not based on LZ77. The main disadvantage of using compressed bitvectors is their redundancy, which is nevertheless sublinear

* Supported by Fondecyt 1-140796, Chile.
** Supported by Academy of Finland grant 268324.
*** This work was carried out while the third author was employed at the University of Melbourne, supported by ARC Grant DP110101743.
† Supported by Academy of Finland grant 258308.

E. Moura and M. Crochemore (Eds.): SPIRE 2014, LNCS 8799, pp. 13–17, 2014.
© Springer International Publishing Switzerland 2014

in the length of the original file. As far as we are aware, such an implementation has never been tried in practice. In this paper we describe an implementation with which we can, e.g., compress 36 *S. cerevisiae* genomes from 464 MB to 11 MB and still support random access to them in under 50 nanoseconds per character, even when the accessed substrings are short. Our theoretical contribution is a compressed representation of RLZ's pointers that is optimized for genomic databases.

2 Relative Lempel-Ziv

Given a collection of similar genomes, RLZ works by selecting or generating a reference genome R, which it leaves uncompressed (or only entropy-compressed), then compressing each of the remaining genomes relative to R. To compress another genome $S[0..n-1]$ relative to R, our implementation of RLZ greedily parses S into phrases such that each phrase consists of a substring of R followed by a single character, called a mismatch character. When the alphabet size is constant this can be done in $\mathcal{O}(n)$ time using, e.g., an FM-index [3] for R (see, e.g., [6]).

Kuruppu et al. originally defined RLZ such that each phrase is either a substring of R or a single character. Deorowicz and Grabowski pointed out, however, that with this definition, single-nucleotide polymorphisms (SNPs) — the most common kind of differences between individuals' genomes — tend to cause two phrase breaks each, instead of only one.

If we want only to compress S, we need only store a sequence $(\ell_0, p_0, c_0), \ldots, (\ell_{z-1}, p_{z-1}, c_{z-1})$ of triples, where z is the number of phrases. Each triple (ℓ_r, p_r, c_r) indicates that the corresponding phrase is $R[p_r..p_r + \ell_r - 1] c_r$, or just c_r if $\ell_r = 0$ (in which case p_r is irrelevant). To decompress S later, we simply replace each triple by its phrase.

2.1 Compressed Bitvectors

A bitvector B is a binary string that supports **access**, **rank** and **select** queries: B.access(i) returns $B[i]$ (and is often written simply $B[i]$); B.rank(i) returns the number of 1s in $B[0..i]$; and B.select(i) returns the position of the ith 1 in B. The best theoretical bound is due to Pătraşcu [11], who showed how B can be stored in $|B|H_0(B) + \mathcal{O}(|B|/\log^a |B|)$ bits, where $H_0(B)$ is the 0th-order empirical entropy of B and a is any constant, such that all three kinds of queries can be answered in $\mathcal{O}(1)$ time.

2.2 Absolute Pointers

If we want to support random access, then we can store a bitvector $B_1[0..n]$ with 1s marking where phrases start in S; an array $P[0..z-1] = [p_0, \ldots, p_{z-1}]$; and an array $C[0..z-1] = [c_0, \ldots, c_{z-1}]$. We set $B_1[0] = 0$ so that B_1.rank(i) is the index of the phrase containing $S[i]$, and we set $B_1[n] = 1$ so that $S[i]$

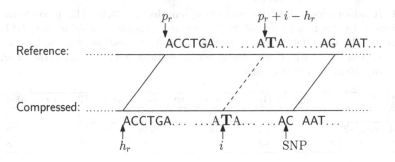

Fig. 1. We can use the relative pointer $P'[r] = p_r - h_r$ instead of the absolute pointer $P[r] = p_r$ because $P'[r] + i = p_r + i - h_r = P[r] + i - h_r$

is a mismatch character if and only if $B_1[i + 1] = 1$. For simplicity, we assume $B_1.\text{select}(0) = 0$.

To access $S[i]$, we

1. find the index $r = B_1.\text{rank}(i)$ of the phrase containing $S[i]$;
2. check whether $B_1[i + 1] = 1$, to see if $S[i]$ is a mismatch character;
3. if so, return $C[r]$;
4. if not, find the starting position $h_r = B_1.\text{select}(r)$ of the rth phrase;
5. return $R[P[r] + i - h_r]$.

We say this approach uses *absolute* pointers because each cell of P contains a direct pointer to the starting position of the appropriate substring of R.

2.3 Relative Pointers

We can simplify our access procedure somewhat if we store *relative* pointers instead of absolute pointers. That is, we store an array $P'[0..z - 1] = [p_0 - h_0, \ldots, p_{z-1} - h_{z-1}]$, where h_r is again the starting position in S of phrase r. Notice that $P'[r] + i = p_r + i - h_r = P[r] + i - h_r$ for any i — see Figure 1 — so we no longer need select to access S. Specifically, to access $S[i]$ we now

1. find the index $r = B_1.\text{rank}(i)$ of the phrase containing $S[i]$;
2. check whether $B_1[i + 1] = 1$, to see if $S[i]$ is a mismatch character;
3. if so, return $C[r]$;
4. if not, return $R[P'[r] + i]$.

2.4 Compressed Pointers

Another benefit of relative pointers is that we can compress P' more easily than P. For example, Kuruppu et al. noted that the difference between phrases' absolute pointers is often the total length of the phrases between them, and used

Table 1. Random access time (for extraction lengths $m = 8, 64, 512, 4096$ characters) for RLZ and GDC for a collection of 36 *S. cerevisiae* genomes, totalling 464 MB before compression. The two GDC rows correspond to different settings for the R-block size and D-block size (see [2]). Times are given in nanoseconds per extracted character averaged over 10 million extractions.

Method	Size (MB)	$m = 8$	$m = 64$	$m = 512$	$m = 4096$
RLZ	11	50	7	3	2
GDC-ra-2^8-2^8	14	500	68	15	8
GDC-ra-2^{12}-2^{12}	10	750	102	19	8

that observation to achieve better compression by discarding pointers that can be computed from earlier pointers and phrases' lengths. They did not support fast random access with that implementation, but they pointed out that it could be reintroduced by sampling the missing pointers, creating a tradeoff between compression and access time.

The most likely explanation for Kuruppu et al.'s observation is that, if phrase r breaks because of an SNP, then usually $p_{r+1} = p_r + \ell_r + 1$, in which case $P'[r + 1] = P'[r]$. To take advantage of this, we run-length compress P' as follows: we store a bitvector $B_2[0..z - 1]$ with 1s marking the relevant pointers in P' that differ from the preceding relevant pointers; we store an array P'' containing the pointers in B_2; and then we discard P'. (Recall that a pointer is irrelevant if its phrase is only a mismatch character, and relevant otherwise.)

We set $P''[0]$ to be the first relevant pointer in P' but we do not mark it in B_2, so that $B_2.\mathsf{rank}(r)$ is the index of the run in P containing the pointer for phrase r. The last step in our access procedure now becomes "if not, return $R[P''[B_2.\mathsf{rank}(r)] + i]$".

3   Experiments

We implemented the above scheme in C++ using the Succinct Data Structure Library (SDSL) [5] version 2.0.1 (available at https://github.com/simongog/sdsl-lite) for the bitvectors. As a baseline we also tested Deorowicz and Grabowski's GDC data structure, which is the best RLZ variant supporting random access of which we are aware.

Setup. We performed experiments on a machine equipped with a 3.16GHz Intel Core 2 Duo CPU with 6144KiB L2 cache and 4GiB of main memory. The machine had no other significant CPU tasks running and only a single thread of execution was used. The OS was Linux (Ubuntu 12.04, 64bit) running kernel 3.2.0. All programs were compiled using g++ version 4.8 with -O3 -static -DNDEBUG options. All reported runtimes are recorded with the C clock function.

Data. We tested our implementation on a collection of 36 *S. cerevisiae* (yeast) genomes, each about 12 MB long, and 464 MB in total.

Mismatch Characters. The genomes were over the alphabet $\{A, C, G, T, N\}$ but there were relatively few Ns in the array C of mismatch characters. Because of this, we stored in a hash table the positions of all the Ns in C; replaced the Ns by As; and packed the mismatch characters into two bits each. We considered C blocks of 20 mismatch characters and stored a binary string in which the ith bit indicated whether the ith block originally contained any Ns. Whenever we read an A from C, we checked the binary string to see if the block containing that A originally contained any Ns and, if so, checked the hash table to see if that particular A was originally an N.

Results. Sizes and times for random access are shown in Table 1. Our implementation of RLZ had very fast extraction times for short substrings. For example, it was over 10 times faster than GDC for substrings of length 8. For longer substrings the gap between the speed of the two approaches narrowed, but even for 4KB substrings RLZ was still more than 4 times faster than GDC.

References

1. Deorowicz, S., Danek, A., Grabowski, S.: Genome compression: A novel approach for large collections. Bioinformatics 29(20), 2572–2578 (2013)
2. Deorowicz, S., Grabowski, S.: Robust relative compression of genomes with random access. Bioinformatics 27(21), 2979–2986 (2011)
3. Ferragina, P., Manzini, G.: Indexing compressed text. J. ACM 52(4), 552–581 (2005)
4. Gagie, T., Gawrychowski, P., Puglisi, S.J.: Faster approximate pattern matching in compressed repetitive texts. In: Asano, T., Nakano, S.-i., Okamoto, Y., Watanabe, O. (eds.) ISAAC 2011. LNCS, vol. 7074, pp. 653–662. Springer, Heidelberg (2011)
5. Gog, S., Beller, T., Moffat, A., Petri, M.: From theory to practice: Plug and play with succinct data structures. In: Gudmundsson, J., Katajainen, J. (eds.) SEA 2014. LNCS, vol. 8504, pp. 326–337. Springer, Heidelberg (2014)
6. Kärkkäinen, J., Kempa, D., Puglisi, S.J.: Lightweight Lempel-Ziv parsing. In: Bonifaci, V., Demetrescu, C., Marchetti-Spaccamela, A. (eds.) SEA 2013. LNCS, vol. 7933, pp. 139–150. Springer, Heidelberg (2013)
7. Kreft, S., Navarro, G.: On compressing and indexing repetitive sequences. Theor. Comp. Sci. 483, 115–133 (2013)
8. Kuruppu, S., Puglisi, S.J., Zobel, J.: Relative Lempel-Ziv compression of genomes for large-scale storage and retrieval. In: Chavez, E., Lonardi, S. (eds.) SPIRE 2010. LNCS, vol. 6393, pp. 201–206. Springer, Heidelberg (2010)
9. Kuruppu, S., Puglisi, S.J., Zobel, J.: Optimized relative Lempel-Ziv compression of genomes. In: Proc. ACSC, pp. 91–98 (2011)
10. Maruyama, S., Tabei, Y., Sakamoto, H., Sadakane, K.: Fully-online grammar compression. In: Kurland, O., Lewenstein, M., Porat, E. (eds.) SPIRE 2013. LNCS, vol. 8214, pp. 218–229. Springer, Heidelberg (2013)
11. Pătraşcu, M.: Succincter. In: Proc. FOCS, pp. 305–313 (2008)
12. Verbin, E., Yu, W.: Data structure lower bounds on random access to grammar-compressed strings. In: Fischer, J., Sanders, P. (eds.) CPM 2013. LNCS, vol. 7922, pp. 247–258. Springer, Heidelberg (2013)
13. Wandelt, S., Leser, U.: FRESCO: Referential compression of highly-similar sequences. IEEE Trans. Comp. Bio. Bioinf. 10(5), 1275–1288 (2013)
14. Ziv, J., Lempel, A.: A universal algorithm for sequential data compression. IEEE Trans. Inf. Theor. 23(3), 337–343 (1977)

Efficient Compressed Indexing
for Approximate Top-k String Retrieval*

Héctor Ferrada and Gonzalo Navarro

Department of Computer Science, University of Chile, Chile
{hferrada,gnavarro}@dcc.uchile.cl

Abstract. Given a collection of strings (called documents), the *top-k document retrieval* problem is that of, given a string pattern p, finding the k documents where p appears most often. This is a basic task in most information retrieval scenarios. The best current implementations require 20–30 bits per character (bpc) and k to $4k$ microseconds per query, or 12–24 bpc and 1–10 milliseconds per query. We introduce a Lempel-Ziv compressed data structure that occupies 5–10 bpc to answer queries in around k microseconds. The drawback is that the answer is approximate, but we show that its quality improves asymptotically with the size of the collection, reaching over 85% of the accumulated term frequency of the real answer already for patterns of length 4–6 on rather small collections, and improving for larger ones.

1 Introduction

Finding the k documents most relevant to a search query is the most basic information retrieval problem. Originally defined on natural language text collections, its generalization to collections of arbitrary strings turns out to be a problem arising naturally in areas like bioinformatics, multimedia databases, software repositories, and so on [11]. For example, one might want to find the genes where a certain motif appears most often (as they may deserve further biological analysis), modules where a function is called most often (to spot cohesion issues in software design), songs containing most occurrences of a certain sequence (to hint plagiarism), and so on. On East Asian languages like Chinese and Korean, classical solutions for Western natural languages are not applicable, and they are usually handled as generic string collections as well.

Our collection will contain D *documents*, which are strings $d_1, \ldots, d_D$, over an alphabet $[1, \sigma]$, of total length $n = \sum |d_i|$. We preprocess them to build an *index*. Later, given a pattern string $p[1, m]$ and a threshold k, we must list the k documents where p appears most often. In natural language searching the measure of relevance can be more sophisticated than just the number of occurrences of p, but frequency is still a key component. Usually even more complex measures are used in a second step, where the top-k documents are further filtered to obtain the final result [14].

---

* Partially funded by Fondecyt grant 1-140796, Chile.

E. Moura and M. Crochemore (Eds.): SPIRE 2014, LNCS 8799, pp. 18–30, 2014.
© Springer International Publishing Switzerland 2014

Hon et al. [4] proposed a first index for this problem, but its space usage is superlinear, $O(n \log n)$ words; their implementation also uses too much space. Later, Hon et al. [6] presented a structure using linear space, that is, $O(n)$ words or $O(n \log n)$ *bits*. They solved queries in $O(m + k \log k)$ time. Navarro and Nekrich [12] reduced the time to the optimal $O(m + k)$. Konow and Navarro [7] implemented this index, obtaining an index that uses 20–30 bits per character (bpc)[1] and answers top-k queries in k to $4k$ microseconds (μsec). Their time complexity is $O(m + (k + \log \log n) \log \log n)$ with high probability, on statistically typical texts [15]. Shah et al. [5] proposed another index that is not yet implemented, but it is likely to perform similarly, and has a time complexity of $O(m + (\log \log n)^6 + k(\log \sigma \log \log n)^{1+\epsilon})$ for any constant $\epsilon > 0$. Navarro and Valenzuela [13] aimed at using less space, reaching 12–24 bpc depending on the compressibility of the collections, but retrieval times are an order of magnitude higher, 1 to 10 milliseconds (the time complexity is upper-bounded by $O(m + k \log^{4+\epsilon} n)$). There are several other theoretical proposals [11] that promise to use much less space than current implementations, but that are most likely to be even slower in practice (as already hinted in current studies [13]).

In this paper we introduce an index that uses much less space and time than current alternatives. It is based on Lempel-Ziv compression, precisely LZ78 [16], which compresses texts by building a dictionary of frequent strings (called phrases) and then parsing the text as a sequence of n' phrases. It holds $n' \leq n/\lg_\sigma n$ for any text, and moreover $n' \lg n = nH_k + O(n(k \log \sigma + \log \log n)/\log_\sigma n)$ for any k, where H_k is the k-th order empirical entropy of the text [8]. This is $n' \lg n = nH_k + o(n \log \sigma)$ for any $k = o(\log_\sigma n)$. Our structure builds on previous LZ78-compressed indexes called LZ-indexes, developed for finding all the occurrences of p [10,2] and for listing all the documents where p occurs [3]. Like these indexes, our structure uses, in practice, $(2 + \epsilon)n' \lg n + O(n' \log \sigma) = (2 + \epsilon)nH_k + o(n \log \sigma)$ bits, and it solves top-k queries in time $O(m + k \log D)$. In practice, the space is around 5–10 bpc to achieve a query time around k μsec. This time/space tradeoff is well below that of previous implementations.

In exchange, our top-k answer is approximate, as we consider only the occurrences of p within phrases. If the text is generated by a stationary source, the occurrences of any pattern p appear regularly, every d positions on average (e.g., $d = \sigma^m$ if the symbols are generated uniformly and independently). On the other hand, since $n' \leq n/\lg_\sigma n$, only $(n/d)m(n'/n) \leq (n/d)m/\lg_\sigma n$ of those occurrences hit a phrase boundary on average. This means that a fraction of $1 - m/\lg_\sigma n$ of the occurrences are within phrases (the fraction improves to $1 - mH_k/\lg n$ on compressible texts). Thus, we are considering asymptotically all of the occurrences of p when building the approximate top-k answers for short enough patterns, $m = o(\log_\sigma n)$. Note that, if $m \geq \lg_\sigma n$, then it occurs $O(1)$ times on average in the collection, and then a plain listing of all the documents where it appears [3] is an appropriate tool to find its top-k documents.

We show that, already on moderate collections of $n = 25$–130 MB, the quality of the answer (measured as the number of occurrences of p on the k retrieved

---

[1] The space results they report [7] are somewhat underestimated, as we show here.

documents as a percentage of the number of occurrences on the actual top-k documents) is always over 85% for short patterns ($m = 4$–6), improving as the collection size grows and as the collection becomes more compressible with LZ78.

2   The LZ-Index

Assume we concatenate the documents $d_1 \cdots d_D$ (each terminated with a special symbol \$, which always marks the end of a phrase) into a text $T[1, n]$ over alphabet $[1, \sigma]$. The LZ78 compression algorithm cuts the text into n' distinct *phrases*, each of which is equal to the longest possible previous phrase plus the following new symbol. Each phrase is then replaced by the *id* of the previous corresponding phrase and the extra symbol. The number of bits output by the compressor is $|\mathsf{LZ78}| = n'(\lg n' + \lg \sigma)$, which converges to the statistical entropy of T [8], and it always holds $n' \leq n/\lg_\sigma n$. The LZ-index [10] is a text index built on the LZ78 parsing of the text, and it supports locating the occurrences of a pattern $p[1, m]$ in T. The index is formed by the following components (among others not relevant for this paper).

1. **LZTrie**: a trie composed of all the phrases produced by the LZ78 parsing. Note that the set of phrases is prefix-closed (every prefix of a phrase is also a phrase), so LZTrie has n' nodes. It stores the phrase identifier of each node.
2. **RevTrie**: a trie storing the reversed phrases. It is not prefix-closed, so there are *empty* nodes not associated with phrases. It contains originally t_{rev} nodes. We contract unary paths of empty nodes to a single edge, after which the trie has $n_{rev} = n' + n_e \leq 2n'$ nodes, where n_e empty nodes remain after contracting. The phrase numbers of the n' nonempty nodes are stored.
3. **Node**: an array mapping from phrase numbers to their preorder in LZTrie.

The modern version [2] of the LZ-index uses $(2+\epsilon)|\mathsf{LZ78}|(1+o(1))$ bits, for any $\epsilon > 0$. The occurrences of pattern p are divided into type 1 (those completely inside a phrase), and types 2 and 3 (those spanning two or more phrases, respectively). Those are found separately at search time. In this paper we are only interested in finding occurrences of type 1. For those, we search for p^r (the reversed pattern) in RevTrie, arriving at node v^r. Each node u^r descending from v^r (including v^r) corresponds to an occurrence of type 1 where p appears at the end of the phrase. The other occurrences of type 1 are the nodes u' that descend from u in LZTrie, where u corresponds to u^r. Thus, for each node u^r that is nonempty, we read the phrase id f_u of u^r, compute $u = Node(f_u)$, and report all the phrase ids in the subtree of u. It takes $O(m + occ)$ time to report the occ type-1 occurrences.

3   An LZ-Index for Approximate Top-k Retrieval

Our top-k retrieval LZ-Index is a modification of the original LZ-Index. The parsing into phrases is changed so as to force phrases to finish when the document terminator \$ is seen, so that no phrase crosses a document border. The LZ-Index is then represented with the following structures, inspired in previous work [3]:

LZTrie: We store only the topology and the documents where the phrases lie, using in total $n' \lg D + O(n')$ bits.

P_{lz}: The LZTrie topology represented with parentheses in a preorder traversal, and made navigable in $O(1)$ time using $2n' + o(n')$ bits (FF [1]).

D_{lz}: An array storing, for each node v of LZTrie in preorder, the document identifier (using $\lceil \lg D \rceil$ bits) where its corresponding phrase lies.

Revtrie: We store the structures needed to carry out searches directly on it, without the help of the LZTrie, using in total $t_{rev} \lg \sigma + O(t_{rev})$ bits. In theory t_{rev} can be as large as n but in practice it is much closer to $n_{rev} \leq 2n'$.

P_{rev}: The tree topology using parentheses and made constant-time navigable, using $2t_{rev} + o(t_{rev})$ bits (FF [1]).

E_{rev}: A bitmap marking empty nodes, in preorder, using t_{rev} bits.

U_{rev}: A bitmap marking empty unary nodes (i.e., contracted), from those that are marked empty in E_{rev}, using $t_{rev} - n'$ bits.

L_{rev}: A sequence of the n_{rev} letters that label the non-contracted edges leading to the nodes, in preorder. Used to find the child nodes at searching.

M_{rev}: A sequence of the $t_{rev} - n_{rev}$ letters that label the contracted edges leading to the nodes, in preorder. Used to check that the characters in the contracted edge match the search pattern.

All the bitmaps are stored with sublinear extra data structures that solve *rank/select* operations in constant time [9]. This allows, for example, finding the jth 0 or 1 in the bitmap in constant time, or count the number of 0s or 1s in any bitmap interval.

Node: This is recast as a mapping from nonempty RevTrie nodes to their LZTrie preorder numbers, using $n' \lg n' + O(n')$ bits.

Top: To solve top-k document retrieval for any $k \leq k^*$, where k^* is a parameter defined at construction time, we will store the top-k^* answers, in decreasing frequency order, for some RevTrie nodes. We use a parameter g to define the RevTrie nodes that will store their top-k^* answer: These will be the (empty or nonempty) nodes representing a string with at least gk^* occurrences of type 1 in T. Empty unary nodes will not store their answer set, as this is the same of its child. The marking will be node for all the k^* values in $[1..D]$ that are a power of 2. Nodes v will store their top-k^* answers for the maximum k^* value for which they are marked. This is implemented with the following additional structures:

B_{top}: A bitmap of size n_{rev} marking which RevTrie nodes have top-k^* answers precomputed, in preorder.

K_{top}: The sequences of k^* most frequent documents where each node marked in B_{top} appears, concatenated in the same order of B_{top}. The identifiers are stored using $\lceil \lg D \rceil$ bits, in decreasing frequency order.

LK_{top}: A bitmap marking the starting positions of the sequences in K_{top}.

A_{top}: Since there may be less than k^* distinct documents where the marked node appears, this bitmap indicates whether a node marked in B_{top} already lists all of the possible documents.

The larger g, the fewer RevTrie nodes store their top-k^* documents: While in theory we might store up to $n_{rev} k^* \lg D$ bits, in practice this is much closer to $(n_{rev}/(gk^*)) k^* \lg D = (n_{rev} \lg D)/g$, which added over the $\lg D$ values for k^* gives $(n_{rev} \lg^2 D)/g$ bits. The other bitmaps use $O(n_{rev})$ bits.

Figure 1 illustrates the main components of our index. The overall space is, in practice, upper bounded by $n'(\lg n' + \lg D + 2 \lg \sigma + 2(\lg^2 D)/g + O(1))$ bits. Thus a value like $g = \Theta(\lg D)$ obtains space similar to the original pattern-matching LZ-index, $(2 + \epsilon)n' \lg n + O(n' \log \sigma)$ bits [2].

Since phrases are cut at the end of documents, there may appear a few repeated phrases across the collection. Therefore, at construction time, we have to consider the special case when two or more documents end with the same phrase. This is handled by storing a short linked list, both in RevTrie and LZTrie, attached to the nodes representing phrases that appear more than once.

Querying. At query time, we find the RevTrie node corresponding to p^r, move to its highest descendant not marked in U_{rev}, v^r, and check if it is marked in B_{top}. If marked and either (1) A_{top} indicates it stores all the possible documents, or (2) LK_{top} indicates that it stores $k' \geq k$ top documents, then we return the first k documents stored for v^r in K_{top} and finish. Otherwise, we need to solve the answer by brute force, by traversing all its type-1 occurrences. By construction, this takes place only if v^r has $k' < k$ answers stored (including the case $k' = 0$). If k^* is the power of 2 next to k, then v^r does not store its top-k^* answers, thus by construction it has less than gk^* occurrences of type 1. Therefore the brute-force process must traverse $O(gk)$ occurrences.

In order to solve v^r by brute force, we use P_{rev} to compute its preorder i_v and its subtree size s_v. Thus all the subtree of v^r has the preorder interval $[i_v, i_v + s_v - 1]$. Then we use $rank$ on E_{rev} to map it to the interval $[i_1, i_2]$ of nonempty preorder values. For each i in this interval, we compute $i_u = Node(i)$, which is the preorder of the corresponding node in LZTrie, and then use P_{lz} to obtain the corresponding node u in LZTrie. Then we compute the size s_u of u and obtain the interval $[i_u, i_u + s_u - 1]$ of all the descendants of u in LZTrie. We process all the document identifiers in $D_{lz}[i_u, i_u + s_u - 1]$, for all the nodes u in LZTrie that correspond to all the RevTrie descendants u^r of v^r, accumulating their frequencies and then choosing the k highest ones. The whole process takes $O(m + gk + k \log k) = O(m + k \log D)$ time.

4   Experimental Results

We use various document collections, following previous work [13,7] and exploring different aspects of statistical compressibility, size, number of documents, and repetitiveness: ClueWiki (English, few large documents), DNA (synthetic, mildly repetitive with 5% mutations among documents), KGS (Go game records), Wiki (more and shorter documents), Proteins (many more documents, almost incompressible), and TodoCL (a snapshot of the Chilean Web, with real queries, used to

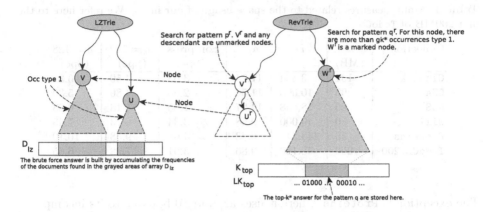

Fig. 1. The main data structures of our index. The search for the pattern q^r reaches node W^r in the RevTrie, which is marked in B_{top}, therefore the answer is retrieved from the vector K_{top} using the marks in the bit array LK_{top}. The search for p^r, instead, reaches node V^r in the RevTrie. Since this is an unmarked node, the answer is computed online by accumulating frequencies from the document array of phases, D_{lz}.

measure quality). Table 1 shows their main characteristics (column "compress" shows how the LZ78-based Unix Compress program compresses them).

Our machine is an Intel Xeon with 8 processors of 2.4GHz and 12MB cache, with 96GB RAM. It runs Linux 2.6.32-46-server, and we use gcc with full optimization and no multithreading. We chose 40,000 patterns of lengths $m = 3$ and $m = 8$ extracted randomly from the collection.

4.1   Time and Space

Table 1 shows the size of our structure with $g = 512$, where it uses almost its minimum possible space, and with $g = 128$, where it achieves around k μsec to solve queries, as we will see. The minimum space ranges from 1.2 to 2.8 times the space of Unix Compress. For this value of g our analysis predicts a factor around 2. On the other hand, our index uses around 5–10 bpc with $g = 128$.

Fig. 2 gives the breakdown of the space obtained by our index on those collections, for increasing values of g. The components are LZTrie (the tree topology and the document identifiers, which dominate), RevTrie (the tree topology and the letters), array Node, and Top (the storage of the best documents for some precomputed nodes). We show the breakdown as cumulative space curves. As g increases, the Top component is reduced and the structure becomes slower.

Now we compare our structure with previous work. We consider search patterns of lengths $m = 3$ in Fig. 3 and $m = 8$ in Fig. 4, and measure the cost to compute the top-10 and top-100 documents, for $g = 512, 256, 128, \ldots$. We denote DCC'13 the existing fast and large structure [7] and denote SEA'12 a choice of relevant space/time tradeoffs of the existing small and slow structure [13]. In most texts, our structure uses 5–7 bpc to solve top-k queries in around k μsec.

Table 1. Main measures related to the space usage of our index. We refer here to the first 200MB of `TodoCL`.

Collection	n (MB)	D	n/n'	compress (bpc)	$g = 512$ (bpc)	$g = 128$ (bpc)
ClueWiki	131	3,334	17.24	3.63	4.50	6.31
DNA	95	10,000	11.50	2.68	4.86	5.30
KGS	25	18,838	14.97	1.85	5.13	6.23
Wiki	80	40,000	9.58	3.34	6.73	7.43
Proteins	56	143,244	6.43	4.61	9.58	10.10
TodoCL.200	200	48,186	9.86	3.91	7.32	6.65

The exception is `Proteins`, where it uses around 10 bpc due to its incompressibility. Except on `Proteins`, where it uses over 20 bpc, structure SEA'12 can use similar or less space than ours, but at the cost of being 4–5 orders of magnitude slower. Even if using much more space, SEA'12 is at least 10 times slower than ours. Structure DCC'13, on the other hand, is 4–50 times slower than ours, and uses 2.5–7 times our space.

4.2 Quality

The drawback is that our structure delivers approximate top-k answers. We present in Fig. 5 two measures of the quality of the answer. On the left we show traditional recall, measured in the following way: for each value $k' \in [1, k]$, we measure how many of the (correct) top-k' documents are reported within the (approximate) top-k results. For example, the point at $k' = 1$ indicates how many times the most relevant document is contained in our top-k answer. The point at $k' = k$ indicates how many of the correct top-k documents are actually returned. This measure is interesting in applications where the top-k answer is postprocessed with a more sophisticated relevance function in order to deliver a final answer of $k' \ll k$ results. The figure shows that, in this scenario, the k' most relevant candidates are frequently in the approximate top-k answer set, for small k'. However, when k' becomes closer to k, the recall degrades, more or less depending on the collection, and faster for $m = 8$ than for $m = 3$. On the other hand, there are no significant differences between the results for $k = 10$ and for $k = 100$. Results are particularly bad for `DNA`, `KGS`, and `Proteins`.

If our index fails to return a top-k document but returns another one with the same frequency, we take it as a hit, as both are equally good. In this sense, recall is a too strict measure of relevance: if the system returns a document with only slightly fewer occurrences than the correct one, it counts as zero. As the frequency is only a rough measure of relevance, a much more precise measure of quality is the sum of the frequencies of the documents in the approximate top-k answer as a fraction of the sum in the correct top-k answer.

Fig. 5 (right) shows the results under this measure of quality. All collections perform very well for $k = 3$, reaching 90%–100% of quality even for $k' = k$. For $k = 8$, collections `ClueWiki` and `KGS` still achieve a reasonable quality over

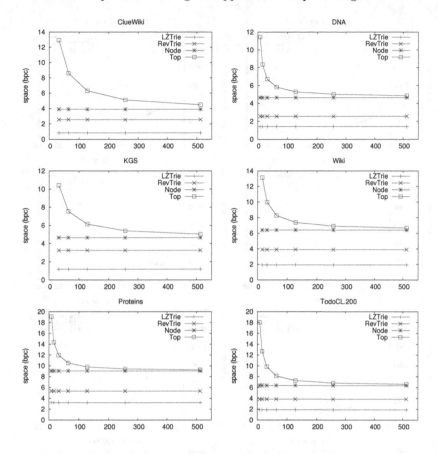

Fig. 2. Space breakdown of our structures for different g values (g is the x-axis)

80%, DNA over 60%, Wiki over 50%, and Proteins only 10%. These differences in quality can be predicted with Table 1: the less compressible the collection, the smaller n/n', and the worse quality obtained for a given pattern length m.

On the other hand, the fact that better quality is obtained for shorter patterns coincides with our probabilistic analysis. Fig. 6 illustrates this effect more closely, for increasing pattern lengths. It can be seen that, for the moderate collection sizes of 25–130 MB we considered, we obtain quality well above 85% for $m = 4$–6, depending on the text type and its n/n' value. Fig. 7 shows the case of real query words (of length > 3 to exclude most stopwords, average length 7.2) and 2-word phrases (average length 8.0), on increasing prefixes of TodoCL converted to lowercase (as case is generally ignored in natural language queries). As predicted, the quality improves with n, from 33%–46% on 200 MB ($n/n' = 10.1$) up to 59%–72% on 1.6 GB ($n/n' = 12.5$).

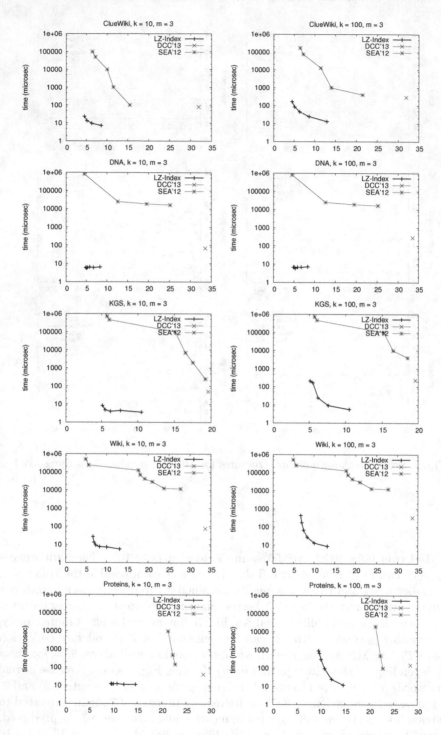

Fig. 3. Space/time comparison for pattern length $m = 3$. Space (bpc) is the x-axis.

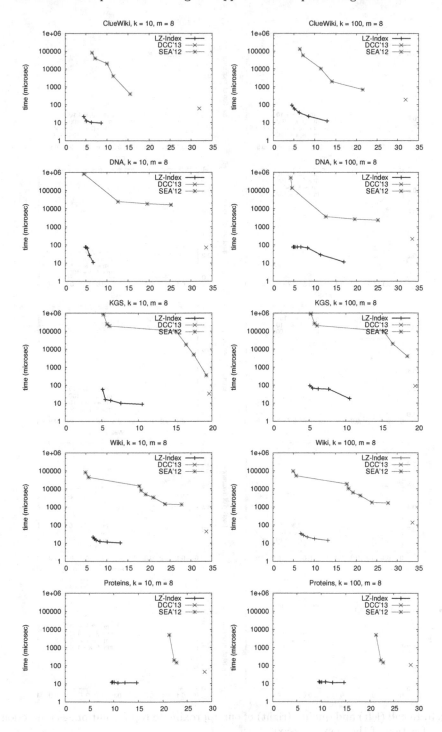

Fig. 4. Space/time comparison for pattern length $m = 8$. Space (bpc) is the x-axis.

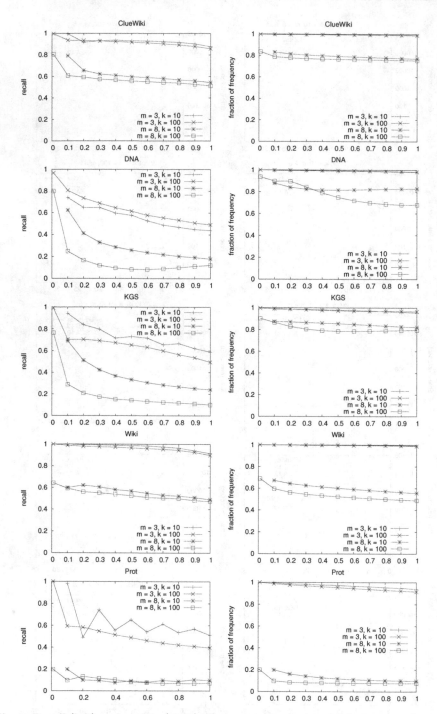

Fig. 5. Recall (left) and quality (right) of our approximate top-k solution, as a function of the fraction of the answer (x-axis)

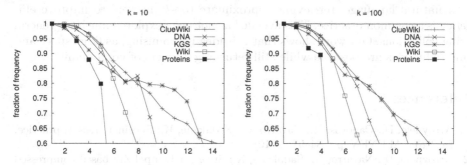

Fig. 6. Quality of our approximate top-k solution, as a function of the pattern length, for top-10 (left) and top-100 (right)

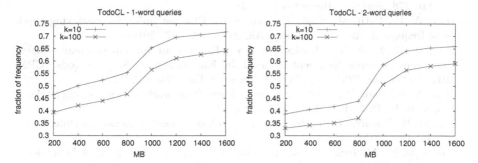

Fig. 7. Quality of our approximate top-k solution, as a function of the prefix size of TodoCL in MB, for words (left) and phrases of 2 words (right)

5   Conclusions

We have introduced a top-k retrieval index for general string collections, based on Lempel-Ziv compression. The index is orders of magnitude faster, and uses much less space, than previous work. In exchange, it delivers approximate top-k answers, which is acceptable in most applications. We analytically show that, under reasonable assumptions on the text distribution, the answers tend to exactness asymptotically, when the collection is large enough compared to the pattern length. Our experiments also show that the quality of the answer is good enough for short patterns already on our moderate-size text collections. The larger the text collection, or the more compressible it is with LZ78, the longer the patterns that can be searched with high quality. In this sense, the index is a very promising alternative to handle the large text collections one aims at in real life.

We obtain good-quality results for real word queries on a moderately large text collection. Our next step is to use our index to find top-k candidate documents for the individual words of multiword queries and then postprocessing the result into weighted conjunctive or disjunctive queries [14].

In natural language, retrieving approximate top-k answers to improve efficiency is a common practice. This avenue has not been explored much for general string collections. Our work shows that this idea is promising, as large space and time reductions are possible while still returning answers of good quality.

References

1. Arroyuelo, D., Cánovas, R., Navarro, G., Sadakane, K.: Succinct trees in practice. In: Proc. ALENEX, pp. 84–97 (2010)
2. Arroyuelo, D., Navarro, G., Sadakane, K.: Stronger Lempel-Ziv based compressed text indexing. Algorithmica 62(1), 54–101 (2012)
3. Ferrada, H., Navarro, G.: A Lempel-Ziv compressed structure for document listing. In: Kurland, O., Lewenstein, M., Porat, E. (eds.) SPIRE 2013. LNCS, vol. 8214, pp. 116–128. Springer, Heidelberg (2013)
4. Hon, W.-K., Patil, M., Shah, R., Wu, S.-B.: Efficient index for retrieving top-k most frequent documents. J. Discr. Alg. 8(4), 402–417 (2010)
5. Hon, W.-K., Shah, R., Thankachan, S.V.: Towards an optimal space-and-query-time index for top-k document retrieval. In: Kärkkäinen, J., Stoye, J. (eds.) CPM 2012. LNCS, vol. 7354, pp. 173–184. Springer, Heidelberg (2012)
6. Hon, W.-K., Shah, R., Vitter, J.: Space-efficient framework for top-k string retrieval problems. In: Proc. FOCS, pp. 713–722 (2009)
7. Konow, R., Navarro, G.: Faster compact top-k document retrieval. In: Proc. DCC, pp. 351–360 (2013)
8. Kosaraju, R., Manzini, G.: Compression of low entropy strings with Lempel-Ziv algorithms. SIAM J. Comp. 29(3), 893–911 (1999)
9. Munro, I.: Tables. In: Proc. FSTTCS, pp. 37–42 (1996)
10. Navarro, G.: Indexing text using the Ziv-Lempel trie. J. Discr. Alg. 2(1), 87–114 (2004)
11. Navarro, G.: Spaces, trees and colors: The algorithmic landscape of document retrieval on sequences. ACM Comp. Surv. 46(4), article 52 (2014)
12. Navarro, G., Nekrich, Y.: Top-k document retrieval in optimal time and linear space. In: Proc. SODA, pp. 1066–1077 (2012)
13. Navarro, G., Valenzuela, D.: Space-efficient top-k document retrieval. In: Klasing, R. (ed.) SEA 2012. LNCS, vol. 7276, pp. 307–319. Springer, Heidelberg (2012)
14. Clarke, C., Büttcher, S., Cormack, G.: Information Retrieval: Implementing and Evaluating Search Engines. MIT Press (2010)
15. Szpankowski, W.: A generalized suffix tree and its (un)expected asymptotic behaviors. SIAM J. Comp. 22(6), 1176–1198 (1993)
16. Ziv, J., Lempel, A.: Compression of individual sequences via variable-rate coding. IEEE Trans. Inf. Theor. 24(5), 530–536 (1978)

Grammar Compressed Sequences
with Rank/Select Support*

Gonzalo Navarro[1] and Alberto Ordóñez[2]

[1] Dept. of Computer Science, Univ. of Chile, Chile
gnavarro@dcc.uchile.cl
[2] Lab. de Bases de Datos, Univ. da Coruña, Spain
alberto.ordonez@udc.es

Abstract. Sequence representations supporting not only direct access to their symbols, but also rank/select operations, are a fundamental building block in many compressed data structures. In several recent applications, the need to represent highly repetitive sequences arises, where statistical compression is ineffective. We introduce grammar-based representations for repetitive sequences, which use up to 10% of the space needed by representations based on statistical compression, and support direct access and rank/select operations within tens of microseconds.

1 Introduction

Given a sequence $S[1,n]$ drawn over an alphabet $\Sigma = [1,\sigma]$, an intensively studied problem in the past few years has been how to represent S space-efficiently while solving operations $\mathrm{rank}_b(S,i)$ (number of occurrences of b in $S[1,i]$), $\mathrm{select}_b(S,i)$ (i-th occurrence of b in S), and $\mathrm{access}(S,i) = S[i]$. The motivation comes from a wide number of applications involving these functionalities: text indexes, document retrieval, data grids, and many others [25].

The most well-known data structure to solve rank/select/access (rsa) queries is the *wavelet tree* (WT) [18] (with several recent improvements for large alphabets [3,12]). These data structures are able to statistically compress the input sequence while efficiently solving rsa queries. However, they are unable to compress S beyond its statistical entropy.

Although statistical compression is appropriate in many contexts, it is unsuitable in various other domains. This is the case of an increasing number of applications that deal with highly repetitive sequences: compressed software repositories, versioned document collections, DNA datasets of individuals of the same species, and so on, which contain many near-copies of the same source code, document, or genome [24]. In this scenario, statistical compressors, or a

---

* Funded in part by Fondecyt Grant 1-140796, Chile, CDTI EXP 000645663/ITC-20133062 (CDTI, MEC, and AGI), Xunta de Galicia (PGE and FEDER) ref. GRC2013/053, and by MICINN (PGE and FEDER) refs. TIN2009-14560-C03-02, TIN2010-21246-C02-01, TIN2013-46238-C4-3-R and TIN2013-47090-C3-3-P and AP2010-6038 (FPU Program).

E. Moura and M. Crochemore (Eds.): SPIRE 2014, LNCS 8799, pp. 31–44, 2014.
© Springer International Publishing Switzerland 2014

compressed WT, do not take a proper advantage of the repetitiveness [20], which is crucial to reduce the size of those usually huge datasets by orders of magnitude.

Grammar- and Lempel-Ziv-based compressors are very efficient at handling repetitive sequences. However, even supporting operation access is difficult on them. Let $S[1, n]$ be compressible into a grammar of size r, so that a grammar-based compressor uses $r \lg(r + \sigma)$ bits. Bille et al. [5] show how to represent S using $O(r \log n)$ bits so that access(S, i) is solved in $O(\log n)$ time. Let z be the number of phrases into which a Lempel-Ziv parser factors S. Then a Lempel-Ziv compressor achieves $z(\lg n + \lg \sigma)$ bits. Gagie et al. [14] show how to represent S using $O(z \log n \log(n/z))$ bits so that access(S, i) can be supported in time $O(\log n)$. Verbin and Yi [32] show that both times are essentially optimal. Note, however, that the spaces are at best proportional to the size of the compressed string, and that operations rank and select are not supported. This is to be contrasted with, for example, alphabet partitioning techniques [3,4] which obtain asymptotically the same space of a kth-order statistical compression of S, support access in $O(1)$ time, select in almost-constant time (or vice versa), and rank in the optimal $O(\log \frac{\log \sigma}{\log w})$ time on a RAM machine of w bits.

Various scenarios require rsa support on repetitive sequences. Some examples are: document retrieval on repetitive sequence collections, to represent the so-called "document array" [28]; XPath queries on versioned XML data, to represent the sequence of tags [2]; simulating positional inverted indexes on repetitive natural language text collections, by representing the sequence of words [6,15]; and bidirectional navigation of Web graphs, to represent adjacency lists [10].

The only current solution to provide rsa support on repetitive sequences is of practical nature [28]. The key idea is that repetitions in the input sequence S should also induce repetitions in the bitmaps of a WT built on it. This is true at least for the first few levels of the WT, since the WT construction algorithm splits such repetitions as we move downward in the tree. Therefore, if S is grammar-compressible, so are the first bitmaps of the WT. These first levels are compressed with an enhanced Re-Pair (a grammar compressor [21]) representation for bitmaps (RPB [28]) that supports rsa queries in $O(\log n)$ time. The remaining levels, which are not grammar-compressible, are compressed with statistical techniques for bitmaps (RRR [29]) or even not compressed at all (CM [9,23]). Thus, the rsa operations are supported in $O(\log n \log \sigma)$ worst-case time.

This solution, dubbed WTRP, has two main drawbacks: (a) Re-Pair compressed bitmaps RPB [28] are in practice orders of magnitude slower than RRR or CM to support rsa operations ($O(\log n)$ vs $O(1)$ time, in theory), what makes the WTRP significantly slower than a regular WT; (b) the WT construction quickly destroys the repetitiveness of S, and thus the size of the WT can be many times larger than the Re-Pair compressed sequence (there is no theoretical guarantee here).

In this paper we propose two new solutions for rsa queries over grammar compressed sequences. The first one, tailored to sequences over small alphabets, is obtained by enhancing and improving the RPB representation for bitmaps [28]. We dub this solution GCC (Grammar Compression with Counters). This may directly apply, for example, to sequences of XML tags. Our second structure combines GCC with alphabet partitioning (AP) [3] and is aimed to sequences with large alphabets. AP splits the sequence S into subsequences over smaller alphabets, what lets us apply GCC on them (or a simpler and faster representation on the subsequences that are not grammar-compressible).

Our experiments on various real-life repetitive sequences show that our new representations use significantly less space, and are an order of magnitude faster, than WTRP, the only current solution [28]. They are still an order of magnitude slower than statistically compressed representations, but they also use an order of magnitude less space on repetitive sequences. We show, as a concrete application, the improvement obtained by plugging our structure to represent the sequence of tags within SXSI, a system that supports XPath queries on compactly represented XML data, when the collections are repetitive.

2   Basic Concepts and Related Work

2.1   Grammar compression of Sequences and Re-Pair

Grammar-compressing a sequence S means to find a context-free grammar that generates (only) S. Finding the smallest grammar that generates a given sequence S is NP-complete [8], but heuristics like Re-Pair [21] perform very well in practice, in linear time and space. This will be our compressor of choice.

Re-Pair finds the most frequent pair of symbols ab in S, adds a rule $X \rightarrow ab$ to a dictionary R, and replaces each occurrence of ab in S by X. This process is repeated (X can be involved in future pairs) until the most frequent pair appears only once. The result is a tuple (R, C), where the dictionary R contains $r = |R|$ rules and C, of length $c = |C|$, is the final reduction of S after all the replacements carried out. Note that C is drawn from an alphabet of size $\sigma + r$, not only σ. Thus, the total output size is $(2r + c) \lg r$ bits. By using the technique of Tabei et al. [31], we represent the dictionary in $r \log r + O(r)$ bits, reducing the total space to $(r + c) \log r + O(r)$ bits. Finally, it is possible to force the grammar to be *balanced*, that is, that the grammar tree is of height $O(\log n)$ [30].

2.2   Bitmap Representations and RPB

Several classical solutions represent a binary sequence $B[1, n]$ with rsa support. Clark and Munro [9,23] (CM) use $o(n)$ bits on top of B and solve all the queries in $O(1)$ time. Raman et al. [29] (RRR) also support the operations in $O(1)$ time, but they statistically compress B to $nH_0(B) + o(n)$ bits, where $H_0(B)$ is the empirical zero-order entropy of B: if B has m 1s, then $H_0(B) = \frac{m}{n} \lg \frac{n}{m} + \frac{n-m}{n} \lg \frac{n}{n-m}$.

The only solution that exploits the repetitiveness of the bitmap was proposed by Navarro et al. [28] (RPB). They Re-Pair compress B with a balanced grammar and enhance the output (R, C) with extra information to solve rsa queries: Let $exp(X)$ be the string of terminals X expands to; then they store, for each rule $X \rightarrow YZ$, $\ell(X) = |exp(X)|$, the length of $exp(X)$, and $z(X) = rank_0(exp(X), \ell(X))$, the number of 0s in $exp(X)$.

Note that both values can be recursively computed as $\ell(X) = \ell(Y) + \ell(Z)$, with $\ell(0) = \ell(1) = 1$; and $z(X) = z(Y) + z(Z)$, with $z(0) = 1, z(1) = 0$. To save space, they store $\ell(\cdot)$ and $z(\cdot)$ only for a subset of nonterminals, and compute the others recursively by partially expanding the nonterminal. Given a parameter δ, they guarantee that, to compute any $\ell(X)$ or $z(X)$, we have to expand at most 2δ rules. The sampled rules are marked in a bitmap $B_d[1, r]$ and the sampled values are stored in two vectors, S_ℓ and S_z, of length $rank_1(B_d, r)$. To obtain $\ell(X)$ we check whether $B_d[X] = 1$. If so, then $\ell(X) = S_\ell[rank_1(B_d, X)]$. Otherwise $\ell(X)$ is obtained recursively as $\ell(Y) + \ell(Z)$. The process for $z(X)$ is analogous.

Finally, every sth position of B is sampled, for a parameter s. An array $S_n[0, n/s]$ stores a tuple (p, o, rnk) at $S_n[i]$, where the expansion of $C[p]$ contains $B[i \cdot s]$, that is, $p = \max\{j, L(j) \leq i \cdot s\}$, where $L(j) = 1 + \sum_{k=1}^{j-1} \ell(C[k])$; $o = i \cdot s - L(p)$ is the offset within that symbol; and $rnk = rank_0(B, L(p) - 1)$. Let $S[0] = (0, 0, 0)$.

To solve $rank_0(B, i)$, let $S_n[\lfloor i/s \rfloor] = (p, o, rnk)$ and set $l = s \cdot \lfloor i/s \rfloor - o$. Then we move forward from $C[p]$, updating $l = l + \ell(C[p])$, $rnk = rnk + z(C[p])$, and $p = p + 1$, as long as $l + \ell(C[p]) \leq i$. When $l \leq i < l + \ell(C[p])$, we have reached the rule $C[p] = X \to YZ$ whose expansion contains $B[i]$. Then, we recursively traverse X as follows. If $l + \ell(Y) > i$, we recursively traverse Y. Otherwise we update $l = l + \ell(Y)$ and $rnk = rnk + z(Y)$, and recursively traverse Z. This is repeated until $l = i$ and we reach a terminal symbol in the grammar. Finally, we return rnk. Obviously, we can also compute $rank_1(B, i) = i - rank_0(B, i)$. Solving $access(B, i)$ is completely equivalent, but instead of returning rnk we return the terminal symbol we reach when $l = i$.

To solve $select_0(B, j)$, we binary search S_n to find $S_n[i] = (p, o, rnk)$ and $S_n[i + 1] = (p', o', rnk')$ such that $rnk < j \leq rnk'$. Then we proceed as for $rank_0$, but iterating as long as $z + z(C[p]) \leq j$, and then traversing by going left (to Y) when $z + z(Y) > j$, and going right (to Z) otherwise. The process for $select_1(B, j)$ is analogous (note X contains $\ell(X) - z(X)$ 1s).

On a balanced grammar, a rule is traversed in $O(\log n)$ time. The time to iterate over C between samples is $O(s)$. Therefore, the total time for rsa is $O(s + \log n)$ and the total space is $O(r \log n + (n/s) \log n) + c \lg(\sigma + r)$ bits. The time is multiplied by δ if we use sampling.

2.3   Sequence Representations

The wavelet tree [18] (WT) is a complete balanced binary tree that represents a sequence S on $\Sigma = [1, \sigma]$. It is able to statistically compress the sequence and solves rsa queries in $O(\log \sigma)$ time. For large alphabets, a variant called wavelet matrix (WM) [12] performs better in practice. Assume we use a plain encoding of symbols in $\lceil \lg \sigma \rceil$ bits, where $a\langle j \rangle$ the jth most significant bit of $a \in \Sigma$. The WM construction algorithm starts with $S_l = S$ at level $l = 1$ and proceeds as follows: (1) build a single bitmap $B_l[1, n]$ where $B_l[i] = S_l[i]\langle l \rangle$; 2) compute $\tilde{z}_{l+1} = rank_0(B_l, n)$; (3) build sequence S_{l+1} such that, for $k \leq \tilde{z}_{l+1}$, $S_{l+1}[k] = S_l[select_0(B_l, k)]$, and for $k > \tilde{z}_{l+1}$, $S_{l+1}[k] = S_l[select_1(B_l, k - \tilde{z}_{l+1})]$; (4) repeat the process until $l = \lceil \log \sigma \rceil$. This is actually a reshuffling of the bits of $S[i]\langle j\rangle$ for all i and j (akin to radix sorting the symbols of S), with $n\lceil\lg\sigma\rceil$ bits in total (plus $\lg n\lg\sigma$ for the $\tilde{z}_l$). The rsa operations are carried out with one binary rsa operation per level of the WM.

By representing the bitmaps B_l with CM [9,23], the total space is $n\lg\sigma(1+o(1))$ bits and the rsa time is $O(\log\sigma)$. By using RRR bitmap representation [29], the time complexity is retained but the space reduces to $nH_0(S) + O(\sigma\log n)$ bits, although the times are higher in practice. Zero-order compression is also obtained, with faster time in practice, by retaining the CM representation but using a tree with Huffman [19] shape instead of a balanced one, which gives $n(H_0(S) + 1)(1 + o(1)) + O(\sigma\log n)$ bits. The results are called WTH (Huffman-shaped WT) or WMH (Huffman-shaped WM [13]).

An alternative solution for rsa queries over large alphabets is alphabet partitioning (AP) [3], which obtains $nH_0(B) + o(n(H_0(B) + 1))$ bits and solves rsa in $O(\log\log\sigma)$ time. The main idea is to partition Σ into several subalphabets Σ_j, and S into the corresponding subsequences S_j over Σ_j. A string $K[1, n]$ indicates the sequence each symbol of S belongs. Then rsa operations on S are translated into rsa operations on K and on some subsequence S_j. Furthermore, the symbols in each Σ_j are of roughly the same frequency, so that using a fast compact (but not compressed) representation of S_j (GMR) [16] yields $O(\log\log\sigma)$ time and does not ruin the statistical compression of S. The actual implementation defines Σ_j as the set of the 2^{j-1}th to the $(2^j - 1)$th most frequent symbols, and uses WT when this alphabet is small, and GMR when it is large.

The mapping to subalphabets is represented in a sequence $M[1,\sigma]$, where $M[a] = j$ iff $a \in \Sigma_j$. In each subsequence S_j, each $a \in \Sigma_j$ is rewritten as $rank_j(M,a)$, so the local alphabet is $[1, 2^{j-1}]$. Now, to find $S[i]$ we compute $j = K[i]$, $v = S_j[rank_j(K,i)]$, and $S[i] = select_j(M,v)$. To find $rank_a(S,i)$, we compute $j = M[a]$, $v = rank_j(M,a)$, $r = rank_j(K,i)$, and $rank_a(S,i) = rank_v(S_j,r)$. Finally, to find $select_a(S,i)$, we compute $j = M[a]$, $v = rank_j(M,a)$, $s = select_v(S_j,i)$, and $select_a(S,i) = select_j(K,s)$.

2.4   Re-Pair Compressed WT

As far as we know, WTRP [28] (or WMRP if implemented on a WM) is the only solution to support rsa on grammar-compressed sequences. The structure is a WT where all the bitmaps at each level l are concatenated, and then the bitmap B_l of each level l is compressed with RPB [28]. The rationale is that the repetitiveness of S is reflected in the bitmaps of the WT, at least for the first levels. That is because the WT construction splits the alphabet at each level, which potentially blurs the repeated substrings into many shorter repetitions.

Therefore, the bitmaps of the first few WT levels are likely to be compressible with Re-Pair, while the remaining ones are not. The authors [28] use at each level l the technique to represent B_l that yields the least space, RPB, RRR, or CM. In case of a highly compressible sequence, the space can be drastically reduced, but the search performance degrades by one or more orders of magnitude compared to using CM or RRR: If all the levels use RPB, the rsa time complexities become

$O(\log \sigma(s + \log n))$. On the other hand, as repetitiveness is destroyed at deeper levels, the total space is far from that of a plain Re-Pair compression of S.

3   Efficient rsa for Sequences on Small Alphabets

Our first proposal, dubbed GCC *(Grammar Compression with Counters)* is aimed at solving rsa queries on grammar-compressed sequences with small alphabets. We generalize the existing solution for bitmaps (RPB, Section 2.2), to sequences with $\sigma > 2$. Besides, we introduce several enhancements that improve its space usage.

Let (R, C) be the result of a balanced Re-Pair grammar compression of S. We store $S_\ell[X] = \ell(X)$ for each grammar rule $X \in R$. In addition, we store a sequence of counters $S_a[X]$ for each symbol $a \in \Sigma$: $S_a[X] = rank_a(exp(X), \ell(X))$ is the number of occurrences of a in $exp(X)$.

The input sequence S is also sampled according to the new scenario: each element (p, o, rnk) of $S_n[1, n/s]$ is now replaced by $(p, o, lrnk[1, \sigma])$, where $lrnk[a] = rank_a(S, L(p) - 1)$ for all $a \in \Sigma$, s being the sampling period.

The rsa algorithms stay practically the same as for RPB; now we use the symbol counter of a for $rank_a$ and $select_a$. The resulting data structure solves rsa in time $O(s + \log n)$ and takes $O(r\sigma \log n + \sigma(n/s) \log n) + c\lg(\sigma + r)$ bits.

The extra space incurred by σ can be reduced by using the same δ-sampling of RPB, which increases the time by a factor δ. In this case we also use the bitmap $B_d[1, r]$ that marks which rules store counters. We further reduce the space by noting that many rules are short, and therefore the values in S_ℓ and S_a are usually small. We represent them using direct access codes (DACs [7]), which store variable-length numbers while retaining direct access to them. The o components of S_n are also represented with DACs for the same reason.

On the other hand, the p and $lrnk[1, \sigma]$ values are not small but increasing. We reduce their space using a two-layer strategy: we sample S_n at regular intervals of length ss. We store $SS_n[j] = S_n[j \cdot ss]$, and then represent the values of $S_n[i] = (p, o, lrnk[1, \sigma])$ in differential form, in array $S'_n[i] = (p', o, lrnk'[1, \sigma])$, where $p' = p - p^*$ and $lrnk'[a] = lrnk[a] - lrnk^*[a]$, with $SS_n[\lfloor i/ss \rfloor] = (p^*, o^*, lrnk^*[1, \sigma])$. The total space for the p and $lrnk[1, \sigma]$ components is $O(\sigma(n/s) \log(s \cdot ss) + (n/(s \cdot ss)) \log n)$ bits. For example, if we use $ss = \lg n$ and $s = \log^{O(1)} n$ (a larger value would imply an excessively high query time), the space becomes $O(r\sigma \log n + \sigma(n/s) \log \log n) + c\lg(\sigma + r)$ bits. This can be reduced to $O((r\sigma + c) \log n)$ bits by sampling regularly C instead of S and using $s = \Theta(\log n)$, but the described sampling works better in practice.

When σ is small, this data srtucture is very space- and time-efficient. It compress better than WTRP [28] since it does not destroy the repetitiveness of S when building the wavelet tree. Besides, it runs faster compared to the $O(\log \sigma \log n)$ time obtained by WTRP: we need just one operation on GCC, not $\log \sigma$ operations on RPB. However, this solution becomes prohibitive when the alphabet becomes large since it has a σ multiplicative term in the space.

4   Efficient rsa for Sequences on Large Alphabets

For large alphabets, our idea is to combine GCC with AP [3] (Section 2.3), which splits $S[1, n]$ into a sequence $K[1, \lg \sigma]$ of classes and $\lg \sigma$ subsequences $S_{[1, \log \sigma]}$. That is, AP partitions the original sequence into subsequences over smaller alphabets, which is the scenario GCC handles well.

Note that, if S is grammar-compressible, then K is grammar-compressible as well, as K consists of a (non-injective) mapping of the symbols of S. It is also reasonable to expect that the subsequences S_j grammar-compress well, at least for the first levels (i.e., the most frequent symbols): If ab is a frequent pair in S, then it is expected that they are frequent individually as well. As a consequence, it is likely that a and b belong to the same first classes. Even for the less frequent symbols, if they appear frequently together, then their individual frequencies are likely to be similar, and thus they have a good chance to be assigned to the same class. If the most frequent pairs of symbols ab are assigned to the same subsequence S_j, then all the space saved by the rule $X \to ab$ is also saved if choosing the same rule when grammar-compressing S_j.

We apply GCC to K and to the first sequences S_j, since they have a small alphabet. For the remaining subsequences we have two choices: (a) represent them using GMR (APRep, recall that subsequences S_j are not statistically compressible [3]); or (b) attempt to grammar-compress them using WMRP (APRep-WMRP). Which is better depends on whether the subsequences on large alphabets (which contain less frequent symbols) are still repetitive or not. While the choice (b) yields higher times than (a), we note that, if queries have the same statistical distribution of the symbols in S, then most queries will refer to more frequent symbols, which will be handled with the fast GCC representation.

5   Experimental Results and Discussion

We used an Intel(R) Xeon(R) E5620 at 2.40GHz with 96GB of RAM memory, running GNU/Linux, Ubuntu 10.04, with kernel 2.6.32-33-server.x86_64. All our implementations use a single thread and are coded in C++. The compiler is g++ version 4.6.3, with -O9 optimization. We implemented our solutions inside LIBCDS (github.com/fclaude/libcds) and use Navarro's implementation of Re-Pair (www.dcc.uchile.cl/gnavarro/software/repair.tgz).

Table 1 shows statistics of interest about the datasets used and their compressibility: length (n), alphabet size (σ), zero-order entropy (H_0), bits per symbol (bps) obtained by Re-Pair (RP, assuming $(2r + c)\lceil \lg(\sigma + r) \rceil$ bits), and bps obtained by p7zip (LZ, www.7-zip.org), a Lempel-Ziv compressor.

5.1   Results for Small Alphabets

To test our structure on small alphabets, we use some DNA datasets (para and escherichia) from *PizzaChili Repetitive Corpus* (pizzachili.dcc.uchile.cl/repcorpus), and influenza from the *SuDS Project* (www.cs.helsinki.fi/group/suds/rlcsa/data/fiwiki.bz2). From *SuDS* we also extract fiwikitags, the sequence of opening and closing tags from a subset of the Finnish Wikipedia.

Table 1. Statistics of the datasets. Length n is measured in millions (and rounded).

dataset	n	σ	H_0	RP	LZ	dataset	n	σ	H_0	RP	LZ
para	429	5	1.12	0.37	0.19	software	37	48	3.23	0.08	0.47
influenza	322	16	1.98	0.23	0.15	einstein	17	8,046	9.91	0.08	0.04
escherichia	113	15	2.00	1.04	0.52	fiwiki	84	99,797	11.04	0.24	0.16
fiwikitags	49	24	3.36	0.11	0.32	indochina	50	685,100	13.94	0.88	0.32

We show results for GCC, using sampling steps $s = 2^{\{10,12,14\}}$ and supersteps $ss = 2^{\{4,6,8\}}$ for C, and $\delta = \{0, 2, 4, 8\}$ for R. We also compare WMRP [28], which takes, for the bitmap of each level, the representation using least space between RPB, RRR (with sampling value 32), and CM (a simple implementation [17] with sampling value 32). We also include in the comparison the WTH (Huffman-compressed WT), as a good statistically compressed solution for rsa. For the WTH bitmaps we use RRR with sampling steps in $\{32, 64, 128\}$.

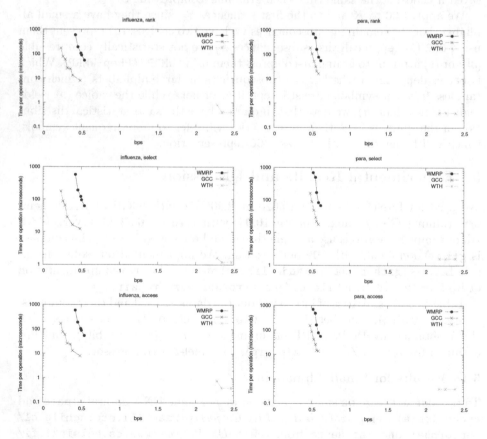

Fig. 1. Space-time tradeoffs for rsa queries over small alphabets: collections influenza and para (note logscale in time)

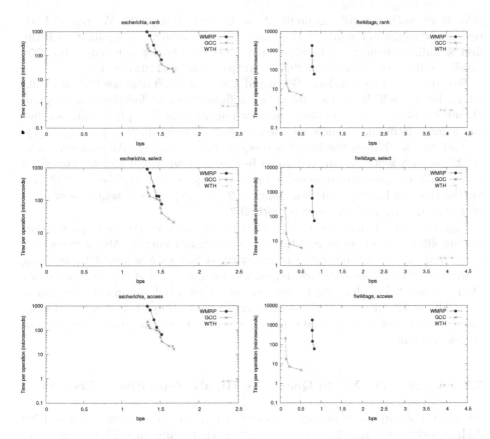

Fig. 2. Space-time tradeoffs for `rsa` queries over small alphabets: collections `escherichia` and `fiwikitags` (note logscale in time)

Figures 1 and 2 show the results for all the operations and collections. Our GCC dominates WMRP both in space and in `rsa` time. The difference in space with WMRP is larger as the sequence is more grammar-compressible (see Table 1). This is because GCC preserves all the repetitiveness of S, while paying a price only in terms of the alphabet size. Instead, WMRP destroys the repetitiveness after a few wavelet tree levels. In terms of `rsa` performance, GCC is up to two orders of magnitude faster than WMRP for the same space usage. Note that the collection in which GCC and WMRP are closest is `escherichia`, the least repetitive one.

On the other hand, the representation that compresses statistically, WTH, is about an order of magnitude faster than GCC, but it also takes many times more space (up to 10 times in case of `fiwikitags`).

5.2 Results for Large Alphabets

For large alphabets, we use collection `einstein` (also from *PizzaChili*), which contains Wikipedia versions of the article about Einstein in German, and `fiwiki` (also from *SuDS*), a 400MB prefix of the Finnish Wikipedia. We regard both texts as sequences of words. A third collection is indochina, a subset with the first 50 million elements of the adajacency lists of the Web graph Indochina2004 (available from the *WebGraph Project*, http://law.dsi.unimi.it).

We study our two solutions, APRep and APRep-WMRP. These use GCC and WMRP internally, for which we use the same configurations as for the case of small alphabets. Besides, we introduce two new parameters: $\beta \in \{2, \ldots, 10\}$, so that the β most frequent symbols are directly stored in K [3], and $f \in \{2, \ldots, 7\}$, so that we use GCC on the first f subsequences, $S_1, \ldots, S_f$. We compare these solutions with WMRP, parameterized as before, and with WMH and AP, two good statistically-compressing representations for large alphabets. We use RRR [29] for the the WMH bitmaps with samplings $\{32, 64, 128\}$. The K sequence of AP is represented with a WT and each S_j with GMR.

Figures 3 and 4 show the results. APRep-WMRP obtains the best space, dominating WMRP in both space and time by a significant margin. APRep takes over when more space is used, being up to twice as fast as APRep-WMRP (yet using twice the space). The statistical representations are, as before, up to an order of magnitude faster than our fastest representations, but use much more space, especially on the most repetitive collections. In those, they are two orders of magnitude faster, but use up to 10 times more space, than our most space-efficient representations.

5.3   Application: XPath Queries on Highly Repetitive Collections

We show the impact of our new representations in the indexing of repetitive XML collections. SXSI [2] is a recent system that represents XML datasets in compact form and solves XPath queries on them. Its query processing strategy uses a tree automaton that traverses the XML data, using several queries on the content and structure to speed up navigation towards the points of interest. SXSI represents the XML data using three separate components: (1) a text index that represents and carries out pattern searches over the text nodes (any compressed full-text index [26] can be used); (2) a balanced parentheses representation of the XML topology that supports navigation using $2 + o(1)$ bits per node (various alternatives exist [1]); and (3) an rsa-capable representation of the sequence of the XML opening and closing tags, using some sequence representation.

When the XML collection is repetitive (e.g., versioned collections like Wikipedia, versioned software repositories, etc.), one can use the RLCSA [22], a full-text index that performs well on a repetitive collection of text nodes, for (1). Components (2) and (3), which are usually less relevant in terms of space, may become dominant if they are represented without exploiting repetitiveness. For (2), we compare GCT, a tree representation aimed at repetitive topologies [27], with a classical representation (FF [1]). For (3), we will use our new repetition-aware sequence representations, comparing them with the alternative proposed in SXSI (MATRIX, using one compressed bitmap per tag) and a WTH representation. All variants will use the RLCSA with no text sampling as their text index.

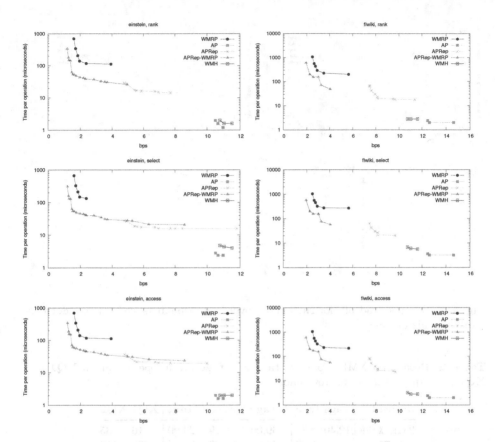

Fig. 3. Space-time tradeoffs for `rsa` queries over large alphabets: collections `einstein` and `fiwiki` (note logscale in time)

We use a repetitive data-centric XML collection of 200MB from a real software repository. Its sequence of XML tags, called `software`, is described in Table 1. We run two XPath queries that make intensive use of the sequence of tags and the tree topology: `XQ1=//class[//methods]`, and `XQ2=//class[methods]`.

Table 2 shows the space in bpe (bits per element) of components (2) and (3). An element here is an opening or a closing tag, so there are two elements per XML tree node. The space of the `RLCSA` is always 2.3 bits per character of the XML document. The table also shows the impact of each component in the total size of the index. Finally, the table shows the time to solve both queries.

The original `SXSI` (`MATRIX+FF`) is very fast but needs almost 14 bpe, which amounts to over 75% of the index space in this repetitive scenario (in non-repetitive text-centric XML, this space is negligible). By replacing the `MATRIX` by a `WTH`, the space drops significantly, to slightly over 4 bpe, yet times degrade by a factor of 3–6. By using our `GCC` for the tags, a new significant space reduction is obtained, to 2.65 bpe, and the times increase by a factor of 4–5, becoming 13–28 times slower than the original `SXSI`. Finally, changing `FF` by `GCT` [27], we can

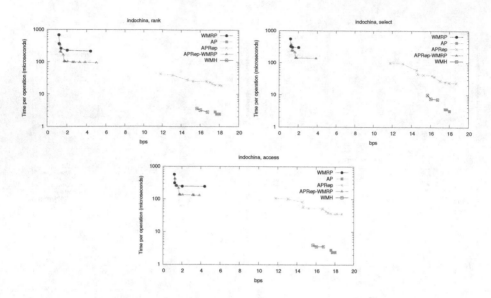

Fig. 4. Space-time tradeoffs for `rsa` queries over large alphabets: collection `indochina` (note logscale in time)

Table 2. Results on XML. Columns **tags** and **tree** are in **bpe**. Columns **XQ1** and **XQ2** show query time in microseconds.

dataset	tags	tree	%tags	%tree	%text	XQ1	XQ2
MATRIX+FF	12.40	1.27	69.00	7.19	23.90	16	35
WTH+FF	2.88	1.27	34.07	15.09	50.84	92	113
GCC+FF	0.37	1.27	6.26	21.45	72.29	442	462
GCC+GCT	0.37	0.19	7.66	3.93	88.42	1,032	3,302

reach as low as 0.56 bpe, 24 times less than the original SXSI, and using less than 12% of the total space. Once again, the price is the time, which becomes 65–95 times slower than the basic SXSI. The price of using the slower GCT is more noticeable on XQ2, which requires more operations on the tree.

While the time penalty is 1–2 orders of magnitude, we note that the gain in space can make the difference between running the index in memory or on disk; in the latter case we can expect it to be up to 6 orders of magnitude slower. On the other hand, the time differences will blur on queries that do not only access the tags and the tree, but also involve the text, as these cost the same in all the representations. Finally, we note that the RLCSA becomes the space bottleneck in GCC+GCT. It is worthwhile to consider even more compressed text representations, for example based on grammars [11] or on LZ77 [20].

6   Final Remarks

Our new ideas permit much more exploration. We have used the same partitioning into sequences given in the alphabet partitioning work [3], with alphabets of doubling sizes. However, other partitionings may be more suitable to our needs, for example building all the subsequences with the same alphabet size ρ, so that alphabet $[1, \rho]$ can be comfortably handled with our basic method for small alphabets. This may induce a hierarchy of classes, instead of two levels as in alphabet partitioning [3]. The result would be indeed a ρ-ary version of the current (2-ary) wavelet-tree based solution [28], which may reduce space and time by increasing the arity. Furthermore, we plan to study heuristics for grouping symbols into classes, aiming to avoid separating symbols that form long repeated substrings, so that fewer repetitions are destroyed when forming the classes.

A more far-fetched goal is to achieve Lempel-Ziv compressed representations that support these operations. Lempel-Ziv is more powerful than grammar compression, but thought to be harder to handle even for supporting direct access.

References

1. Arroyuelo, D., Cánovas, R., Navarro, G., Sadakane, K.: Succinct trees in practice. In: Proc. ALENEX, pp. 84–97 (2010)
2. D. Arroyuelo, F. Claude, S. Maneth, V. Mäkinen, G. Navarro, K. Nguyễn, J. Sirén, and N. Välimäki. Fast in-memory xpath search over compressed text and tree indexes. In: Proc. 26th ICDE, pp. 417–428 (2010)
3. Barbay, J., Claude, F., Gagie, T., Navarro, G., Nekrich, Y.: Efficient fully-compressed sequence representations. Algorithmica 69(1), 232–268 (2014)
4. Belazzougui, D., Navarro, G.: New lower and upper bounds for representing sequences. In: Epstein, L., Ferragina, P. (eds.) ESA 2012. LNCS, vol. 7501, pp. 181–192. Springer, Heidelberg (2012)
5. Bille, P., Landau, G., Raman, R., Sadakane, K., Rao Satti, S., Weimann, O.: Random access to grammar-compressed strings. In: Proc. 22nd SODA, pp. 373–389 (2011)
6. Brisaboa, N., Fariña, A., Ladra, S., Navarro, G.: Implicit indexing of natural language text by reorganizing bytecodes. Inf. Retr. 15(6), 527–557 (2012)
7. Brisaboa, N., Ladra, S., Navarro, G.: DACs: Bringing direct access to variable-length codes. Inf. Proc. Manag. 49(1), 392–404 (2013)
8. Charikar, M., Lehman, E., Liu, D., Panigrahy, R., Prabhakaran, M., Sahai, A., Shelat, A.: The smallest grammar problem. IEEE Trans. Inf. Theor. 51(7), 2554–2576 (2005)
9. Clark, D.: Compact Pat trees. PhD thesis, Univ. of Waterloo, Canada (1998)
10. Claude, F., Navarro, G.: Extended compact web graph representations. In: Elomaa, T., Mannila, H., Orponen, P. (eds.) Ukkonen Festschrift 2010. LNCS, vol. 6060, pp. 77–91. Springer, Heidelberg (2010)
11. F. Claude and G. Navarro. Improved grammar-based compressed indexes. In *Proc. 19th SPIRE*, LNCS 7608, pages 180–192, 2012.
12. Claude, F., Navarro, G.: The wavelet matrix. In: Calderón-Benavides, L., González-Caro, C., Chávez, E., Ziviani, N. (eds.) SPIRE 2012. LNCS, vol. 7608, pp. 167–179. Springer, Heidelberg (2012)

13. Claude, F., Navarro, G., Ordóñez, A.: The wavelet matrix: An efficient wavelet tree for large alphabets. Information Systems (to appear, 2014)
14. Gagie, T., Gawrychowski, P., Kärkkäinen, J., Nekrich, Y., Puglisi, S.J.: LZ77-based self-indexing with faster pattern matching. In: Pardo, A., Viola, A. (eds.) LATIN 2014. LNCS, vol. 8392, pp. 731–742. Springer, Heidelberg (2014)
15. Gagie, T., Navarro, G., Puglisi, S.J.: New algorithms on wavelet trees and applications to information retrieval. Theor. Comp. Sci. 426-427, 25–41 (2012)
16. Golynski, A., Munro, I., Rao, S.: Rank/select operations on large alphabets: a tool for text indexing. In: Proc. 17th SODA, pp. 368–373 (2006)
17. González, R., Grabowski, S., Mäkinen, V., Navarro, G.: Practical implementation of rank and select queries. In: Poster Proc. 4th WEA, pp. 27–38 (2005)
18. Grossi, R., Gupta, A., Vitter, J.: High-order entropy-compressed text indexes. In: Proc. 14th SODA, pp. 841–850 (2003)
19. Huffman, D.A.: A method for the construction of minimum-redundancy codes. Proceedings of the I.R.E. 40(9), 1098–1101 (1952)
20. Kreft, S., Navarro, G.: On compressing and indexing repetitive sequences. Theor. Comp. Sci. 483, 115–133 (2013)
21. Larsson, J., Moffat, A.: Off-line dictionary-based compression. Proc. of the IEEE 88(11), 1722–1732 (2000)
22. Mäkinen, V., Navarro, G., Sirén, J., Välimäki, N.: Storage and retrieval of highly repetitive sequence collections. J. Comp. Biol. 17(3), 281–308 (2010)
23. Munro, I.: Tables. In: Proc. 16th FSTTCS, pp. 37–42 (1996)
24. Navarro, G.: Indexing highly repetitive collections. In: Smyth, B. (ed.) IWOCA 2012. LNCS, vol. 7643, pp. 274–279. Springer, Heidelberg (2012)
25. Navarro, G.: Wavelet trees for all. J. Discr. Alg. 25, 2–20 (2014)
26. Navarro, G., Mäkinen, V.: Compressed full-text indexes. ACM Comp. Surv. 39(1), article 2 (2007)
27. Navarro, G., Ordóñez, A.: Faster compressed suffix trees for repetitive text collections. In: Gudmundsson, J., Katajainen, J. (eds.) SEA 2014. LNCS, vol. 8504, pp. 424–435. Springer, Heidelberg (2014)
28. Navarro, G., Puglisi, S.J., Valenzuela, D.: Practical compressed document retrieval. In: Pardalos, P.M., Rebennack, S. (eds.) SEA 2011. LNCS, vol. 6630, pp. 193–205. Springer, Heidelberg (2011)
29. Raman, R., Raman, V., Srinivasa Rao, S.: Succinct indexable dictionaries with applications to encoding k-ary trees, prefix sums and multisets. ACM Transactions on Algorithms 3(4), article 43 (2007)
30. Sakamoto, H.: A fully linear-time approximation algorithm for grammar-based compression. J. Discr. Alg. 3(2-4), 416–430 (2005)
31. Tabei, Y., Takabatake, Y., Sakamoto, H.: A succinct grammar compression. In: Fischer, J., Sanders, P. (eds.) CPM 2013. LNCS, vol. 7922, pp. 235–246. Springer, Heidelberg (2013)
32. Verbin, E., Yu, W.: Data structure lower bounds on random access to grammar-compressed strings. In: Fischer, J., Sanders, P. (eds.) CPM 2013. LNCS, vol. 7922, pp. 247–258. Springer, Heidelberg (2013)

Algorithms for Jumbled Indexing, Jumbled Border and Jumbled Square on Run-Length Encoded Strings

Amihood Amir[1,*], Alberto Apostolico[2], Tirza Hirst[3], Gad M. Landau[4,**],
Noa Lewenstein[5], and Liat Rozenberg[6]

[1] Bar-Ilan University and Johns Hopkins University
[2] Georgia Tech and IASI, CNR
[3] Machon Tal
[4] University of Haifa and NYU
[5] Netanya College
[6] University of Haifa

Abstract. *Jumbled Indexing*, the problem of indexing a text for histogram queries, has been of much interest lately. In this paper we consider jumbled indexing for run-length encoded texts. We refute a former conjecture and show an algorithm for general sized alphabets. We also consider *Jumbled Borders*, the extension of borders to jumbled strings. Borders are the basis for various algorithms. Finally, we consider *Jumbled Squares*, strings which are of the form $x\bar{x}$, where $\bar{x}$ is a jumbling of x. We show efficient algorithms for these problems.

1 Introduction

In this paper we investigate jumbled (abelian) versions of three classical strings problems: pattern matching, string borders and string squares.

Jumbled Pattern Matching and Jumbled Indexing: The **Jumbled Pattern Matching** problem is an important extension of the standard pattern matching problem. The problem asks to decide whether any permutation of a pattern P occurs in a string S. In other words, whether there is a substring of size $|P|$ in S, where each alphabet letter appears the same number of times as in P. This problem has many applications in computational biology, such as interpretation of mass spectrometry data [6], alignment [4], SNP discovery [5], repeated pattern discovery [13] and metabolic network analysis [19].

While the case of a single query can be easily and efficiently solved using a sliding window approach, the *indexed* version of the problem, where the string S is fixed and we need to answer many queries on S, is much more difficult, even for a binary alphabet.

In the last few years many papers have focused on the indexed version of the problem on binary alphabet strings. In [10], the authors proposed an algorithm that constructs a data structure of size $O(n)$ in $O(n^2)$-time, where n is the size of S, and answer queries in $O(1)$ time. Later, Moosa and Rahman [20] and Burcsi *et al.* [6] independently improved the construction time to $O(n^2/\log n)$ (see also [7,8]). Then, Moosa

---

* Partly supported by NSF grant CCR-09-04581, ISF grant 347/09, and BSF grant 2008217.
** Partially supported by the National Science Foundation Award 0904246, Israel Science Foundation grants 347/09 and 571/14,Yahoo, Grant No. 2008217 from the United States-Israel Binational Science Foundation (BSF) and DFG.

E. Moura and M. Crochemore (Eds.): SPIRE 2014, LNCS 8799, pp. 45–51, 2014.
© Springer International Publishing Switzerland 2014

and Rahman [21] further improved it to $O(n^2/(\log n)^2)$-time in the RAM model. Recently, Hermelin *et al.* [17] gave an $n^2/2^{\Omega(\log n/\log\log n)^{1/2}}$ time solution. Badkobeh *et al.* [3] used the run-length encoded string of S (see Section 2) to construct a data structure L of size $min(n, r^2)$, where r is the number of runs in S, in time $O(r^2 \log r)$, and query time $O(\log|L|)$. Recently, Giaquinta and Grabowski [16] proposed some time and space tradeoffs based on Badkobeh *et al.* data structure and Cicalese *et al.* [11] proposed an $O(n)$-space index with $O(\log n)$ query time. In [14] a grammar-based construction has been proposed. In [11,15] the binary jumbled pattern matching was investigated on trees and tree-like structures.

Lately, there has been some progress also for non-binary alphabets. Kociumaka et al. [18] presented a solution for jumbled indexing for any constant-sized alphabet Σ that uses $O(\frac{n^2 \log^2\log n}{\log n})$ preprocessing time and space and answers queries in $O((\frac{\log n}{\log\log n})^{2|\Sigma|-1})$ time. Amir et al. [1] proposed a solution for constant-sized alphabets that preprocesses in $O(n^{1+\epsilon})$ time and answers queries in $\tilde{O}(m^{\frac{1}{\epsilon}})$ time, where m is the sum of the Parikh vector elements. In an even newer paper, Durocher et al. [12] considered alphabet size $|\Sigma| = o((\frac{\log n}{\log\log n})^2)$ and showed how to construct an index in $O(|\Sigma|(\frac{n}{\log_{|\Sigma|} n})^2)$ time and answer queries in $O(n^\epsilon + |\Sigma|)$ time, where $\epsilon > 0$ is an arbitrary small constant.

This still leaves us in a sad state of affairs. In all the (exact) solutions mentioned for $|\Sigma| \geq 3$ the time complexity of preprocessing or the time complexity of querying is always within polylogarithmic factors of one of the above two naive algorithms.

The question that has troubled the community in these last few years is whether jumbled indexing could be solved with $O(n^{2-\epsilon})$ preprocessing time and $O(n^{1-\delta})$ for some constants $\epsilon, \delta > 0$. In a recent result [2] it was shown that for alphabets of $\omega(1)$ size this is impossible under a 3SUM-hardness assumption. They further show that for any constant alphabet size $r \geq 3$ there exist describable fixed constants ϵ_r and δ_r such that jumbled indexing requires $\Omega(n^{2-\epsilon_r})$ preprocessing time or $\Omega(n^{1-\delta_r})$ query time under a stronger 3SUM-hardness assumption.

In this paper we propose an algorithm for the *general* alphabet case (as opposed to the binary case of Badkobeh *et al.* [3]) that uses the run-length encoded string of S to construct a data structure of size $O(r^2|\Sigma|)$ in time $O(r^2(\log r + |\Sigma|\log|\Sigma|)$, and answer queries in $O(|\Sigma|^3 \log r)$-time. Note that, as opposed to former algorithms, we can support queries that return all matches rather than only answering an existential query. We do so in $O(|\Sigma|^3 \log r + occ)$-time where occ is the number of answers. In the full version of this paper, we will outline a solution which works without the $|\Sigma|\log|\Sigma|$ overhead as well.

We also refute a conjecture that was left open in [3].

In addition to the Jumbled Indexing problem, we also studied the **Jumbled Borders** and **Jumbled Squares** problems. Due to a lack of space, the discussion of these problems is left to the full journal version.

2   Definitions and Notations

Let $S[1..n]$ be a string of length n over alphabet Σ, and $S[i..j]$ substring of $S[1..n]$ that starts at index i and ends at index j. We denote by $|S[i..j]|_\sigma$ the number of occurrences of the letter $\sigma \in \Sigma$ in substring $S[i..j]$. An alphabet histogram of S is a $|\Sigma|$-length array $< a_1, a_2, \ldots, a_{|\Sigma|} >$ such that $a_i = |S|_{\sigma_i}$. Two strings S and T are said to *jumble match* if they have the same alphabet histogram.

Let $S'[1..r]$ be the *run-length encoded string* of $S[1..n]$. S' is a sequence of ordered pairs (σ, i), denoted also as a symbol σ^i, where σ is an alphabet letter and i is a positive integer. Each symbol σ^i corresponds to a *run* in S consisting of i consecutive occurrences of the letter σ. For example, if $S = aaabbaacbbbcc$, its run-length encoded string $S' = < (a,3), (b,2), (a,2), (c,1), (b,3), (c,2) >$. For simplicity of reading, for the examples we will use compressed notation, e.g. $S = a^3 b^2 a^2 c^1 b^3 c^2$. Note that r, the size of the run-length encoded string, can be significantly shorter than the size of the original string.

The definition of $|S[i..j]|_\sigma$ is not as useful for run-length encoded strings. We specifically give alternative definitions for run-length encoded strings. We start with a definition for a run k. Then we generalize it to contiguous runs (k to l).

$$C^{[k]}(\sigma) = \begin{cases} q & \text{if } S'[k] = (\sigma, q) \\ 0 & \text{o/w, (i.e. } \sigma \notin \text{ run } k) \end{cases}$$

$$C^{[k,\ell]}(\sigma) = \Sigma_{i=k}^{\ell} C^{[i]}(\sigma)$$

3   The Jumbled Indexing Problem

The *jumbled indexing* problem is defined as follows.

Input: A string $S[1..n]$ in run-length format $S'[1..r]$ to be indexed.
Query input: A pattern $P[1..m]$, represented by its histogram array $< a_1, a_2, \ldots, a_{|\Sigma|} >$
Query output: All substrings $S[i,j]$ of S such that $S[i,j]$ and P jumble-match.

We will show a couple of results in this section. First of all, in [3] an algorithm for the binary alphabet was given and it was suggested that this may be optimal. We show a counterexample to this conjecture. Secondly, we propose an algorithm for a general alphabet.

3.1   Counterexample to Conjecture of [3]

In [3] a data structure was constructed for a binary string that is represented in run-length encoding. The data structure is based upon a function defining a collection of points, which we describe now.

Let S be a string of length n over binary alphabet $\{a, b\}$. Consider $S'[1 \ldots r]$, the run length representation of S. Let *max-b(i)* denote the maximal number of b's over all substrings of S that contain exactly i a's.

Example 1. Let $S = aabbabbaabbbab$. In this case:

$$max\text{-}b(0) = 3$$
$$max\text{-}b(1) = 4$$
$$max\text{-}b(2) = 5$$
$$max\text{-}b(3) = 7$$
$$max\text{-}b(4) = 8$$
$$max\text{-}b(5) = 8$$
$$max\text{-}b(6) = 8$$

It can easily be seen that the function *max-b* is a non-decreasing function.

In [3] the authors defined 2D points, based upon the minimal *max-b* values. That is in our example we have 2D points $(0,3), (1,4), (2,5), (3,7), (4,8)$. In [3] an algorithm was presented that runs as a near-linear function of the number of these points. The number of points is easily shown to be $O(\min(n, r^2))$. The authors point out in the introduction that "preliminary investigations have indicated that $|L|$ (the number of points) may often be close to r". We show a counterexample to this conjecture, showing a lower bound matching the upper bounds, i.e. $\Omega(\min(n, r^2))$.

Claim. The image of *max-b* (i.e. all the possible answers) can be of size $\Omega(r^2)$ (and can be $\Omega(|S|)$).

Proof. Let $S = (ba)^n (b^n a^n)^n$. It is straightforward to verify that $max\text{-}b(i) = n + i$. This happens because the image of *max-b* is: $\{k \mid n \le k \le n^2 + n\}$ which is of size n^2 whereas the text is of size $|S| = 2n^2 + 2n$ and the number of runs is $r = 4n$.   □

3.2   Background for the Algorithm

Say we are examining the matches starting in run k and ending in run ℓ. Then for any character σ the number of σ's is (obviously) upper-bounded by $C^{[k,\ell]}(\sigma)$ and lower-bounded by $C^{[k+1,\ell-1]}(\sigma)$. This yields the following range.

$$R^{[k,\ell]}(\sigma) = [C^{[k+1,\ell-1]}(\sigma), C^{[k,\ell]}(\sigma)]$$

The following *vectors array* contains the ranges for all σ.

$$V^{[k,\ell]} = < R^{[k,\ell]}(\sigma_1), R^{[k,\ell]}(\sigma_2), \cdots, R^{[k,\ell]}(\sigma_{|\Sigma|}) >.$$

Observe that if $C^{[k]}(\sigma) = 0$ and $C^{[\ell]}(\sigma) = 0$ then $C^{[k,\ell]} = C^{[k+1,\ell-1]}$ and, in turn, $R^{[k,\ell]}(\sigma)$ is a *singleton*. Hence, for every k and l such that $k \le \ell$, there are either one or two ranges in $V^{[k,\ell]}$ that are not singletons (specifically, the letters of runs k and ℓ). Lets define $V'^{[k,\ell]}$ to be the vector $V^{[k,\ell]}$ with the non-singleton ranges replaced with the value -1.

For example consider $S = a^3 c^2 b^2 a^5 d^8 b^4 e^7 c^2 b^3 c^2 a^1 c^2 b^4$. Now consider runs $k = 2$ and $\ell = 11$, emphasized by $S = a^3$ **c^2** $b^2 a^5 d^8 b^4 e^7 c^2 b^3 c^2$ **a^1** $c^2 b^4$. Now $V^{[2,11]} = < [5,6], [9,9], [4,6], [8,8], [7,7] >$ which can be shorthanded to $V^{[2,11]} = < [5,6], 9, [4,6], 8, 7 >$. Therefore, $V'^{[2,11]} = < -1, 9, -1, 8, 7 >$.

3.3   Algorithm Outline

A jumbled indexing query P is given as a histogram array, $< a_1, \ldots, a_{|\Sigma|} >$. P and a substring starting within run k and ending within run ℓ can jumble-match if each a_j, the number of σ_j's in P, satisfies $a_j = R^{[k,\ell]}(\sigma_j)$, for singleton values, or $a_j \in R^{[k,\ell]}(\sigma_j)$, for non-singleton ranges.

If we adapt the histogram array of P by replacing the two ranges that correspond to the letters of runs k and ℓ with -1, then a comparison between the adapted histogram array and $V'^{[k,\ell]}$ checks all the singleton values.

For the two non-singleton ranges, assume σ_j is the letter of run k and $\sigma_{j'}$ is the letter of run ℓ, then we need to verify whether $a_j \in R^{[k,\ell]}(\sigma_j)$ and $a_{j'} \in R^{[k,\ell]}(\sigma_{j'})$. This is equivalent to a classical computational geometry problem of planar point location, which is defined as: given a set of rectangles in the plane and a 2D point q, find all rectangles that contain q. This kind of queries are also known as point enclosure, or stabbing queries. Chazelle [9] proposed a data structure for this problem that requires $O(n)$ space (for n rectangles) and supports queries in time $O(\log n + t)$, where t is the size of the output. See Figure 1 for an example of this data structure.

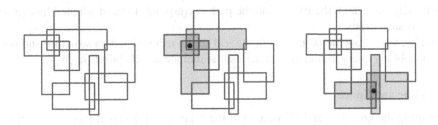

Fig. 1. (a) A collection of rectangles, (b) and (c) rectangles stabbed (at two example points)

In our case, we can ask whether the 2D point $(a_j, a_{j'})$ lies within the rectangle $R^{[k,\ell]}(\sigma_j) \times R^{[k,\ell]}(\sigma_{j'})$. Note that in the case of one non-singleton range we get a line, which is a special case of a rectangle.

Therefore, our algorithm consists of constructing an index that supports efficient comparison of singleton values and answering stabbing queries. We now detail the index construction and then detail the query.

3.4 Vectors Index

In a preprocessing stage we first compute all vectors $V^{[k,\ell]}$ and $V'^{[k,\ell]}$ for all pairs of runs k and ℓ ($k \leq \ell \leq r$). Note that two or more pairs of runs will share the same V' vector, if (a) the non-singleton ranges are on the same letters, and (b) all other letters have the same singleton values. For example, consider $S = a^3 c^2 b^2 a^5 b^4 a^1 c^2 b^3$, then $V^{[1,5]} = < [5,8], [2,6], 2 >$, $V^{[4,7]} = < [1,6], [4,7], 2 >$, and $V'^{[1,5]} = V'^{[4,7]} = < -1, -1, 2 >$.

This leads to the following index idea. Generate a trie, which we call the *vectors index* of all V' vectors, where each vector is represented by one string of length $|\Sigma|$ from root to leaf. Then, generate a data structure to represent all V vectors that share the same path in the trie, or more specifically represent the two non-singleton ranges of these vectors. That means that each data structure includes a set of vectors that share exactly the same values for the same $|\Sigma| - 2$ letters, and may differ only in the non-singleton ranges of the other two letters. Assume the two non-singleton ranges are on letters σ_j and σ'_j, we define for each vector in the set, $V^{[k,\ell]}$, a rectangle $R^{[k,\ell]}(\sigma_j) \times R^{[k,\ell]}(\sigma_{j'})$. Then we generate a data structure of these rectangles for stabbing queries as the one proposed by Chazelle [9]. Finally, we connect each leaf in the trie to the corresponding rectangles data structure.

3.5   Answering the Query

For a query P, we perform three steps. First, we adapt for each pair of letters $(\sigma_j, \sigma_{j'})$ in P, a new histogram $h'^{(j,j')}$. This histogram is identical to the input histogram $h = <a_1, \ldots, a_{|\Sigma|}>$, except of the values $(a_j, a_{j'})$, which are set to -1. For example, for query $P = aabbabc$ and input histogram $h = < 3, 3, 1 >$ we get the following three adapted histograms: $h'^{(1,2)} = < -1, -1, 1 >$, $h'^{(1,3)} = < 3, -1, -1 >$ and $h'^{(2,3)} = < 3, -1, -1 >$.

Secondly, we search the trie to find the path corresponds to each adapted histogram $h'^{(j,j')}$, if such exists.

Finally, we follow the pointer from the leaf to the rectangles data structure and perform a stabbing query to find all rectangles that include the 2D point (a_j, a'_j).

3.6   Time and Space

Computing the $O(r^2)$ V and V' vectors in the preprocessing stage, takes $O(r^2)$ time. The vectors index maintains a trie over the vectors which are "strings" of length $|\Sigma|$ each. Moreover, there are $O(r^2)$ vectors in the index. The stabbing query data structures are bounded by $O(r^2)$, because each vector appears in only one data structure. Generating the stabbing query data structures for each set of vectors that share the same V' vector, will require $O(k \log k)$ time and $O(k)$ space if there are k vectors in the set. Hence, the preprocessing and index constructing time is $O(r^2 \log r)$. Note that the time to generate the trie is $O(r^2|\Sigma| \log |\Sigma|)$. Hence, the space of the data structure is $O(r^2|\Sigma|)$ and the total time is $O(r^2(\log r + |\Sigma| \log |\Sigma|))$. The query time is dominated by the stabbing queries data structure and, hence, is $O(|\Sigma|^3 \log r + occ)$, where occ is the number of answers.

We note that in the full version of this paper we will show how to remove the $|\Sigma|$ factor in the space by maintaining a data structure without the trie. That is we carefully generate hashing codes on the vectors directly to the set of vectors that share the same V' vector (and the stabbing structure). The $|\Sigma| \log |\Sigma|$ factor of time will also be spared. Hence, we will be able to achieve $O(r^2 \log r)$ preprocessing time, $O(r^2)$ preprocessing space, and $O(|\Sigma|^3 \log r + occ)$ query time.

References

1. Amir, A., Butman, A., Porat, E.: On the relationship between histogram indexing and block-mass indexing. Philosophical Transactions A (to appear)
2. Amir, A., Chan, T.M., Lewenstein, M., Lewenstein, N.: On hardness of jumbled indexing. In: Esparza, J., Fraigniaud, P., Husfeldt, T., Koutsoupias, E. (eds.) ICALP 2014. LNCS, vol. 8572, pp. 114–125. Springer, Heidelberg (2014)

3. Badkobeh, G., Fici, G., Kroon, S., Lipták, Z.: Binary jumbled string matching for highly run-length compressible texts. Information Processing Letters 113(17), 604–608 (2013)
4. Benson, G.: Composition alignment. In: Benson, G., Page, R.D.M. (eds.) WABI 2003. LNCS (LNBI), vol. 2812, pp. 447–461. Springer, Heidelberg (2003)
5. Böcker, S.: Simulating multiplexed snp discovery rates using base-specific cleavage and mass spectrometry. Bioinformatics 23(2), 5–12 (2007)
6. Burcsi, P., Cicalese, F., Fici, G., Lipták, Z.: On table arrangements, scrabble freaks, and jumbled pattern matching. In: Boldi, P. (ed.) FUN 2010. LNCS, vol. 6099, pp. 89–101. Springer, Heidelberg (2010)
7. Burcsi, P., Cicalese, F., Fici, G., Lipták, Z.: Algorithms for jumbled pattern matching in strings. International Journal of Foundations of Computer Science 23(02), 357–374 (2012)
8. Burcsi, P., Cicalese, F., Fici, G., Lipták, Z.: On approximate jumbled pattern matching in strings. Theory of Computing Systems 50(1), 35–51 (2012)
9. Chazelle, B.: Filtering search: A new approach to query-answering. SIAM Journal on Computing 15(3), 703–724 (1986)
10. Cicalese, F., Fici, G., Lipták, Z.: Searching for jumbled patterns in strings. In: Proceedings of the Prague Stringology Conference, pp. 105–117 (2009)
11. Cicalese, F., Gagie, T., Giaquinta, E., Laber, E.S., Lipták, Z., Rizzi, R., Tomescu, A.I.: Indexes for jumbled pattern matching in strings, trees and graphs. In: Kurland, O., Lewenstein, M., Porat, E. (eds.) SPIRE 2013. LNCS, vol. 8214, pp. 56–63. Springer, Heidelberg (2013)
12. Durocher, S., Munro, J.I., Mondal, D., Thankachan, S.V.: Jumbled pattern matching over large alphabets. Manuscript, Personal Communication (2014)
13. Eres, R., Landau, M.G., Parida, L.: Permutation pattern discovery in biosequences. Journal of Computational Biology 11(6), 1050–1060 (2004)
14. Gagie, T.: Grammar-based construction of indexes for binary jumbled pattern matching. CoRR, abs/1210.8386 (2012)
15. Gagie, T., Hermelin, D., Landau, G.M., Weimann, O.: Binary jumbled pattern matching on trees and tree-like structures. CoRR, abs/1301.6127 (2013)
16. Giaquinta, E., Grabowski, S.: New algorithms for binary jumbled pattern matching. Information Processing Letters 113(14-16), 538–542 (2013)
17. Hermelin, D., Landau, G.M., Rabinovich, Y., Weimann, O.: Binary jumbled pattern matching via all-pairs shortest paths. CoRR, abs/1401.2065 (2014)
18. Kociumaka, T., Radoszewski, J., Rytter, W.: Efficient indexes for jumbled pattern matching with constant-sized alphabet. In: Bodlaender, H.L., Italiano, G.F. (eds.) ESA 2013. LNCS, vol. 8125, pp. 625–636. Springer, Heidelberg (2013)
19. Lacroix, V., Fernandes, C.G., Sagot, M.-F.: Motif search in graphs: Application to metabolic networks. IEEE/ACM Trans. Comput. Biology Bioinform. 3(4), 360–368 (2006)
20. Moosa, T.M.: Rahman M. S. Indexing permutations for binary strings. Information Processing Letters 110(18-19), 795–798 (2010)
21. Moosa, T.M., Rahman, M.S.: Sub-quadratic time and linear space data structures for permutation matching in binary strings. Journal of Discrete Algorithms 10, 5–9 (2012)

Relative FM-Indexes

Djamal Belazzougui[1,2,*], Travis Gagie[1,2,**], Simon Gog[3,***],
Giovanni Manzini[4], and Jouni Sirén[5,†]

[1] University of Helsinki
[2] Helsinki Institute for Information Technology
[3] Karlsruhe Institute of Technology
[4] University of Eastern Piedmont
[5] University of Chile

Abstract. Intuitively, if two strings S_1 and S_2 are sufficiently similar and we already have an FM-index for S_1 then, by storing a little extra information, we should be able to reuse parts of that index in an FM-index for S_2. We formalize this intuition and show that it can lead to significant space savings in practice, as well as to some interesting theoretical problems.

1 Introduction

FM-indexes [4] are core components in most modern DNA aligners (e.g., [8–10]) and have thus played an important role in the genomics revolution. Medical researchers are now producing databases of hundreds or even thousands of human genomes, so bioinformatics researchers are working to improve FM-indexes' compression of sets of nearly duplicate strings. As far as we know, however, the solutions proposed so far (e.g., [3, 11]) index the concatenation of the genomes, so we can search the whole database easily but searching only in one specified genome is more difficult. In this paper we consider how to index each of the genomes individually while still using reasonable space and query time.

Our intuition is that if two strings S_1 and S_2 are sufficiently similar and we already have an FM-index for S_1 then, by storing a little extra information, we should be able to reuse parts of that index in an FM-index for S_2. More specifically, it seems S_1's and S_2's Burrows-Wheelers Transforms [2] (BWTs) should also be fairly similar. Since BWTs are the main component of FM-indexes, it is natural to try to take advantage of such similarity to build an index for S_2 that "reuses" information already available in S_1's FM-index.

Among the many possible similarities one can find and exploit in the BWTs, in this paper we consider the longest common subsequence (LCS). The BWT sorts

* Partially supported by Academy of Finland under grant 250345 (CoECGR).
** Funded by Academy of Finland grant 268324.
*** This work was carried out while the third author was employed at the University of Melbourne, supported by ARC Grant DP110101743.
† Funded by the Jenny and Antti Wihuri Foundation, Finland, and Fondecyt Grant 1-140796, Chile.

E. Moura and M. Crochemore (Eds.): SPIRE 2014, LNCS 8799, pp. 52–64, 2014.
© Springer International Publishing Switzerland 2014

the characters of a string into the lexicographic order of the suffixes following those characters. For example, if

$$S_1 = \text{GCACTTAGAGGTCAGT}, \qquad S_2 = \text{GCACTAGACGTCAGT};$$

then

$$\text{BWT}(S_1) = \text{TCTGCGTAAAAGGTGC}, \qquad \text{BWT}(S_2) = \text{TGCTCGTAAAACGCG};$$

whose LCS TCTCGTAAAAGG is nearly as long as either BWT.

We introduce the concept of *BW-distance* $\text{BWD}(S_1, S_2)$ between S_1 and S_2 defined as $|S_1| + |S_2| - 2|\text{LCS}(\text{BWT}(S_1), \text{BWT}(S_2))|$. Note that this coincides with the edit distance between $\text{BWT}(S_1)$ and $\text{BWT}(S_2)$ when only insertions and deletions are allowed. We prove that, if we are willing to tolerate a slight increase in query times, we can build an index for S_2 using an unmodified FM-index for S_1 and additional data structures whose total space in words is asymptotically bounded by $\text{BWD}(S_1, S_2)$ (Theorem 1).

This first result is the starting point for our investigation as it generates many challenging issues. First, since we are interested in indexing whole genomes, we observe that finding the LCS of strings whose length is of the order of billions is outside the capabilities of most computers. Thus, in Section 3.1 we show how to approximate the LCS of two BWTs, using combinatorial properties of the BWT to align the sequences. In the same section we also discuss and test several practical alternatives for building the index for S_2 given the one for S_1 and we analyze their time/space trade-offs.

If we need an index not only for counting queries but also for locating and extracting, we must enrich it with suffix array (SA) samples. Such samples usually take significantly less space than the main index. However, we may still want to take advantage of the similarities between S_1 and S_2 to "reuse" SA samples from S_1 for S_2's index. In Section 4 we show that this is indeed possible if, instead of considering the LCS between the BWTs, we use a common subsequence with the additional constraint of being *BWT-invariant* (Theorem 2). This result motivates the problem of finding the longest BWT-invariant subsequence, which unfortunately turns out to be NP-hard (Theorem 3). We therefore devise a heuristic to find a "long" BWT-invariant subsequence in $\mathcal{O}(|S_1| \log |S_1|)$ time.

We have tested our approach in practice by building an FM-index for the genomes of two human individuals, "reusing" an FM-index of the human reference genome. The reference genome is 3096 million base pairs, the individual genomes are 3002 million and 3036 million base pairs, and we found common subsequences of 2935 million and 2992 million base pairs, respectively. Our index is 3.8–5.0 times or 2.2–2.9 times smaller than a standard implementation of a stand-alone FM-index, depending on the encoding of the stand-alone index. On the other hand, queries to our index take about 11 times or 1.9 times longer, respectively. Since our index is compressed relative to the underlying index for the reference, we call it a relative FM-index.

2   Review of the FM-Index Structure

The core component of an FM-index for a string $S[1..n]$ is a data structure supporting rank queries on the Burrows-Wheeler Transform $\mathsf{BWT}(S)$ of S. This transform permutes the characters in S such that $S[i]$ comes before $S[j]$ in $\mathsf{BWT}(S)$ if $S[i+1..n]$ is lexicographically less than $S[j+1..n]$.

If the lexicographic range of suffixes of S starting with β is $[i..j]$, then the range of suffixes starting with $a\beta$ is

$$\left[\mathsf{BWT}(S).\mathsf{rank}_a(i-1) + 1 + \sum_{a' \prec a} S.\mathsf{rank}_{a'}(n).. \right.$$
$$\left. \mathsf{BWT}(S).\mathsf{rank}_a(j) + \sum_{a' \prec a} S.\mathsf{rank}_{a'}(n) \right]$$

It follows that, if we have precomputed an array storing $\sum_{a' \prec a} S.\mathsf{rank}_{a'}(n)$ for each distinct character a (i.e., the number of characters in S less than a), then we can find the range of suffixes starting with a pattern $P[1..m]$ — and, thus, count its occurrences — using $\mathcal{O}(m)$ rank queries.

If the position of $S[i]$ in $\mathsf{BWT}(S)$ is j, then the position of $S[i-1]$ is

$$\mathsf{BWT}(S).\mathsf{rank}_{S[i]}(j) + \sum_{a \prec S[i]} \mathsf{BWT}(S).\mathsf{rank}_a(n) .$$

It follows that, if we have also precomputed a dictionary storing the position of every rth character of S in $\mathsf{BWT}(S)$ with its position in S as satellite information, then we can find a character's position in S from its position in $\mathsf{BWT}(S)$ using $\mathcal{O}(r)$ rank and membership queries. Therefore, once we know the lexicographic range of suffixes starting with P, we can locate each of its occurrences using $\mathcal{O}(r)$ rank queries.

Finally, if we have also precomputed an array storing the position of every rth character of S in $\mathsf{BWT}(S)$, in order of appearance in S, then given i and j, we can extract $S[i..j]$ using $\mathcal{O}(r+j-i)$ rank queries.

3   BW-Distance and Relative FM-Indices

Given two strings $S_1[1..n_1]$ and $S_2[1..n_2]$ we define the *BW-distance* $\mathsf{BWD}(S_1, S_2)$ between S_1 and S_2 as

$$\mathsf{BWD}(S_1, S_2) = n_1 + n_2 - 2|\mathsf{LCS}(\mathsf{BWT}(S_1), \mathsf{BWT}(S_2))|. \tag{1}$$

Note that the BW-distance is nothing but the edit distance between $\mathsf{BWT}(S_1)$ and $\mathsf{BWT}(S_2)$ when only insertions and deletions are allowed [13] (also known as the shortest edit script or indel distance), and is thus at most twice their normal edit distance. We now show how to support counting queries on S_2 using an FM-index for S_1 and some auxiliary data structures taking $\mathcal{O}(\mathsf{BWD}(S_1, S_2))$

words of space. Specifically, we consider how we can support rank queries on $BWT(S_2)$ and partial-sum queries on the distinct characters' frequencies.

Let C denote a LCS of $BWT(S_1)$ and $BWT(S_2)$ with $|C| = m$. Let $C = c_1 \cdots c_m$, and for $i = 1, \ldots, m$, let α_i (resp. β_i) be the position of c_i in $BWT(S_1)$ (resp. $BWT(S_s)$) with $\alpha_1 < \cdots < \alpha_m$ (resp. $\beta_1 < \cdots < \beta_m$). Define

- bitvector $B_1[1..n_1]$ with 0s in positions $\alpha_1, \ldots, \alpha_m$,
- bitvector $B_2[1..n_2]$ with 0s in positions of $\beta_1, \ldots, \beta_m$,
- subsequence D_1 of $BWT(S_1)$ marked by 1s in B_1; D_1 is the complement of C in $BWT(S_1)$,
- subsequence D_2 of $BWT(S_2)$ marked by 1s in B_2; D_2 is the complement of C in $BWT(S_2)$.

We claim that if we can support fast rank queries on $BWT(S_1)$, B_1, B_2, D_1 and D_2 and fast select_0 queries on B_1, then we can support fast rank queries on $BWT(S_2)$. To see why, notice that

$$BWT(S_2).\text{rank}_X(i) = C.\text{rank}_X(B_2.\text{rank}_0(i)) \\ + D_2.\text{rank}_X(B_2.\text{rank}_1(i))$$

and, by the same reasoning,

$$C.\text{rank}_X(j) = BWT(S_1).\text{rank}_X(B_1.\text{select}_0(j)) \\ - D_1.\text{rank}_X(B_1.\text{rank}_1(B_1.\text{select}_0(j))).$$

Therefore,

$$BWT(S_2).\text{rank}_X(i) = BWT(S_1).\text{rank}_X(k) \\ - D_1.\text{rank}_X(B_1.\text{rank}_1(k)) \\ + D_2.\text{rank}_X(B_2.\text{rank}_1(i))$$

where $k = B_1.\text{select}_0(B_2.\text{rank}_0(i))$.

For example, for the strings

$$S_1 = \text{GCACTTAGAGGTCAGT}, \qquad S_2 = \text{GCACTAGACGTCAGT}$$

given as an example in Section 1,

$$BWT(S_1) = \text{TCTGCGTAAAAGGTGC}, \qquad BWT(S_2) = \text{TGCTCGTAAAACGCG};$$

and $LCS(BWT(S_1), BWT(S_2)) = \text{TCTCGTAAAAGG}$ so

$$B_1 = 0001000000000111 \qquad\qquad D_1 = \text{GTGC}$$
$$B_2 = 010000000001010 \qquad\qquad D_2 = \text{GCC}.$$

Suppose we want to compute $BWT(S_2).\text{rank}_C(13)$. Since $B_1.\text{select}_0(B_2.\text{rank}_0(13)) = 12$,

$$\text{BWT}(S_2).\text{rank}_C(13) = \text{BWT}(S_1).\text{rank}_C(12) - D_1.\text{rank}_C(B_1.\text{rank}_1(12))$$
$$+ D_2.\text{rank}_C(B_2.\text{rank}_1(13)) = 3.$$

Observing that the number of 1s in B_1 and B_2 is $\mathcal{O}(\max(n_1, n_2) - m) = \mathcal{O}(\text{BWD}(S_1, S_2))$, we can store data structures for B_1, B_2, D_1 and D_2 in $\mathcal{O}(\text{BWD}(S_1, S_2))$ space such that the desired rank/select queries take $\mathcal{O}(\log \text{BWD}(S_1, S_2))$ time.

The only other component required for an FM-index for S_2 for counting, is a data structure for computing $\sum_{a' \prec a} S_2.\text{rank}_{a'}(n)$ for each character a. Notice that $\text{BWD}(S_1, S_2)$ is at least the number of distinct characters whose frequencies in S_1 and S_2 differ. It follows that in $\mathcal{O}(\text{BWD}(S_1, S_2))$ space we can store

- a $\mathcal{O}(\log \text{BWD}(S_1, S_2))$-time predecessor data structure storing those distinct characters,
- an array storing $\sum_{a' \prec a} S_2.\text{rank}_{a'}(n_2)$ for each such distinct character a.

For any distinct character b, we can find the preceding distinct character a whose frequencies in S_1 and S_2 differ and compute

$$\sum_{a' \prec b} S_2.\text{rank}_{a'}(n_2) = \sum_{a' \prec b} S_1.\text{rank}_{a'}(n_1) - \sum_{a' \prec a} S_1.\text{rank}_{a'}(n_1) + \sum_{a' \prec a} S_2.\text{rank}_{a'}(n_2)$$

using $\mathcal{O}(\log \text{BWD}(S_1, S_2))$ time. Summing up:

Theorem 1. *If we have an FM-index for S_1, we can store a relative FM-index for S_2 using $\mathcal{O}(\text{BWD}(S_1, S_2))$ words of extra space. Counting queries on the relative FM-index take time an $\mathcal{O}(\log \text{BWD}(S_1, S_2))$ factor larger than on S_1.*

3.1   A Practical Implementation

A longest common sequence of $\text{BWT}(S_1)$ and $\text{BWT}(S_2)$ can be computed in $\mathcal{O}(n_1 n_2 / w)$ time, where w is the word size [12]. Since we are mainly interested in strings with a small BW-distance, a better alternative could be the algorithms whose running times are bounded by the number of differences between the input sequences (see eg [7, 13]). Unfortunately none of these algorithms is really practical when working with such very large files as the complete genomes we considered in our tests. Hence, to make the construction of a relative FM-index practical, we approximate the LCS of the two Burrows-Wheeler transforms, using the combinatorial properties of the BWT to align the sequences.

Let S_1 be a random string of length n over alphabet Σ of size σ, and let string S_2 differ from it by s insertions, deletions, and substitutions. In the expected case, the lexicographic rank of each suffix of S_1 is determined by a prefix of length $\mathcal{O}(\log_\sigma n)$ of that suffix. Thus, the s edit operations are expected to affect the relative lexicographic order of $\mathcal{O}(s \log_\sigma n)$ suffixes [11], possibly causing the characters immediately preceding those suffixes to appear in different positions in $\mathsf{BWT}(S_1)$ and $\mathsf{BWT}(S_2)$. The edits can also change the characters immediately preceding at most s suffixes. If we remove the characters preceding the affected suffixes from $\mathsf{BWT}(S_1)$ and $\mathsf{BWT}(S_2)$, we have a common subsequence of length $n - O(s \log_\sigma n)$ in the expected case.

Assume that we have partitioned the BWTs according to the first k characters of the suffixes, for $k \geq 0$. For all $x \in \Sigma^k$, let $\mathsf{BWT}_x(S_1)$ and $\mathsf{BWT}_x(S_2)$ be the substrings of the BWTs corresponding to the suffixes starting with x. If we remove the suffixes affected by the edit operations, as well as the suffixes where string x covers an edit, we have a common subsequence BWT'_x of $\mathsf{BWT}_x(S_1)$ and $\mathsf{BWT}_x(S_2)$. If we concatenate the sequences BWT'_x for all x, we get a common subsequence of $\mathsf{BWT}(S_1)$ and $\mathsf{BWT}(S_2)$ of length $n - O(s(k + \log_\sigma n))$ in the expected case. This suggests that we can find a long common subsequence of $\mathsf{BWT}(S_1)$ and $\mathsf{BWT}(S_2)$ by partitioning the BWTs, finding an LCS for each partition, and concatenating the results.

In practice, we partition the BWTs by variable-length strings. We use backward searching on the BWTs to traverse the suffix trees of S_1 and S_2, selecting a partition when either the length of $\mathsf{BWT}_x(S_1)$ or $\mathsf{BWT}_x(S_2)$ is at most 1024, or the length of the pattern x reaches 32. For each partition, we use the greedy LCS algorithm [13] to find the longest common subsequence of that partition. To avoid hard cases, we stop the greedy algorithm if it would need diagonals beyond ± 50000, and match only the most common characters for that partition. We also predict in advance the common cases where this happens (the difference of the lengths of $\mathsf{BWT}_x(S_1)$ and $\mathsf{BWT}_x(S_2)$ is over 50000, or $x = \mathsf{N}^{32}$ for DNA sequences, where N is the "any base" symbol), and match the most common characters in that partition directly.

We implemented the counting structure of the relative FM-index using the SDSL library [5], and compared its performance to a regular FM-index.[1] To encode the BWTs and sequences D_1 and D_2, we used Huffman-shaped wavelet trees with either plain or entropy-compressed (RRR) [15] bitvectors. For marking the positions of the LCS in $\mathsf{BWT}(S_1)$ and $\mathsf{BWT}(S_2)$, we used either entropy-compressed or sparse [14] bitvectors.

The implementation was written in C++ and compiled on g++ version 4.7.3. We used a system with 32 gigabytes of memory and two quad-core 2.53 GHz Intel Xeon E5540 processors, running Ubuntu 12.04 with Linux kernel 3.2.0. Only one CPU core was used in the experiments.

For our experiments, we used the 1000 Genomes Project assembly of the human reference genome as sequence S_1.[2] As sequence S_2, we used the genome of a Han Chinese male from the YanHuang project[3], and the genome of the 1000 Genomes Project individual NA12878 (Utah female, maternal haplotype) [16]. The properties of the datasets can be seen in Table 1. As our pattern set, we

---

[1] The implementation is available at http://jltsiren.kapsi.fi/relative-fm
[2] GRCh37, ftp://ftp-trace.ncbi.nih.gov/1000genomes/ftp/technical/reference/
[3] ftp://public.genomics.org.cn/BGI/yanhuang/fa/

Table 1. Properties of the datasets. The length of the sequence and the common subsequence of the BWTs; the number of matching patterns and the total number of occurrences for those patterns.

Dataset	Length	LCS	Matches	Occurrences
Reference	3096M	–	–	–
YanHuang	3002M	2935M	1.14M	5.49M
NA12878	3036M	2992M	1.21M	5.67M

Table 2. Experiments with human genomes. Dataset, bitvector used in the wavelet trees (WT) and for the LCS; time and space requirements for building the relative FM-index; time required for counting queries and index size for a regular FM-index and a relative FM-index. The query times are averages over five runs.

Dataset	WT	LCS	Construction		Regular		Relative	
			Time [s]	Space [MB]	Time [s]	Size [MB]	Time [s]	Size [MB]
YanHuang	Plain	RRR	708	9124	56.45	1090	621.47	288
YanHuang	Plain	Sparse	711	9124	56.45	1090	1162.47	290
YanHuang	RRR	RRR	5898	7823	328.86	628	1637.44	256
YanHuang	RRR	Sparse	5882	7823	328.86	628	1994.89	257
NA12878	Plain	RRR	589	9124	57.31	1090	619.81	218
NA12878	Plain	Sparse	575	9124	57.31	1090	1058.75	199
NA12878	RRR	RRR	5454	7823	325.49	636	1614.56	192
NA12878	RRR	Sparse	5412	7823	325.49	636	1921.92	173

used 3.68 million reads of length 108 from the 1000 Genomes Project individual HG00122 (British female). The results of the experiments can be seen in Table 2.

The fastest variant of the relative FM-index uses plain bitvectors in the wavelet trees and RRR bitvectors for the LCS. It is 3.8–5.0 times smaller and 11 times slower than a regular FM-index using plain bitvectors, and 2.2–2.9 times smaller and 1.9 times slower than a regular index using RRR bitvectors. Switching to compressed bitvectors in the wavelet trees does not yield a good trade-off. Using sparse bitvectors for the LCS is a slightly better option on the NA12878 dataset, making the relative index 1.1 times smaller and 1.7 times slower. On the YanHuang dataset, sparse bitvectors require more space than RRR bitvectors, because the sequence is too different from the reference.

Bitvectors B_1 and B_2 take 70% to 85% of the total size of the relative index, so improving their compression may be the best way to make the index smaller. Hybrid bitvectors using different encodings for different parts of the bitvector [6] could be one option, but the existing implementation does not work with vectors longer than 2^{31} bits. It should be noted that the size difference between NA12878 and the reference is mostly due to the inclusion of chromosome Y (59 million base pairs) in the reference. Therefore we can expect the relative FM-index to work significantly better with male genomes than female genomes.

Building a relative FM-index out of regular FM-indexes for two human genomes takes 10–12 minutes. Using RRR bitvectors for the wavelet trees increases this to 90–98 minutes, as extracting substrings from the wavelet trees becomes the bottleneck. Decompressing the regular FM-index of S_2 from the regular index of S_1 and the relative index of S_2 should be even faster. As a comparison, building BWT for a human genome takes 19–20 minutes and 25–26 gigabytes of memory using libdivsufsort 2.0.1[4], depending on the sequence. The space usage of relative FM-index construction has not been optimized, and it can probably be improved significantly.

4 Relative FM-Indices Supporting Locating and Extracting

As mentioned in Section 2, an FM-index for S_1 usually has an SA sample that takes an only slightly sublinear number of bits. This sample has two parts: the first consists of a bitvector R with 1s marking the positions in $\mathsf{BWT}(S_1)$ of every rth character in S_1, and an array A storing a mapping from the ranks of those characters' positions in $\mathsf{BWT}(S_1)$ to their positions in S_1; the second is an array storing a mapping from the ranks of those characters' positions in S to their positions in $\mathsf{BWT}(S_1)$. With these, given the position of a sampled character in $\mathsf{BWT}(S_1)$, we can find its position in S_1, and vice versa.

These parts are used for locating and extracting queries, respectively, and the worst-case query times are proportional to r. On the other hand, the size of the sample in words is proportional to the length of S divided by r. For details on how the sample works, we direct the reader to the full description of FM-indexes [4]. We note only that if we sample irregularly, then the worst-case query times for locating and extracting are proportional to the maximum distance in S between two consecutive sampled characters. We leave consideration of extracting for the full version of the paper — it is nearly symmetric to locating — so we do not discuss the second part of the sample here.

Let $G = S_1[i_1] \cdots, S_1[i_\ell]$ denote a length-ℓ common subsequence of S_1 and S_2 (not their BWTs). That is, we have $i_1 < \cdots < i_\ell$ and there exists $j_1 < \cdots < j_\ell$ such that

$$S_1[i_1] = S_2[j_1], \ldots, S_1[i_\ell] = S_2[j_\ell].$$

Since there is a one-to-one correspondence between the characters in a text and in its BWT, we can define the indexes $v_1, \ldots, v_\ell$ (resp. $w_1, \ldots, w_\ell$) such that for $k = 1, \ldots, \ell$, $\mathsf{BWT}(S_1)[v_k]$ is the character corresponding to $S_1[i_k]$ (resp. $\mathsf{BWT}(S_2)[w_k]$ is the character corresponding to $S_2[j_k]$). We say that the common subsequence G is *BWT-invariant* if there exists a permutation $\pi : \{1, \ldots, \ell\} \to \{1, \ldots, \ell\}$ such that we have simultaneously

$$v_{\pi(1)} < v_{\pi(2)} < \cdots < v_{\pi(\ell)}, \quad \text{and} \quad w_{\pi(1)} < w_{\pi(2)} < \cdots < w_{\pi(\ell)}. \tag{2}$$

---

[4] https://code.google.com/p/libdivsufsort/

In other words, when we go from the texts to the BWTs the elements of G are permuted in the same way in S_1 and S_2.

An immediate consequence of (2) is that the sequence

$$G' = \text{BWT}(S_1)[v_{\pi(1)}]\,\text{BWT}(S_1)[v_{\pi(2)}] \cdots \text{BWT}(S_1)[v_{\pi(\ell)}]$$

is a common subsequence of $\text{BWT}(S_1)$ and $\text{BWT}(S_2)$. We can therefore generalize (1) and define

$$\text{BWD}_G(S_1, S_2) = \max(n_1, n_2) - |G|$$

and repeat the construction of Theorem 1 with BWD replaced by BWD_G. However, since G is BWT-invariant it is now possible to reuse the the SA samples from S_1 relative to positions in G for the string S_2 provided that we have

- bitvector $M_1[1..n_1]$ with 0s in positions $i_1, \ldots, i_\ell$, supporting fast rank queries,
- bitvector $M_2[1..n_2]$ with 0s in positions $j_1, \ldots, j_\ell$, supporting fast select$_0$ queries.

Summing up, we have (proof idea in the appendix):

Theorem 2. *For any BWT-invariant subsequence G, if we already have an FM-index for S_1, then we can store $\mathcal{O}(\text{BWD}_G(S_1, S_2))$ extra words such that the time bounds for locating and extracting queries on S_2 are an $\mathcal{O}(\log \text{BWD}_G(S_1, S_2))$ factor larger than on S_1.*

In view of the above theorem, it is certainly desirable to find the longest common subsequence of S_1 and S_2 which is BWT-invariant. Unfortunately, this problem is NP-hard as shown by the following result.

Theorem 3. *It is NP-complete to determine whether there is an LCS of S_1 and S_2 which is BWT-invariant, even when the strings are over a ternary alphabet.*

Proof. Clearly we can check in polynomial time whether a given subsequence of S_1 and S_2 has this property, so the problem is in NP. To show that it is NP-complete, we reduce from the NP-complete problem of permutation pattern matching [1], for which we are given two permutations π_1 and π_2 over n and $m \le n$ elements, respectively, and asked to determine whether there is a subsequence of π_1 of length m such that the relative order of the elements in that subsequence is the same as the relative order of the elements in π_2. For example, if $\pi_1 = 6, 3, 2, 1, 4, 5$ and $\pi_2 = 4, 2, 1, 3$, then $6, 2, 1, 4$ is such a subsequence. Specifically, we set

$$S_1 = \text{AB}^{\pi_1[1]}\text{AB}^{\pi_1[2]} \cdots \text{AB}^{\pi_1[n]}, \qquad S_2 = \text{AC}^{\pi_2[1]}\text{AC}^{\pi_2[2]} \cdots \text{AC}^{\pi_2[m]},$$

so the unique LCS of S_1 and S_2 is A^m. For our example,

$$S_1 = \text{AB}^6\text{AB}^3\text{AB}^2\text{ABAB}^5 = \text{ABBBBBBABBBABBABABBBBB}$$
$$S_2 = \text{AC}^4\text{AC}^2\text{ACAC}^3 = \text{ACCCCACCACACCC}.$$

The BWT sorts the m copies of A in S_2 according to π_2 and sorts any subsequence of m copies of A in S_1 according to the corresponding subsequence of π_1. Therefore, there is an LCS of S_1 and S_2 such that the relative order of its characters is $\mathsf{BWT}(S_1)$ and $\mathsf{BWT}(S_2)$ is the same, if and only if there is a subsequence of π_1 of length m such that the relative order of the elements in that subsequence is the same as the relative order of the elements in π_2.   □

In view of the above result, for large inputs we cannot expect to find the longest possible BWT-invariant subsequence, so, as for the LCS, we have devised the following fast heuristic for computing a "long" BWT-invariant subsequence.

We first compute the suffix array SA_{12} for the concatenation $S_1 \# S_2$ and we use it to define the array A of size $n_1 \times 2$ as follows

- $A[i][1] = j$ iff $S_1[i] = S_2[j]$ and suffix $S_2[j+1, n_2]$ immediately follows suffix $S_1[i+1, n_1]$ in SA_{12}. If no such j exists $A[i][1]$ is undefined.
- $A[i][2] = j$ iff $S_1[i] = S_2[j]$ and suffix $S_2[j+1, n_2]$ is the lexicographically largest suffix of S_2 preceding suffix $S_1[i+1, n_1]$ in SA_{12}. If no such j exists $A[i][2]$ is undefined.

Next, we compute the longest subsequence $1 \leq i_1 < i_2 < \cdots < i_\ell \leq n_1$ such that there exist $b_1, \ldots, b_\ell$, with $b_k \in \{1, 2\}$ and the sequence

$$A[i_1][b_1] < A[i_2][b_2] < \cdots < A[i_\ell][b_\ell]$$

is the longest possible (every $A[i_k][b_k]$ must be defined). The values $i_1, \ldots, i_\ell$ and $b_1, \ldots, b_\ell$ can be computed in $\mathcal{O}(n_1 \log n_1)$ time using a straightforward modification of the dynamic programming algorithm for the longest increasing subsequence. Setting, for $k = 1, \ldots, \ell$, $j_k = A[i_k][b_k]$ we get that

$$G \;=\; S_1[i_1]S_1[i_2] \cdots S_1[i_\ell] \;=\; S_2[j_1]S_2[j_2] \cdots S_2[j_\ell]$$

is a common subsequence of S_1 and S_2.

Lemma 1. *The subsequence G is BWT-invariant.*

Proof. Let $v_1, \ldots, v_\ell$ (resp. $w_1, \ldots, w_\ell$) such that for $k = 1, \ldots, \ell$, $\mathsf{BWT}(S_1)[v_k]$ is the character corresponding to $S_1[i_k]$ (resp. $\mathsf{BWT}(S_2)[w_k]$ corresponds to $S_2[j_k]$). It suffices to prove that for any pair h, k, with $1 \leq h, k \leq \ell$, the inequality $v_h < v_k$ implies $w_h < w_k$. Let $\prec$ denote the lexicographic order. By construction, and by the properties of the BWT, we have $v_h < v_k$ iff the suffix $S_1[i_h + 1, n_1] \prec S_1[i_k + 1, n_1]$ and we must prove that this implies $S_2[j_h + 1, n_2] \prec S_2[j_k + 1, n_2]$.

Since $j_h = A[i_h][b_h]$ and $j_k = A[i_k][b_k]$, the proof follows considering the four possible cases: $b_h, b_k \in \{1, 2\}$. We consider the case $b_h = 1$, $b_k = 2$ leaving the others to the reader. If $j_h = A[i_h][1]$ and $j_k = A[i_k][2]$ then $S_2[j_h + 1, n_2]$ immediately follows $S_1[i_h + 1, n_1]$ in SA_{12}. At same time $S_2[j_k + 1, n_2]$ precedes $S_1[i_k + 1, n_1]$ but there are no other suffixes from S_2 between them. Since $j_h \neq j_k$ the only possible ordering of the suffixes in SA_{12} is

$$S_1[i_h + 1, n_1] \prec S_2[j_h + 1, n_2] \prec S_2[j_k + 1, n_2] \prec S_1[i_k + 1, n_1]$$

implying $S_2[j_h + 1, n_2] \prec S_2[j_k + 1, n_2]$ as claimed.   □

Table 3. Comparison between $|G|$ and $|\mathsf{LCS}|$. The normalizing factor n is the length of sequence 273614N.

	322134S	378604X	BC187	DBVPG1106
$|\mathsf{LCS}|/n$	0.9341	0.9669	0.9521	0.9590
$|G|/n$	0.8694	0.8655	0.8798	0.8800

To evaluate whether the subsequence G derived from the above procedure is still able to capture the similarity between S_1 and S_2, we have compared the length of G with the LCS length for pairs of *S.cerevisiae* genomes from the Saccharomyces Genome Resequencing Project.[5] In particular we compared the 273614N sequence with sequences 322134S, 378604X, BC187, and DBVPG1106. For each sequence we report in Table 3 the ratio between the length of G and $\mathsf{LCS}(\mathsf{BWT}(S_1), \mathsf{BWT}(S_2))$ and the length of sequence 273614N (roughly 11.9 MB). We see that in all cases more than 85% of BWT positions are in G which roughly indicates that more than 85% of the SA samples from 273614N could be reused as SA samples for the other sequences.

5   Conclusions

In this paper we have considered the problem of building an index for a string S_2 given an FM-index for a similar string S_1. We have shown how to build such a "relative" index using space bounded by the BW-distance between S_1 and S_2. The BW-distance is simply the edit distance between $\mathsf{BWT}(S_1)$ and $\mathsf{BWT}(S_2)$ when only insertions and deletions are allowed. We have also introduced the notion of BWT-invariant subsequence and shown that it can be used to determine a set of S_1 suffix array samples that can be easily "reused" for an index for S_2.

We have tested our approach by building a relative index for a Han Chinese individual and a 1000 Genomes Project individual with respect to an FM-index of the human reference genome. We leave as a future work the development of these ideas and the complete implementation of a relative FM-index supporting locating and extracting. We also leave as future work proving bounds on the BW-distance and the length of the longest BWT-invariant subsequence in terms of the edit distance of the strings.

References

1. Bose, P., Buss, J.F., Lubiw, A.: Pattern matching for permutations. Inf. Process. Lett. 65(5), 277–283 (1998)
2. Burrows, M., Wheeler, D.J.: A block sorting lossless data compression algorithm. Technical Report 124, Digital Equipment Corporation (1994)
3. Ferrada, H., Gagie, T., Hirvola, T., Puglisi, S.J.: Hybrid indexes for repetitive datasets. Phil. Trans. Royal Society A 372, 2014 (2016)

---

[5] https://www.sanger.ac.uk/research/projects/genomeinformatics/sgrp.html

4. Ferragina, P., Manzini, G.: Indexing compressed text. Journal of the ACM 52(4), 552–581 (2005)
5. Gog, S., Beller, T., Moffat, A., Petri, M.: From theory to practice: Plug and play with succinct data structures. In: Gudmundsson, J., Katajainen, J. (eds.) SEA 2014. LNCS, vol. 8504, pp. 326–337. Springer, Heidelberg (2014)
6. Kärkkäinen, J., Kempa, D., Puglisi, S.J.: Hybrid compression of bitvectors for the FM-index. In: Proc. 2014 IEEE Data Compression Conference, DCC 2014, pp. 302–311 (2014)
7. Landau, G.M., Vishkin, U., Nussinov, R.: An efficient string matching algorithm with k differences for nucleotide and amino acid sequences. Nucleic Acids Research 14(1), 31–46 (1986)
8. Langmead, B., Trapnell, C., Pop, M., Salzberg, S.L.: Ultrafast and memory-efficient alignment of short DNA sequences to the human genome. Genome Biology, 10:R25 (2009)
9. Li, H., Durbin, R.: Fast and accurate short read alignment with Burrows-Wheeler transform. Bioinformatics 25(14), 1754–1760 (2009)
10. Li, R., Yu, C., Li, Y., Lam, T.-W., Yiu, S.-M., Kristiansen, K., Wang, J.: SOAP2: An improved ultrafast tool for short read alignment. Bioinformatics 25(15), 1966–1967 (2009)
11. Mäkinen, V., Navarro, G., Sirén, J., Välimäki, N.: Storage and retrieval of highly repetitive sequence collections. Journal of Computational Biology 17(3), 281–308 (2010)
12. Myers, E.W.: A fast bit-vector algorithm for approximate string matching based on dynamic programming. Journal of the ACM 46(3), 395–415 (1999)
13. Myers, E.W.: An O(ND) difference algorithm and its variations. Algorithmica 1(2), 251–266 (1986)
14. Okanohara, D., Sadakane, K.: Practical entropy-compressed rank/select dictionary. In: Proc. Ninth Workshop on Algorithm Engineering and Experiments (ALENEX 2007), pp. 60–70. SIAM (2007)
15. Raman, R., Raman, V., Rao Satti, S.: Succinct indexable dictionaries with applications to encoding k-ary trees, prefix sums and multisets. ACM Transactions on Algorithms 3(4), 43 (2007)
16. Rozowsky, J., Abyzov, A., Wang, J., Alves, P., Raha, D., Harmanci, A., Leng, J., Bjornson, R., Kong, Y., Kitabayashi, N., Bhardwaj, N., Rubin, M., Snyder, M., Gerstein, M.: AlleleSeq: Analysis of allelespecific expression and binding in a network framework. Molecular Systems Biology 7, 522 (2011)

Appendix: Reusing an SA Sample

Consider the strings S_1, S_2 used as example in Section 3, and the corresponding LCS $C = \mathsf{LCS}(\mathsf{BWT}(S_1), \mathsf{BWT}(S_2))$ and bitvectors B_1 and B_2. The characters of $\mathsf{BWT}(S_1)[1..16]$ and $\mathsf{BWT}(S_2)[1..15]$ are mapped to their positions by the BWT from

$$S_1[16, 2, 6, 8, 13, 1, 12, 3, 7, 9, 14, 10, 15, 5, 11, 4]$$
$$S_2[15, 7, 2, 5, 12, 1, 11, 8, 3, 6, 13, 9, 14, 4, 10]$$

respectively. Notice the lists of indices are just the SAs of $S_1\$$ and $S_2\$$ with each value decremented by one. Therefore, if $r = 3$ then

$$R = 1000110010010001, \qquad A[1..6] = [16, 13, 1, 7, 10, 4]$$

(see beginning of Section 4 for the definition of R and A).

Comparing R and $B_1 = 0001000000000111$ we see that the sampled characters $\mathsf{BWT}(S_1)[1, 5, 6, 9, 12]$ that are in C, are C's 1st, 4th, 5th, 8th and 11th characters. From $B_2 = 010000000001010$ we see that the 1st, 4th, 5th, 8th and 11th characters in C in $\mathsf{BWT}(S_2)$ are $\mathsf{BWT}(S_2)[1, 5, 6, 9, 13]$, which are mapped to their positions by the BWT from $S_2[15, 12, 1, 3, 14]$.

The relative order $5, 3, 1, 2, 4$ of the positions $15, 12, 1, 3, 14$ in S_2 of these characters, is *almost* the same as the relative order $5, 4, 1, 2, 3$ of the positions $16, 13, 1, 7, 10$ in S_1 of the sampled characters in $\mathsf{BWT}(S_1)$ that are in C.

We can get rid of the "almost" if instead of C we consider a subsequence G' derived from a BWT-invariant sequence G. For example, we can choose

$$G' = \mathsf{TCTCGTAAAGG}$$
$$B_1' = 0001000001010101 \qquad\qquad B_2' = 010000010001010$$
$$D_1' = \mathsf{GAGTC} \qquad\qquad\qquad\quad D_2' = \mathsf{GACC}$$

Clearly G' is not an LCS of $\mathsf{BWT}(S_1)$ and $\mathsf{BWT}(S_2)$ and, thus, our data structures for supporting rank in $\mathsf{BWT}(S_2)$ are slightly larger. However, now the characters in $\mathsf{BWT}(S_1)$ and $\mathsf{BWT}(S_2)$ that are in G', are mapped to their positions by the BWT from

$$S_1[16, 2, 6, 13, 1, 12, 3, 7, 14, 15, 11], \qquad S_2[15, 2, 5, 12, 1, 11, 3, 6, 13, 14, 10]$$

and the relative order $11, 2, 4, 8, 1, 7, 3, 5, 9, 10, 6$ of the indices in those two lists is *exactly* the same. Now suppose we store yet another pair of bitvectors

$$M_1 = 0001100111000000, \qquad M_2 = 000100111000000$$

with 1s marking the positions in S_1 and S_2 of characters that are not mapped into G' in $\mathsf{BWT}(S_1)$ and $\mathsf{BWT}(S_2)$ (that is, the characters that are not in the BWT-invariant subsequence G). We claim that if we can support fast rank queries on B_2', R and M_1, fast access to A and fast select_0 queries on B_1' and M_2, then we can support fast access to a (possibly irregular) SA sample for S_2 with as many sampled characters as there are in G' in $\mathsf{BWT}(S_1)$. More specifically, if $\mathsf{BWT}(S_2)[i]$ is in G' and $R[B_1'.\mathsf{select}_0(B_2'.\mathsf{rank}_0(i))] = 1$ — meaning the corresponding character in G' in $\mathsf{BWT}(S_1)$ is sampled — then $\mathsf{BWT}(S_2)[i]$ is mapped to its position by the BWT from

$$S_2\left[M_2.\mathsf{select}_0\left(M_1.\mathsf{rank}_0\left(A\left[R.\mathsf{rank}_1\left(B_1'.\mathsf{select}_0\left(B_2'.\mathsf{rank}_0(i)\right)\right)\right]\right)\right)\right].$$

We leave a detailed explanation to the full version of this paper. We note, however, that this approach works for any sample rate r, and even if the SA sample for S_1 is irregular itself.

In our example, since $\mathsf{BWT}(S_2)[10]$ is in G', $B_1'.\mathsf{select}_0(B_2'.\mathsf{rank}_0(10)) = 9$ and $R[9] = 1$, we know $\mathsf{BWT}(S_2)[10]$ is mapped to its position by the BWT from position $M_2.\mathsf{select}_0\left(M_1.\mathsf{rank}_0\left(A[R.\mathsf{rank}_1(9)]\right)\right) = 6$ in S_2.

Efficient Indexing and Representation of Web Access Logs*

Francisco Claude[1], Roberto Konow[1,2], and Gonzalo Navarro[2]

[1] Escuela de Informática y Telecomunicaciones, Universidad Diego Portales
fclaude@recoded.cl
[2] Department of Computer Science, University of Chile
{rkonow,gnavarro}@dcc.uchile.cl

Abstract. We present a space-efficient data structure, based on the Burrows-Wheeler Transform, especially designed to handle web sequence logs, which are needed by web usage mining processes. Our index is able to process a set of operations efficiently, while at the same time maintains the original information in compressed form. Results show that web access logs can be represented using 0.85 to 1.03 times their original (plain) size, while executing most of the operations within a few tens of microseconds.

1 Introduction

Web Usage Mining (WUM) [14] is the process of extracting useful information from web server access logs, which allows web site administrators, designers and engineers to understand the users' interaction with their web site. This process is used to improve the layout of the web site to better suit their users, or to analyze the performance of their systems in order to apply smart prefetching techniques for faster response, among other applications.

One particular WUM task is to predict the path of web pages that the user is going to traverse within a website. Accurately predicting the web user access behavior can minimize the user perception of latency, which is an important measure of the website quality of service [5,6,28]. This is achieved by fetching the web page *before* the user requests it. Another application is as a recommendation technique [19,29]: the prediction can be displayed to the user, giving an insight of what the user might be looking for, therefore improving the user's experience. Other relevant mining operations include determining how frequently the path has been followed, which users have followed the path, and so on.

The prediction problem can be formalized as follows. Access logs obtained from web servers are used to extract the user's web site visit path as an ordered sequence of web pages $S_u = \langle v_1, v_2, v_3, \ldots, v_m \rangle$ (several sessions of the same user u might be concatenated into S_u). Therefore the system records the set

---

* This work was partially supported by the Conicyt PhD Scholarship, by Fondecyt Iniciación Grant 11130104, and by Millennium Nucleus Information and Coordination in Networks ICM/FIC P10-024F.

E. Moura and M. Crochemore (Eds.): SPIRE 2014, LNCS 8799, pp. 65–76, 2014.
© Springer International Publishing Switzerland 2014

$\mathcal{S} = \{S_1, S_2, \ldots, S_n\}$ of the accesses of each user. Given a new visit sequence (or path) P that is currently being performed by a user, the predicting task has to predict which page will be visited next. One näive approach to this problem is to first return the k web pages that have been visited most commonly by users after following the same path P, that is, we consider each time P appears as a substring of some S_u, and pick the most common symbols following those occurrences of P. After this process is done, more complex recommendation or machine learning algorithms [16] can be employed to accurately predict the next web page that is going to be visited. One particular challenge is that this operation needs to be done in an *on-line* manner, that is, we have to efficiently update our results as new requests are appended at the end of P. The system will eventually add those requests S' to the corresponding S_u sequences in $\mathcal{S}$, via periodic updates. At query time, the set $\mathcal{S}$ can be taken as static. The other mining operations are defined analogously.

Another interesting operation coming from WUM and general data mining is to retrieve the top-k most frequent sequences [12,22] of a certain length. These are commonly used in retailing, add-on sales, customer satisfaction and in many other fields.

A typical WUM system faces two challenges: On the one hand, it has to manage huge amounts of data, that comes directly from the web access logs that store the records of all the interactions between the web server and the users. With the increasing amount of users and content on the Internet, handling this amount of data is a non-trivial task. On the other hand it has to provide accurate results. This is usually performed via a two stage process [16]: The first stage is a fast and simple filtration procedure that returns few hundreds or thousands of candidates from possibly millions of alternatives. During the second stage, more complex data mining techniques are performed to reduce the preliminary results to just a few high-quality results. In this paper we focus on improving the space consumption and time to perform queries on the web access sequences obtained from the web server logs used during the first step, thus freeing resources for the second stage and therefore increasing the performance of the process.

We present a space-efficient data structure in the Word-RAM model for representing web access logs, based on the Burrows-Wheeler Transform (BWT) [2]. Our index is able to efficiently process various queries of interest, while representing the data in compressed form. In this paper we focus on the following key operations; others are described in the Conclusions.

- *Access(u, i)* : Access the i-th web page visited by user u.
- *UserPath(u)* : Return the complete path done by user u.
- *Count(P)* : Count how many times path P has been performed in the collection.
- *MostCommonNextPage(P, k)* : Return the k most common web pages visited after path P.
- *ListUsers(P, k)* : Return k distinct users that have followed path P.
- *MostFrequentPath(k, q)* : Return the k most frequent paths of length q done by the users.

Our experimental results show that our index is able to represent the web access logs using 0.85–1.03 of their plain representations, thereby *replacing* them by a representation that uses about the same space but efficiently answers various queries, within microseconds in most cases. Our index can be easily deployed in other types of applications that handle ordered sequences, such as GPS trajectories, stock price series, customer buying history, and so on.

To our knowledge, this is the first compressed representation of logs that answers queries specific of WUM applications.

2   Basic Concepts

2.1   Rank and Select

Two basic operations used as building blocks for space-efficient data structures are *rank* and *select*. Given a bitmap B of length n, $rank_b(i)$ computes the number of bits b up to position i. The operation $select_b(j)$ retrieves the position where the j-th bit b appears. Munro [17] and Clark [3] obtained constant-time solutions for both operations while using $o(n)$ bits of space on top of B. Raman et al. [23] managed to compress the space to $nH_0(B) + o(n)$ bits[1] while still supporting both operations in constant time.

The wavelet tree [11] of a sequence S of length n over an alphabet of size σ extends the results for *rank* and *select* to general sequences, by decomposing the sequence hierarchically alphabet-wise in the form of a balanced tree. Internal nodes T_v store a binary string B_v. The root contains n bits, one per symbol in the sequence, and they are set to 0 or 1, depending on whether the corresponding symbol of S belongs to the lower or higher half of the alphabet. The left/right subtree is built for the subsequence of elements that have a 0/1 on the root. This decomposition continues, halving the alphabet, until the leaves, which correspond to a single symbol. All bitvectors are processed to handle binary rank and select queries in $O(1)$ time. This data structure accesses any $S[x]$ and solves $rank_b(S, x)$ and $select_b(S, x)$ in $O(\lg \sigma)$ time, using $n \lg \sigma(1 + o(1))$ bits (which is close to the space a plain representation of S would require).

2.2   Range Minimum Queries

A range minimum query asks for the position of the minimum element in a given range (i, j) of an integer array A of length n, that is, $RMQ_A(i, j) = argmin_{i \leq k \leq j} A[k]$. This query can be solved in constant time [8], after building a Cartesian tree over the array, convert it into a general tree using the usual bijection, and representing the general tree with a compact tree representation [27] that answers *lowest common ancestor (lca)* and other queries in constant time. This data structure requires $2n + o(n)$ bits.

The Cartesian tree of array $A[1, n]$ is a binary tree whose root corresponds to the minimum position i in A, its left child is the Cartesian tree of $A[1, i - 1]$, and its right child is the Cartesian tree of $A[i + 1, n]$.

---

[1] $H_0(B)$ is the zero-order entropy of the bitmap B.

2.3   Burrows-Wheeler Transform and the SSA Index

The Succinct Suffix Array (SSA) [20] is a compressed index that builds on the Burrows-Wheeler transform (BWT) of a text [2]. The main idea is to represent the BWT of a text T in compressed space and support pattern matching.

Given a text T of length N, ending with a unique symbol $ smaller than the rest, the BWT corresponds to a permutation of the symbols in T that is reversible. A simple way to describe the transformation is to imagine the $N \times N$ matrix of the N cyclic rotations of the text T, sort the rows lexicographically, and then keep the last column of the matrix, $L[1, N] = BWT(T)$. The first column, formed by all the characters of T in order, is called $F[1, N]$. Note that any $F[i]$ is preceded by $L[i]$ in T.

It has been shown [15] that local compressors tend to handle the BWT of a text much better than the text itself, since the transformation tends to cluster together occurrences of the same symbol. This is not surprising, since the symbols are actually arranged according to their context (symbols appearing after it). An interesting operation that allows to support the ones we are interested in is known as the LF-mapping. $LF(i)$ tells where does $L[i]$ appear in $F[i]$, this way allowing us to retrieve the text that precedes it in T in backward form: $L[i]$, $L[L[i]]$, and so on. The LF operation can be computed as $LF(i) = rank_c(BWT(T), i) + occ[c]$, where $occ[c]$ corresponds to the number of symbols lexicographically smaller than c in T. By representing $BWT(T)$ with a wavelet tree [11], the LF operation takes $O(\lg \sigma)$ time, the same as accessing any position in $BWT(T)$.

The *backward search* operation [7] returns in $O(m \lg \sigma)$ time the range $L[sp, ep]$ from where all the occurrences of a given pattern P of length m can be located. It is called backward search since its procedure seeks the pattern in reverse order. Backward search is sufficient to compute the number of occurrences of P, as $ep - sp + 1$; this procedure is called *count*.

A Succinct Suffix Array (SSA) enhances the BWT representation with a sampling of some suffix array entries. This sampling is used to locate the actual positions where P occurs in T from the range $L[sp, ep]$. Given a *sampling factor* s_a, which requires $O((N/s_a) \lg N)$ further bits of space, the SSA *locates* the position in T of any of the $ep - sp + 1$ occurrences of P in $O(s_a \lg \sigma)$ time. With a similar sampling, the SSA can also *extract* any desired substring $T[l, r]$ in $O((s_a + r - l) \lg \sigma)$ time. By choosing any $s_a = \omega(\lg_\sigma N)$, the SSA index can be represented using $NH_k(T) + o(N \lg \sigma)$ bits of space, where $NH_k(T)$ is the k-th order entropy of the text T.

3   Indexing Web Access Sequences

3.1   Construction

We start by concatenating all ordered web access sequences $\mathcal{S} = \{S_1, S_2, \ldots, S_n\}$ from all users into a sequence $\mathcal{T}(\mathcal{S})$ of size $N = \sum_{i=1}^{n} |S_i|$ over an alphabet of size σ, where σ is the amount of distinct web pages visited by any user in the log file. Instead of building the BWT to index $\mathcal{T}(S)$, we construct the index over $\mathcal{T}(S)^R$, that is, $\mathcal{T}(S)$ has each S_i reversed. This simple trick allows us to maintain a range of elements that match the sequence of requests so far, while performing backward search, and thus allowing us to add new arriving requests by just performing one more step. Having the BWT of $\mathcal{T}(S)^R$ is not enough to reconstruct the information from the log. We also need to store the user identifier associated with each position in the sequence. To do so without spending $N \lg n$ bits to associate a user id to each position in the BWT of $\mathcal{T}(S)^R$, we construct a bitmap B of length N and mark with a 1 the positions where each S_i ends in $\mathcal{T}(S)^R$. We later index B to solve $rank$ and $select$ queries in constant time using compressed space [23]. This is enough to obtain the user associated to a location in the BWT of the sequence, by locating its original position p in the sequence and then performing $rank_1(B, p)$.

For listing the distinct users (strings) where path P occurs, we implement Muthukrishnan's document listing algorithm [18], as compressed by Sadakane [26]. We construct a temporary array $U[i] = rank_1(B, i)$, for $1 \leq i \leq N$, that stores the user ids and then permute the values so that the ids are aligned to the $BWT(\mathcal{T}(S)^R)$ sequence. Another integer array C is constructed by setting $C[i] = select_{U[i]}(U, rank_{U[i]}(U, i) - 1)^2$ for all $0 \leq i \leq N$, and then build a RMQ data structure on C. We keep this structure and discard C and U.

The RMQ data structure requires $2N + o(N)$ bits. The SSA index requires $NH_k(\mathcal{T}(S)) + o(N \lg \sigma)$ bits. The representation of bitmap B takes $H_0(B) + o(N) \leq N + o(N)$ bits. Note that we do not store the users, nor the frequencies in an explicit way.

3.2 Queries

Access(u, i). To obtain the i-th web page visited by user u within the sequence, we need to locate the position in the original sequence $\mathcal{T}(S)^R$ where the user's session begins. We can do this by computing $p = select_1(B, u + 1) - 1 - i$ and then applying $extract\ \mathcal{T}(S)^R[p, p]$ on the SSA index, in $O(s_a \lg \sigma)$ time.

UserPath(u). To obtain the path done by user u, we compute $p_1 = select_1(B, u)$ and $p_2 = select_1(B, u + 1) - 1$ and then $extract\ \mathcal{T}(S)^R[p_1, p_2]$ using the SSA index. This takes $O((\ell + s_a) \lg \sigma)$ time, where $\ell = p_2 - p_1$ is the length of the extracted path.

Count(P). Given a path P of length m we can count its occurrences, by just performing $Count(P)$ on the SSA index in $O(m \lg \sigma)$ time.

MostCommonNextPage(P, k). We describe this operation incrementally. Assume we have already processed the sequence of requests $P = r_1, r_2, \ldots, r_{m-1}$. Our invariant is that we know the interval $[sp, ep]$ corresponding to path P, and

---

^{2} To avoid corner cases, we define $select_{U[i]}(U, 0) = -1$.

$\mathcal{T}(\mathcal{S})^R$	a	a	a	b	c	a	d	$	a	b	c	d	a	$	a	a	a	b	c	d	a	$	a	a	a	$	0
B	1	0	0	0	0	0	0	0	1	0	0	0	0	0	1	0	0	0	0	0	0	0	1	0	0	0	1

BWT	$	a	a	a	d	a	d	d	a	$	0	$	a	a	a	a	$	c	a	a	a	b	b	b	a	c	c
U	4	4	3	2	1	4	3	2	4	3		2	1	3	1	3	1	1	1	3	2	1	3	2	1	3	2
C	-1	0	-2	-3	-4	1	2	3	5	6		7	4	9	12	13	14	16	17	15	11	18	19	20	21	22	23

RMQ

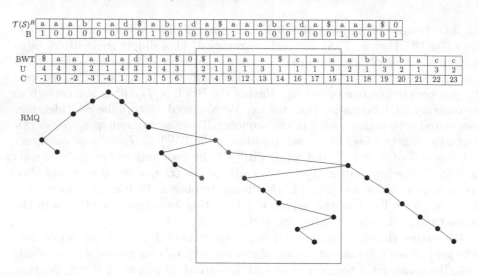

Fig. 1. Layout of our index organization using example sequences: $S_1 = dacbaaa, S_2 = adcba, S_3 = adcbaaa$ and $S_4 = aaa$. The dollar ($) symbol is used to represent the end of each sequence and the zero symbol is used to mark the end of the concatenation of the sequences $\mathcal{T}(\mathcal{S})^R$. At the bottom we show the topology of the Cartesian tree built over the C array representing the RMQ data structure. Recall that arrays U and C are not represented and are only shown for guidance.

this is sufficient to answer query $MostCommonNextPage(P, k)$. Now a new request r_m arrives at the end of P. Then we proceed as follows:

1. Update the range $[sp, ep]$ in $BWT(\mathcal{T}(\mathcal{S}))$ using r_m, in $O(\lg \sigma)$ time [7].
2. Retrieve the k most frequent symbols in $BWT[sp, ep]$, which are precisely those preceding the occurrences of P^R in $\mathcal{T}(\mathcal{S})^R$, or following P in $\mathcal{T}(\mathcal{S})$.

The second step is done with the heuristic proposed by Culpepper et al. [4] to retrieve the k most frequent symbols in a range of a sequence represented with a wavelet tree. It is a greedy algorithm that starts at the wavelet tree root and maps the range (in constant time, using $rank$ on the bitmap of the wavelet tree node) to its children, until reaching the leaves. The traversal is prioritized by visiting the longest ranges first, and reporting the symbols corresponding to the first k leaves found in the process. The worst-case performance of this algorithm is bounded by the number of different symbols present in the string range. This is smaller than both σ and the size of the range. By using more sophisticated data structures (that nevertheless do not add much space) [13, 21], a worst case of $O(k + \text{polylog} n)$ time can be guaranteed. Note that, if we want to list *all* the request that have followed P, we can use an optimal algorithm based on depth-first traversal of the wavelet tree [9].

ListUsers(P, k). To list k (or all) distinct users that have followed path P we first locate the starting and ending points $[sp, ep]$ for the given path, using the

SSA index in $O(m \lg \sigma)$ time (we can also proceed incrementally as in operation *MostCommonNextPage*). Then we apply the optimal document listing algorihtm [18, 26]. Each value $C[p] < sp$, for $sp \leq p \leq ep$, signals a distinct value of $U[p]$ in $U[sp, ep]$. Recall that we do not have C or U anymore, but the procedure for extracting the list of users can be emulated with the RMQ data structure over array C and the bitmap B. The procedure for extracting the list of all users works as shown in Algorithm 1. Function *locate* takes a position in the BWT and maps it to the corresponding position in the original $\mathcal{T}(\mathcal{S})^R$, in $O(s_a \lg \sigma)$ time. Listing k distinct users for path P takes $O((k \cdot s_a + p) \lg \sigma)$ time.

Fig. 1 shows an example of listing users. The framed region represents the range sp, ep. Red nodes in the RMQ tree represent the position of the retrieved users, while the blue node represents the last visited node before returning.

Algorithm 1. – UserListing(sp, ep, users = ∅)

$p \leftarrow RMQ_C(sp, ep)$
if $ep < sp$ then
    return *users*
end if
$u \leftarrow rank_1(B, locate(p))$
if $u \notin users$ then
    $users \leftarrow users \cup u$
    $UserListing(sp, p - 1, users)$
    $UserListing(p + 1, ep, users)$
end if

MostFrequentPath(k, q). We want to retrieve the k most frequent paths of a certain length q done by the users in the system. We start by pushing into a priority queue all ranges obtained by performing a backward-search of paths of length 1 for each possible symbol (σ at most). Now, we extract the biggest range from the priority queue as well as the path that created that range. We execute the same procedure again, creating new paths by appending to the extracted path one further symbol (trying the σ possible ones in the worst case) and we push the ranges obtained by performing the backward search on these new paths into the priority queue. When we extract a path of length q, we report it and remove it from the priority queue. The procedure ends when k paths are reported or when the priority queue is empty. In the worst case, this operation can take $O(\sigma^q)$.

This method may perform poorly when the alphabet σ is large. An optimization is to avoid trying out all the σ characters to extend the current path, but just those symbols that do appear in the current range. Those are found by traversing the wavelet tree from the root towards all the leaves that contain some symbol in the current range [9].

4   Experiments and Results

Setup and Implementations. We used dedicated server with 16 processors Intel Xeon E5-2609 at 2.4GHz, with 256 GB of RAM and 10 MB of cache.

Table 1. Space usage, in bytes, of the data structures used in our index. Plain corresponds to the sum of the space usage of plain representations of the sequence and users. Ratio corresponds to the total index size divided by the plain representation size.

Data Structure	Msnbc	Kosarak	Spanish
SSA	3,175,504	12,500,964	75,667,016
RMQ	1,770,690	2,807,570	39,538,238
Users Bitmap	608,044	777,804	9,413,732
Total	**5,554,246**	**16,086,346**	**124,618,994**
Plain	**5,782,495**	**18,884,286**	**120,613,202**
Ratio	**0,96**	**0,85**	**1,03**

The operating system is Linux with kernel 3.11.0-15 64 bits. We used g++ compiler version 4.8.1 with full optimizations (-O3) flags.

We implemented the SSA index using the public available wavelet tree implementation obtained from libcds (http://www.github.com/fclaude/libcds) and developed the heuristic proposed by Culpepper et al. [4] on top of that implementation. The wavelet tree needed for the SSA index uses a RRR [23] compressed bitmap representation. The bitmap B needed to retrieve the users is used in plain form [10]. We implemented the RMQ data structure based on compact tree representations [1], which in practice requires $2.38n$ bits. Our implementation is available at https://gitlab.com/fclaude/wum-index/.

Experimental Data. We used web access sequences from the public available Msnbc, Kosarak, and Spanish datasets. The Msnbc dataset comes from Internet Information Services log files of msnbc.com for a complete day of September, 28 of 1999. It contains web access sequences from $989,818$ users with an average of 5.7 web page categories visits per sequence, the alphabet size of this dataset is $\sigma = 17$. The Kosarak dataset contains the click-stream data obtained from a Hungarian on-line news portal. It contains sequences of $990,000$ users with an average of 8.1 web page visits per sequence and an alphabet size $\sigma = 41,270$. Finally, the Spanish dataset contains the visitors' click-stream obtained from a Spanish on-line news portal during September 2012. This dataset consists of $9,606,228$ sequences of news-categories that were visited, with an average of 12.3 categories visited per sequence and an alphabet size of $\sigma = 42$.

Space Usage. Table 1 shows the space usage of each data structure for each dataset. Row "Plain" shows the space required to represent the original sequence $\mathcal{T}(\mathcal{S})^R$ using an array of $N \lg \sigma$ bits plus an array to represent the users using $n \lg N$ bits. The table shows that our index is able to compress the sequence by up to 15%, on the Kosarak dataset, while requiring only 3% extra space at most, on the Spanish dataset. Within this space, we are able to support the aforementioned operations, while at the same time can reconstruct the original sequence and the users information.

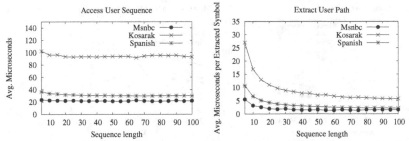

Fig. 2. On the left, average microseconds to perform $Access(u, i)$; on the right, average microseconds per extracted symbol for the operation $UserPath(u)$

We evaluated alternatives to the SSA, such as the *Compressed Suffix Array (CSA)* [24] and the *Compressed Suffix Tree (CST)* [25], using the ones provided by the sdsl-lite library (https://github.com/simongog/sdsl-lite). Table 2 compares their space usage to our SSA at representing the sequence. The SSA is a better choice in this case, using up to 33% less space than the others. The CST could be used to compute some of the operations, since it is naturally a faster alternative. In fact, we evaluated this alternative for counting the occurrences of a sequence, and it is in practice four to twenty times faster than the SSA. We discarded it since the space requirement makes it unpractical for massive datasets. In fact, the CST needs to be augmented in order to support all the operations presented in this paper, which would increase its memory usage.

Time Performance. To evaluate the main operations using our proposed data structure, we generate query paths by choosing uniformly at random a position in the original sequences, and then extracting a path of the desired length.

Fig. 2 (left) shows the time to access positions chosen uniformly at random from users whose traversal log has length 1 to 100. The time does not depend on the length, but directly on $s_a \lg \sigma$. Fig. 2 (right) shows the time per symbol extracted, when we access the whole sequence associated with a user. We can see that for users with short interactions the SSA sampling has a greater effect, and this is amortized when accessing longer sequences, that is, the term $O(s_a \lg \sigma)$ is spread among more extracted symbols and the time converges to $\lg \sigma$.

Table 2. Comparison of the space consumption of the SSA, CSA, and CST for representing sequence $\mathcal{T}(\mathcal{S})^R$. We show the ratio of each index space over the plain representation of the sequence without the user's information by using $N \lg \sigma$ bits.

Data Structure	Msnbc	Kosarak	Spanish
SSA	1,08	0,77	0,85
CSA	1,61	0,87	1,13
CST	3,82	1,43	2,96

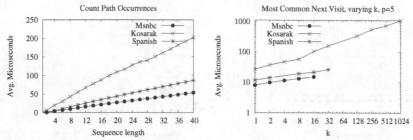

Fig. 3. On the left, average microseconds to perform *Count(P)*; on the right, average microseconds for the *MostFrequentPath* operation with varying k using patterns of fixed length p = 5

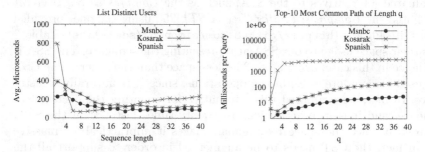

Fig. 4. On the left, average microseconds to list distinct users that traversed paths with varying lengths. On the right, the time required to perform top-10 most common path operation for varying pattern lengths.

We then measured the time to count the number of times a certain path appears in the access sequence. Fig. 3 (left) shows the time per query. As expected, it grows linearly with the length of the path being counted, and the $\lg \sigma$ term determines the slope of the line. On Fig. 3 (right) we show the results for the *MostCommonNextPage* operation. The x-axis and y-axis are in log scale. For datasets containing small alphabets such as Msnbc ($\sigma = 17$) and Spanish ($\sigma = 42$) this operation is performed in under 30 microseconds for all possible values of k (note it is impossible to obtain more than σ distinct symbols), and shows a logarithmic behavior. We also note that the slope of the logarithm depends on the value of σ, as shown in the Kosarak dataset. The operation, however, is still reasonably fast, taking less than 1 millisecond for $k = 1024$.

Fig. 4 (left) shows the time for retrieving the set of users that followed a given access pattern in the system. For shorter sequences, the index has to retrieve a bigger set of users, as these sequences are more likely to appear. As the sequences grow in length, the time decreases, since the resulting set is also smaller. The behavior for Kosarak, which after a certain point starts increasing in time per query, can be explained by the fact that determining the range $[sp, ep]$ grows linearly with the length of the pattern, and at some point it dominates the query time. This is also expected, at a later point, for Msnbc and Spanish.

Finally, Fig. 4 (right) shows the query time for operation $MostFrequentPath(k, q)$. Our first implementation tried following all symbols at every step of the algorithm. This worked quite well for small alphabets but had a very bad performance on the Kosarak dataset. This plot shows the implementation traversing the wavelet tree at each step to only follow symbols that do appear in the range. This gives a slightly worse performance for small alphabets, but a considerable speedup (1–2 orders of magnitude) for larger ones.

5 Discussion and Future Work

We introduced a new data structure for handling web access sequences in compressed form that fully replaces the original dataset and also supports useful operations for typical WUM processes. Our experiments show that our index uses about the same size of the plain representation of the data, and within this space supports various relevant operations within tens of microseconds. This is competitive; consider that in most common scenarios the systems have to reply over a network, which have considerable latency and transfer time. Ours is the first compressed representation tailored for this scenario.

We have not yet fully explored other possible operations of interest in log mining that can be supported with our arrangements. For example, we can count the number of users that followed some path in constant time using $2n + o(n)$ bits using document counting [26], compute the k users that have followed a path most frequently using top-k document retrieval [13], and others.

Our index can be easily adapted to custom scenarios by adding satellite information, such as duration of the visit, actions (buy, login/logout, comment, etc.), browser information, location, and others, to each event in the log and include the information by mapping it to the BWT transform for later processing. This enables our index to be applied in other scenarios involving ordered sequences, such as GPS trajectories, stock price series, customer buying history, and so on.

References

1. Arroyuelo, D., Cánovas, R., Navarro, G., Sadakane, K.: Succinct trees in practice. In: Proc. 11th ALENEX, pp. 84–97 (2010)
2. Burrows, M., Wheeler, D.J.: A block-sorting lossless data compression algorithm. Tech. rep., Digital Equipment Corporation (1994)
3. Clark, D.: Compact Pat Trees. Ph.D. thesis, Univ. of Waterloo, Canada (1996)
4. Culpepper, J.S., Navarro, G., Puglisi, S.J., Turpin, A.: Top-k ranked document search in general text databases. In: de Berg, M., Meyer, U. (eds.) ESA 2010, Part II. LNCS, vol. 6347, pp. 194–205. Springer, Heidelberg (2010)
5. Domènech, J., Gil, J.A., Sahuquillo, J., Pont, A.: Web prefetching performance metrics: A survey. Perform. Eval. 63(9), 988–1004 (2006)
6. Dongshan, X., Junyi, S.: A new markov model for web access prediction. Computing in Science and Eng. 4(6), 34–39 (2002)

7. Ferragina, P., Manzini, G.: Indexing compressed texts. J. ACM 52(4), 552–581 (2005)
8. Fischer, J., Heun, V.: Space-efficient preprocessing schemes for range minimum queries on static arrays. SIAM J. Comp. 40(2), 465–492 (2011)
9. Gagie, T., Navarro, G., Puglisi, S.: New algorithms on wavelet trees and applications to information retrieval. Theor. Comp. Sci. 426-427, 25–41 (2012)
10. González, R., Grabowski, S., Mäkinen, V., Navarro, G.: Practical implementation of rank and select queries. In: Proc. Posters 4th WEA, pp. 27–38 (2005)
11. Grossi, R., Gupta, A., Vitter, J.S.: High-order entropy-compressed text indexes. In: Proc. 14th SODA, pp. 841–850 (2003)
12. Han, J., Cheng, H., Xin, D., Yan, X.: Frequent pattern mining: current status and future directions. Data Mining Knowl. Disc. 15(1), 55–86 (2007)
13. Hon, W.K., Shah, R., Vitter, J.: Space-efficient framework for top-k string retrieval problems. In: Proc. 50th FOCS, pp. 713–722 (2009)
14. Hussain, T., Asghar, S., Masood, N.: Web usage mining: A survey on preprocessing of web log file. In: Proc. ICIET, pp. 1–6 (2010)
15. Manzini, G.: An analysis of the Burrows-Wheeler transform. J. ACM 48(3), 407–430 (2001)
16. Mobasher, B.: Data mining for web personalization. In: Brusilovsky, P., Kobsa, A., Nejdl, W. (eds.) Adaptive Web 2007. LNCS, vol. 4321, pp. 90–135. Springer, Heidelberg (2007)
17. Munro, J.I.: Tables. In: Chandru, V., Vinay, V. (eds.) FSTTCS 1996. LNCS, vol. 1180, pp. 37–42. Springer, Heidelberg (1996)
18. Muthukrishnan, S.: Efficient algorithms for document retrieval problems. In: Proc. 13th SODA, pp. 657–666 (2002)
19. Nadi, S., Saraee, M., Davarpanah-Jazi, M.: A fuzzy recommender system for dynamic prediction of user's behavior. In: Proc. ICITST, pp. 1–5 (2010)
20. Navarro, G., Mäkinen, V.: Compressed full-text indexes. ACM Comp. Surv. 39(1) (2007)
21. Navarro, G., Valenzuela, D.: Space-efficient top-k document retrieval. In: Klasing, R. (ed.) SEA 2012. LNCS, vol. 7276, pp. 307–319. Springer, Heidelberg (2012)
22. Pei, J., Han, J., Mortazavi-Asl, B., Zhu, H.: Mining access patterns efficiently from web logs. In: Terano, T., Liu, H., Chen, A.L.P. (eds.) PAKDD 2000. LNCS, vol. 1805, pp. 396–407. Springer, Heidelberg (2000)
23. Raman, R., Raman, V., Rao, S.: Succinct indexable dictionaries with applications to encoding k-ary trees, prefix sums and multisets. ACM Trans. Alg. 3(4), art. 43 (2007)
24. Sadakane, K.: New text indexing functionalities of the compressed suffix arrays. J. Alg. 48(2), 294–313 (2003)
25. Sadakane, K.: Compressed suffix trees with full functionality. Theor. Comp. Sys. 41(4), 589–607 (2007)
26. Sadakane, K.: Succinct data structures for flexible text retrieval systems. J. Discr. Alg. 5(1), 12–22 (2007)
27. Sadakane, K., Navarro, G.: Fully-functional succinct trees. In: Proc. 21st SODA, pp. 134–149 (2010)
28. Su, Z., Yang, Q., Lu, Y., Zhang, H.: Whatnext: A prediction system for web requests using n-gram sequence models. In: Proc. 1st WISE, pp. 214–224 (2000)
29. Sumathi, C., Valli, R.P., Santhanam, T.: Automatic recommendation of web pages in web usage mining. Intl. J. Comp. Sci. Eng. 2, 3046–3052 (2010)

A Compressed Suffix-Array Strategy
for Temporal-Graph Indexing*

Nieves R. Brisaboa[2], Diego Caro[1], Antonio Fariña[2], and M. Andrea Rodríguez[1]

[1] Dept. Comp. Sci., University of Concepción, Chile
{diegocaro,andrea}@udec.cl
[2] Database Lab., University of A Coruña, Spain
{brisaboa,fari}@udc.es

Abstract. Temporal graphs represent vertexes and binary relations that change over time. In this paper we consider a temporal graph as a set of 4-tuples (v_s, v_e, t_s, t_e) indicating that an edge from a vertex v_s to a vertex v_e is active during the time interval $[t_s, t_e)$. Representing those tuples involves the challenge of not only saving space but also of efficient query processing. Queries of interest for these graphs are both direct and reverse neighbors constrained by a time instant or a time interval. We show how to adapt a Compressed Suffix Array (CSA) to represent temporal graphs. The proposed structure, called Temporal Graph CSA ($TGCSA$), was experimentally compared with a compact data structure based on compressed inverted lists, which can be considered as a fair baseline in the state of the art. Our experimental results are promising. $TGCSA$ obtains a good space-time trade-off, owns wider expressive capabilities than other alternatives, obtains reasonable space usage, and it is efficient even when performing the most complex temporal queries.

1 Introduction

There is an increasing need to handle large graphs that change over time and where not only the current state but also the past state is of interest. For example, consider the evolution of friendship relations when a user adds or removes friends in online social networks, how the citation network grows when new scientific articles are published, how connectivity between mobile devices evolves through time when their base station changes, or how links appear and disappear on the Web graph. The compact representation of temporal graphs is then a relevant problem since direct/reverse-neighboring queries constrained by time instant/interval could benefit from keeping as much data as possible in main memory.

* Founded in part by Fondef [D09I1185], Fondecyt [1140428] and a CONICYT doctoral fellowship (for the Chilean group); and, for the Spanish group, by MINECO (PGE and FEDER) [TIN2013-46238-C4-3-R, TIN2013-47090-C3-3-P]; CDTI, AGI, MINECO [CDTI-00064563/ITC-20133062]; ICT COST Action IC1302; and by Xunta de Galicia (co-founded with FEDER) [GRC2013/053].

E. Moura and M. Crochemore (Eds.): SPIRE 2014, LNCS 8799, pp. 77–88, 2014.
© Springer International Publishing Switzerland 2014

A temporal graph can be seen as a set of 4-tuples of the form (v_s, v_e, t_s, t_e), indicating that an edge from a vertex v_s to a vertex v_e is active during the interval $[t_s, t_e)$. A compact representation for this set of 4-tuples, called *EdgeLog* uses an adjacency list to represent edges and lists of time points indicating when each edge is turned on and off. *EdgeLog* can use existing compact representations for inverted lists [16], d-gaps, or k^2-trees [3].

This paper proposes *Temporal Graph CSA* (*TGCSA*), a novel compact data structure based on the Compressed Suffix Array (*CSA*) [14]. We discuss how *TGCSA* opens new opportunities for the application of suffix arrays that are worth exploring both for temporal and general graphs.

Previous work in this area is still incipient. In [7], they represented a temporal graph as several static graphs (or snapshots), storing the active edges for each time point in the lifetime of the graph. Its main drawback is the amount of space used even if the state of an edge (active or not) does not vary for a long time. Storing differences between some snapshots (carefully chosen) saves space but requires processing them at query time [13,9]. This has the advantage of storing only what changes between consecutive time points and of answering queries about the active direct and reverse neighbors of a vertex. Data structures for temporal graphs based on adjacency lists [4] and on distributed environments [8,10] exist; however, they have focused on improving time performance neglecting space cost. Recently, we found a preliminary effort to define efficient compact structures for temporal graphs [2]. However, their results apply only to medium size graphs and show that there is much work to do in this area.

The structure of this paper is as follows. Section 2 presents preliminary concepts and the *EdgeLog* as a baseline to compare with the proposed structure. Section 3 describes *TGCSA*, which is followed in Section 4 by the experimental evaluation using real and synthetic data. Conclusions and future research directions are given in Section 5.

2   Preliminary Concepts

Temporal Graph Definition. Formally, a temporal graph is a set C of contacts between a set of vertexes V during a set of time points T representing the *lifetime* of the graph. A *contact* of an edge $(u, v) \in E \subseteq V \times V$ is a 4-tuple $c = (u, v, t, t')$, where $[t, t') \in T \times T$ is the time interval when the edge (u, v) is active [11]. We say that an edge (u, v) is *active* at time t if there exists a contact $(u, v, t_s, t_e) \in C$ such that $t \in [t_s, t_e)$. We refer to an *aggregated graph* as the static graph composed by all edges that have been active during the lifetime of the temporal graph.

For the purpose of this paper, we define four operations on the temporal graph for a given time point t: (1) *Edge existence at t* checks if an edge is active at t. (2) *Direct neighbors at t* returns the adjacent active neighbors at a given time point t. (3) *Reverse neighbors at t* gives the active reverse adjacent vertexes at time t. (4) *Snapshot at t* returns all active edges at a time point t. For example, given Figure 1a, the snapshot at $t = 5$ corresponds to the edges $\{(b, c), (b, e), (d, b), (e, d)\}$.

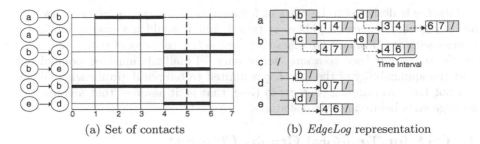

(a) Set of contacts   (b) *EdgeLog* representation

Fig. 1. A temporal graph of 5 vertexes. The dashed line corresponds to the time point $t = 5$. Reverse aggregated graph is omitted in (b). (Figure adapted from [11]).

EdgeLog: *Baseline Representation.* A simple temporal graph representation [4] stores the aggregated graph as $|V|$ adjacency lists, one per vertex, with a sorted list of time intervals attached to each neighbor indicating when that edge is active. Figure 1b shows a conceptual example.

To check if an edge (u, v) is active at time t, we first check if v appears within the adjacency list of vertex u. If the edge is found, then we need to check if t falls into one of the time intervals related to (u, v) that are represented in the time-point list of that edge. Direct neighbors of vertex u at time t are recovered similarly. For each neighbor v in the adjacency list of u, we check if t is within the time intervals of the edge (u, v).

This simple representation has two main drawbacks: (1) it uses much space; and (2) reverse neighbors operation requires traversing all adjacency lists. Both issues are overcome in what we call *EdgeLog*. (1) Since both the adjacency lists and the time-interval lists are sorted (i.e., they are of the form $\langle t_1, t_2, t_3, ..., t_l \rangle$, with $t_i < t_{i+1}$), they can be represented as d-gaps $\langle t_1, t_2 - t_1, t_3 - t_2, ..., t_l - t_{l-1} \rangle$ and compressed using variable-length encoding for the differences (e.g., PForDelta [17], S16 [15]). Also, (2) to avoid traversing all adjacency lists in reverse-neighbor queries, *EdgeLog* stores the reverse aggregated graph containing an adjacency list with the reverse neighbors of each vertex. Therefore, to get the reverse neighbors of vertex v at time t, we first use the reverse adjacency list to obtain the candidate reverse neighbors of v. Then, for each candidate reverse neighbor u, we move to v in its adjacency list and finally check if the edge (u, v) is active at time t (using the time-interval list of the edge).

Strong and Weak Points in *EdgeLog*. Although *EdgeLog* is a simple structure using well-known technology, it is expected to be extremely space-efficient when the temporal graph has a low number of edges per vertex and a large number of contacts per edge. In the opposite way, a low number of contacts per edge will have a negative impact on the compression achieved by *EdgeLog* (as d-gaps become large). Note also that, even with the reverse aggregated graph to find reverse neighbors, the performance is expected to be poor if the number of edges per vertex is high because all their adjacency lists will have to be checked.

EdgeLog is designed to be efficient for queries of the type *edge existence at t* and direct and reverse neighbors at time t, but it could not answer efficiently queries such as: *"Find all the edges that have active contacts at time t"* or *"Find all the edges that have been only active once"*. Finally, it must be pointed out that the applicability of the *EdgeLog* is limited to temporal graphs where edges can not have overlapping contacts in time; that is, it assumes that a contact of an edge ends before another one starts.

3   CSA for Temporal Graphs (*TGCSA*)

Our *Temporal Graph CSA* (*TGCSA*) is an adaptation of Sadakane's Compressed Suffix Array (*CSA*)[14]. More precisely, it is based on the *integer-based CSA* (*iCSA*) that allows *CSA* to deal with large (integer-based) alphabets (see [6] for details). Recall that *CSA* consists of three main elements to support searches: i) The symbols of the source alphabet S; ii) a bitmap D of size n to mark the positions of the suffix array A where the first symbol of the suffixes pointed to changes; and iii) an array Ψ such that $\Psi[i]$ indicates, for each position i in A, the position $z = \Psi[i]$ such that $A[z]$ points to the position $A[i] + 1$.

There is an important difference between the standard *CSA* and our implementation that we conceptually describe here. Let us assume that all the terms in a contact are made up from four disjoint alphabets $\Sigma_1, \Sigma_2, \Sigma_3,$ and Σ_4 such that $\Sigma_1 \prec \Sigma_2 \prec \Sigma_3 \prec \Sigma_4$ ($\prec$ indicates lexicographic order). Note that alphabets Σ_1 and Σ_2 represent the set of nodes V, while Σ_3 and Σ_4 represent the set of time points $\mathcal{T}$. Our procedure starts by creating an ordered list of n contacts, so that the contacts are sorted by their first term, then (if they have the same first term) by the second component, and so on. Now, those sorted contacts are regarded as a sequence of integers with $4n$ elements, and a suffix array $A[1, 4n]$ is built over it. Since the values in the four disjoint alphabets are ordered ($\Sigma_i \prec \Sigma_j$ $\forall i < j$), the first 25% entries in A ($A[1, n]$) will point to the first terms of all the contacts, the next n entries ($A[n + 1, 2n]$) to the second terms, and so on. Consequently, the first 25% entries of Ψ ($\Psi[1, n]$) will point to positions in the range $[n + 1, 2n]$, because, in the indexed sequence, each symbol $u \in \Sigma_1$ is followed by a symbol $v \in \Sigma_2$, and so on.

Note that, in the standard CSA, if $A[i], (i \in [3n + 1, 4n])$ points to the last term of the jth contact, then $\Psi[i]$ would store the position in A pointing to the first term of the following $(j+1)^{th}$ contact in the ordered list ($A[i]+1 = A[\Psi[i]]$), which would be in the range $[1, n]$. However, we modified those pointers in the last 25% of Ψ, because we want that, instead of pointing to the position $x = A[\Psi[i]]$ corresponding to the first term of the following contact, we want them to point to the first term of the same contact. This implies modify Ψ such that $A[\Psi[i]] = x - 1$ for $x \neq 1$, and $A[\Psi[i]] = n$ otherwise.

By starting at any entry i in Ψ and following the pointers $\Psi[\Psi[\Psi[\Psi[i]]]]$, all the elements of the current contact can be retrieved, but no entry from any other tuple will be reached. With our modification, it is not possible to traverse the whole *CSA* just using Ψ because consecutive applications of Ψ will cyclically obtain the four elements of the same contact.

3.1 Detailed Construction of the *TGCSA*

As indicated above, the first step to build a *TGCSA* is to create a sequence S with the ordered n contacts from $\mathcal{C}$. Hence we obtain, $S[1, 4n] = \langle u^1, v^1, t_s^1, t_e^1, u^2, v^2, t_s^2, t_e^2, \ldots, u^n, v^n, t_s^n, t_e^n \rangle$.[1]

Let us assume we have $\nu = |V|$ different vertexes and $\tau = |\mathcal{T}|$ periods of time. It is possible to define a reversible mapping function that maps the terms of any original contact $c = (u, v, t_s, t_e)$ into $c' = (u, v+\nu, t_s+2\nu, t_e+2\nu+\tau)$. To achieve this, we define an array $gaps[1, 4] \leftarrow [0, \nu, 2\nu, 2\nu+\tau]$ and $c'[i] \leftarrow c[i]+gaps[i]\ \forall i = 1\ldots 4$. This mapping defines four ranges of entries in an alphabet Σ' for both vertexes and times such that $|\Sigma'| = 2\nu + 2\tau$. Note that vertex i is mapped to either the integer $i + gaps[1]$ or $i + gaps[2]$ depending on whether it is the source or target vertex of an edge. Similarly, the time instant t is mapped to either $t + gaps[3]$ or $t + gaps[4]$. This will permit us to distinguish between starting/ending vertexes/times by simply checking the range where their value falls in.

Note that even though vertex i always exists in the temporal graph, either source vertex $u' = i + gaps[1] = i$ or target vertex $v' = i + gaps[2]$ may not actually be used. Similarly a time t' could not occur as an initial or as an ending time of a contact, yet we could be interested in retrieving all the edges that are active at that time t'.

To overcome the existence of holes in the alphabet Σ', a bitmap $B[1, 2\nu+2\tau]$ is used. We set $B[i] \leftarrow 1$ if the symbol i from Σ' occurs in a contact, and $B[i] \leftarrow 0$ otherwise. Therefore, each of the four terms within a contact (u, v, t_s, t_e) will correspond to a 1 in B. Now an alphabet Σ of size $\sigma = rank_1(B, 4n)$[2] is created containing the positions in B where the bits set to 1 occurs. For each symbol $i \in \Sigma'$ a *mapID(i)* function assigns an integer $id \in \Sigma$ to i, so that $id \leftarrow mapID(i) = rank_1(B, i)$ if $B[i] = 1$, and $0 \leftarrow mapID(i)$ if $B[i] = 0$. The reverse mapping is provided via a $unmapID(id) = select_1(B, id)$ function[3].

At this point, a sequence of ids $Sid[1, 4n]$ can be created by setting $Sid[i] \leftarrow mapID\ (S[i] + gaps[((i-1) \mod 4) + 1])\forall i = 1\ldots 4n$.

Indeed, being *type* $= 1, 2, 3, 4$, respectively, the types of source vertex, target vertex, starting time and ending time from the source sequence S, any source symbol i from S can be mapped into Sid as $id = getmap(i, type) \leftarrow rank_1(B, i+gaps[type])$. Similarly, the reverse mapping obtains $i = getunmap\ (id, type) \leftarrow select_1(B, id) - gaps[type]$.

Finally, an *iCSA* is built over Sid.[4] Note that since the vocabularies of the ids associated with the four terms of any contact are disjoint, the corresponding suffix array A will have four ranges of length n so that $A[(j-1)n+1, jn], j = 1..4$. Pointers in each range point to suffixes starting with a source vertex, a

---

[1] Note that the ordering is not relevant as we have a set of contacts. Therefore, we will assume contacts are sorted by the first term, then by the second one, and so on.

[2] $rank_1(B, i)$ returns the number of 1s in $B[1, i]$.

[3] $select_1(B, i)$ computes the position of the ith 1 in B.

[4] We actually added four integers set to *zero* that make up a dummy contact $(0,0,0,0)$ at the beginning of Sid. This is required to avoid limit-checks at query time.

target vertex, a starting time, or an ending time, respectively. Similarly, values in $\Psi[1, n]$ in the range of source vertexes, will point to the range of target vertexes $[n+1, 2n]$. Values in $\Psi[n+1, 2n]$ will point to the range of initial times $[2n+1, 3n]$. Those in $\Psi[2n + 1, 3n]$ will point to the range of ending times $[3n + 1, 4n]$. And finally, those in $\Psi[3n + 1, 4n]$ will point to the range of source vertexes $[1, n]$. Indeed, if $A[3n+1]$ points to the ending time of the kth contact of the collection, $z \leftarrow \Psi[3n+1]$ will indicate the position such that $A[z]$ points to the source vertex (first term) of the $(k + 1)^{th}$ contact. This is how Ψ works in a regular $iCSA$.

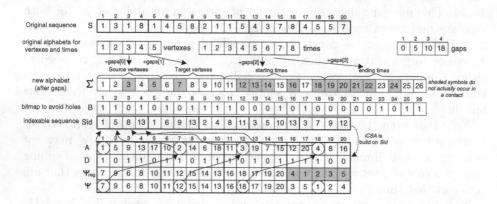

Fig. 2. Structures involved in the creation of a $TGCSA$ for the graph in Example 1

As discussed above, we modified the Ψ array in our $TGCSA$ to allow Ψ to move circularly from one term to the next one within the same contact. To do this, we simply have to modify the values in the regular Ψ so that, $\forall i = 3n + 1 \ldots 4n$, $\Psi[i] \leftarrow ((\Psi[i] - 2) \mod n) + 1$. This small change brings the interesting property of enabling to perform a query for any term of a contact in the same way. We use the $iCSA$ to binary search for any term of a contact obtaining a range $A[l, r]$, and then by circularly applying Ψ up to three times, we can retrieve the other terms of each contact pointed in $A[l, r]$.

Example 1. Let us assume we have a temporal graph with $|v| = 5$ vertexes numbered $1 \ldots 5$ and $|\tau| = 8$ time instants numbered $1 \ldots 8$. The graph contains the following five contacts: $(1, 3, 1, 8)$, $(1, 4, 5, 8)$, $(2, 1, 1, 5)$, $(4, 3, 7, 8)$, and $(4, 5, 5, 7)$. Figure 2 depicts all the structures involved in the creation of a $TGCSA$ that represents the temporal graph.   □

To sum up, the $TGCSA$ representation consists of a bitmap B, and the structures D and Ψ from the $iCSA$. B is compressed with Raman *et al.* strategy [12]. For D we used a faster bitmap from [6] using $1.375|D|$ bits.

3.2 Performing Queries in $TGCSA$

We can take advantage of the $iCSA$ capabilities at search time to solve all the typical queries in a temporal graph regarding *direct* and *reverse* vertexes from contacts that are active at a given time point t. Basically, we binary search the range in $A[l, r]$ for the given source or target vertex, and for each position $i \in [l, r]$, we apply Ψ circularly up to the third or fourth term where we can check if the starting-time and ending-time constraints either hold or not. In Figure 3 we show the pseudocode of the algorithm to obtain direct neighbors.

DirectNeighbors $(vrtx, t)$ //neighbors of $vrtx$ in contact $(vrtx, v, t_1, t_2)$ s.t. $t_1 \leq t < t_2$
(1) $u \leftarrow$ **getmap**$(vrtx, typeVertex = 1)$;   // map into final alphabet without holes
(2)   **if** $u = 0$ **then return** $\emptyset$;   // vertex does not appear as a source vertex
(3) $neighbors \leftarrow \emptyset$;
(4) $t_s \leftarrow$ **getmap**$(t, typeStartTime = 3)$; $t_e \leftarrow$ **getmap**$(t, typeEndTime = 4)$;
(5) $[lu, ru] \leftarrow$ **CSA_binSearch**(u);   // range $A[lu, ru]$ for vertex u
(6) $[lt_s, rt_s] \leftarrow$ **CSA_binSearch**(t_s);   // range $A[lt_s, rt_s]$ for starting time t_s
(7) $[lt_e, rt_e] \leftarrow$ **CSA_binSearch**(t_e);   // range $A[lt_e, rt_e]$ for ending time t_e
(8)   **for** $i \leftarrow lu$ **to** ru // checks time intervals for each occurrence of u
(9) $x \leftarrow \Psi[i]$; $y \leftarrow \Psi[x]$;
(10)   **if** $(y \leq rt_s)$ **then**
(11) $z \leftarrow \Psi[y]$;
(12)   **if** $(z > rt_f)$ **then**
(13) $neighbors \leftarrow neighbors \cup \{$**getunmap**$(x, typeRevVertex = 2)\}$;
(14) **return** $neighbors$;

Fig. 3. Obtaining the direct neighbors of a vertex in a contact that is active at time t

Edge operations consisting in checking if an edge (u, v) is active at time t are expected to be faster than direct neighbor queries as we can binary search for a phrase $u \cdot v$ rather than by a unique vertex u, hence returning a much shorter initial range. Finally, to solve *snapshot* queries returning the set of active contacts (u, v, t_1, t_2) such that $t_1 \leq t < t_2$, we can binary search $[lt_s, rt_s] \leftarrow CSA\_binSearch(getmap(t, 3))$ and $[lt_f, rt_f] \leftarrow CSA\_binSearch(getmap(t, 4))$. All the contacts pointed by $A[2n + 1, rt_i]$ hold $t_s \leq t$, and those in $A[rt_f + 1, 4n]$ hold $t_2 > t$. Therefore, $\forall i \in [2n + 1, rt_s]$, if $\Psi[i] > rt_f$ we recover the source and target vertexes as $\Psi[\Psi[i]]$ and $\Psi[\Psi[\Psi[i]]]$. The original values are obtained via *getunmap()*.

3.3   Strengths and Weak Points in *TGCSA*

One advantage of *TGCSA* with respect to other representations such as those in [2], or our baseline *EdgeLog* is that it actually represents the whole set of 4-tuples. Therefore, it has the same (strong) expressive power as if the set is stored in a relational database. Note that *TGCSA* can represent temporally overlapping contacts for one edge with no limitations.

Another important property is that *TGCSA* can answer queries over any component of a contact with the same mechanism. That is, searching for the contacts of a source vertex u is done in the same way as searching for the contacts starting at a specific time t. First, a binary search is performed over one of the four sectors of A depending on the term of the contact that is searched for (bounded in the query) to locate the range $A[l, r]$ associated with that value. Then, for each of the entries in that range, Ψ is applied three times to recover the other of terms of each contact. Note that other data structures are designed to answer efficiently some types of queries but they are not efficient at others, whereas $TGCSA$ has a more regular behavior.

Note also that inside the section devoted to any given symbol, in any of the four sectors of Ψ, all the pointers are always growing, which is a good property to allow compression. Unfortunately, this becomes a weakness of $TGCSA$ when the vocabularies are huge and symbols occur few times. In this case, Ψ will not be highly compressible. As shown in our experiments, compression in some synthetic datasets we created is poor when the relative number of contacts per time instant is low, or when the number of edges per vertex is low. In those cases, the increasing areas of Ψ are small and the gaps between pointers are compressed poorly.

4   Experimental Evaluation

We evaluated $TGCSA$[5] on real and synthetic datasets. We compared its space needs with *gzip* compressor and with the baseline *EdgeLog*.[6] We also included a comparison in both space and time for some query types. Table 1 describes the experimental datasets. The "base size" is the size of all the uncompressed graphs, representing terms in the contacts with 32-bit integers (128 bits/contact).

The real *Flickr-Days* and *Flickr-Seconds* datasets are well-known temporal graphs where contacts indicate the time-points when two people become friends. Note that each edge has only one contact that ends at the end of the lifetime of the temporal graph. Flickr-Days [5] has a granularity of time in *days*, with a lifetime of 135 days. Flickr-Seconds captures time in *seconds*.

For the synthetic datasets we created an aggregated graph with a uniform degree distribution (following Erdõs-Rényi model [1]), and then assigned a fixed number of contacts to each edge. We used different combinations of the parameters (number of edges per vertex and number of contacts per edge and per instant time) to understand how they affected the compression and behavior of both $TGCSA$ and *EdgeLog*. We present a summary of our results.

The tests were run on a machine with processors Intel(R) Core(TM) i7-3820 CPU @ 3.60GHz, quad-core, and 64GB DDR3 RAM. The operating system was Ubuntu 12.04 and the compiler gcc 4.6.3 (option -O3).

4.1   Space Comparison

In Table 2, we show three configurations of $TGCSA$ (Ψ_{16} corresponds to a dense sampling and Ψ_{256} to a sparser one) and compare them with *gzip* as a baseline to show the compressibility of the source dataset. Even tough an *iCSA*-based self-index built on English text typically reached the compression of *gzip* [6], the

---

[5] We used three different settings of $TGCSA$ that differ in the sample-rate for Ψ.

[6] *EdgeLog* was configured to use PForDelta with $b = \{32, 128\}$.

Table 1. Temporal graph datasets

Dataset	Vertexes (×1000)	Edges (×1000)	Lifetime (×1000)	Contacts (×1000)	average contacts/ vertex	edges/ vertex	contacts/ edge	Base size (MB)
Flickr-Secs	6,204	71,346	167,944	71,346	12	12	1	1,089
Flickr-Days	2,586	33,140	0.135	33,140	13	13	1	506
Erdos1	1,000	10,002	1,000	10,002	10	10	1	153
Erdos5	1,000	10,002	1,000	50,008	50	10	5	763
Erdos50	1,000	10,002	1,000	500,079	500	10	50	7,631
Erdos50R10	1,000	50,001	1	500,079	500	50	10	7,631

Table 2. Comparison on space usage. Space in bits per contact.

Dataset	gzip default	Edgelog Pfor32	Edgelog Pfor128	TGCSA Ψ_{16}	TGCSA Ψ_{64}	TGCSA Ψ_{256}
FlickrSecs	61.51	161.57	161.53	96.30	87.65	85.48
FlickrDays	30.19	102.06	101.13	60.34	51.06	48.71
Erdos1	83.43	185.91	187.63	105.29	99.07	97.51
Erdos5	63.02	70.32	82.09	94.09	86.52	84.63
Erdos50	53.71	36.76	35.67	89.87	80.93	78.65
Erdos50R10	38.37	24.69	24.10	72.62	62.67	60.19

compressibility of temporal graphs is not so good. Actually, the large number of 1-runs that appeared in Ψ when dealing with text is now much smaller in the *TGCSA*, and we are not able to reach the compression levels of *gzip*.

Focusing on *EdgeLog*, we see that it is completely unsuccessful when the number of contacts per edge is very small. However, when there are few edges and the number of contacts per edge grows, it becomes very successful as its inverted lists become highly compressible. *TGCSA* shows a more regular behavior, and reasonable space needs in most cases. It does not require as much space as *EdgeLog* when the number of contacts per edge is small, but it cannot cope with many contacts per edge because Ψ is irregular, as discussed above.

4.2 Performing Queries

We chose the real *Flickr-Secs* and the synthetic *Erdos50R10* datasets to show the main features of *TGCSA* when answering typical temporal-graph queries such as retrieving the active direct and reverse neighbors at a given time t, checking if an edge is active at time t,[7] and recovering all the source contacts that are active at a given time instant.

Results in Table 3 show that *TGCSA* is very fast at retrieving direct-neighbors, and the time to recover a contact is close to 1 μsec when using a dense sampling

---

[7] We used average times for 2000 queries with random values of t.

Table 3. Comparison on query performance. CPU-user times in μsec/contact reported.

Dataset	Edgelog Pfor32	Edgelog Pfor128	TGCSA Ψ_{16}	TGCSA Ψ_{64}	TGCSA Ψ_{256}	contactsReported
FlickrSecs.DirNei	0.02	0.02	1.48	4.26	9.44	960,364
FlickrSecs.RevNei	9.03	8.50	0.91	1.35	3.35	799,273
FlickrSecs.Edge	5.43	5.17	8.12	13.49	43.20	2,000
FlickrSecs.Snapshot	0.02	0.02	1.22	1.63	3.64	71,345,977
Erdos50R.DirNei	0.61	0.57	49.76	112.23	350.18	10,973
Erdos50R.RevNei	21.61	19.24	23.41	43.47	126.22	9,847
Erdos50R.Edge	4.24	4.12	3.48	5.32	14.48	2,000
Erdos50R.Snapshot	0.45	0.41	3.67	4.53	7.90	5,437,058

in Ψ (Ψ_{16}) and many contacts are retrieved. Yet, the performance of the snap-shot operation degrades if many contacts (to check) start before the last time but only a few of them are active at the query time. Note that *EdgeLog* is much faster when answering direct-neighbors as it is designed for these queries. It is also very fast at the snapshot operation. However, its advantage is reduced dras-tically on edge operations (where the *TGCSA* is able to binary search directly the range where the time intervals of the queried edge occurs in A).

As expected, reverse neighbors is the worst case for *EdgeLog*. In particular, its performance degrades when many reverse neighbors need to be checked. Yet, even for this type of queries, in the synthetic collection with less that 50 edges per node, *EdgeLog* was still faster than *TGCSA*.

It is interesting to point out that *TGCSA* is faster when performing reverse neighbor queries than the direct neighbors. For reverse queries, we binary search A for a target vertex v (the second term of the contact), and a single application of Ψ permits us to reach the corresponding starting time of that contact (the third term of the contact). With an additional access to Ψ, we can also obtain the ending time. However, when we perform direct-neighbor queries, we start at the first term of the contact, and we need to access Ψ twice and three times to reach the starting and ending time of the contact respectively.

Flexibility to Support Special Queries. *TGCSA* can give support to other query types that could be interesting in some domains. In particular, queries about exact time-instants or edges can benefit from searching more than one term in the initial binary search in the *TGCSA*. For example, in a temporal graph representing phone calls from a given user to another, starting and ending at a given time, it could be interesting to perform queries such as: *i)* *"who phoned user A exactly at time t_s?"*, or *ii)* *"who received a phone call from B that started at time t_s and ended at t_e?"*. They could be implemented in the *TGCSA* as an initial binary search for $(A \cdot t_s)$ and $(t_s \cdot t_e \cdot B)$, respectively. Then, for the entries in the returned ranges $A[l, r]$, two or one accesses to Ψ, respectively, would be needed to retrieve the caller for the first query, and the receiver in the latter one. Note also that, in the second query, the initial binary search would report a unique entry in A, hence the query is answered almost instantaneously.

5 Conclusions and Future Work

The experimental results showed that $TGCSA$ has reasonable space usage, and succeeds when performing queries that filter out many contacts from the dataset with a single binary search in the $TGCSA$. This avoids the need for sequentially checking a large number of contacts. In particular, our best trade-off between space and query performance was obtained in the real *Flicker-Days* dataset. In general, space needs are between 50-100 bits per contact, and most queries are solved in less than 1 millisecond per contact reported.

As future work, we want to try more Ψ compression alternatives to those in [6]. Since Ψ represents around 80-90% of the size of $TGCSA$, it is almost the only way to reduce space needs. We are also interested in studying the applicability of other self-indexes to the scope of this paper.

References

1. Albert, R., Barabási, A.L.: Statistical mechanics of complex networks. Rev. Modern Physics 74, 47–97 (2002)
2. Bernardo, G.D., Brisaboa, N.R., Caro, D., Rodriguez, M.A.: Compact Data Structures for Temporal Graphs. In: Proc. DCC 2013, p. 477 (2013)
3. Brisaboa, N., Ladra, S., Navarro, G.: Compact representation of web graphs with extended functionality. Inf. Systems 39(1), 152–174 (2014)
4. Buin-Xuan, B.M., Ferreira, A., Jarry, A.: Computing shortest, fastest, and foremost journeys in dynamic networks. Int. J. Found. Comput. Sci. 14(02), 267–285 (2003)
5. Cha, M., Mislove, A., Gummadi, K.P.: A measurement-driven analysis of information propagation in flickr social network. In: Proc. WWW 2009, pp. 721–730 (2009)
6. Fariña, A., Brisaboa, N., Navarro, G., Claude, F., Places, A., Rodríguez, E.: Word-based self-indexes for natural language text. ACM TOIS 30(1), article 1 (2012)
7. Ferreira, A., Viennot, L.: A Note on Models, Algorithms, and Data Structures for Dynamic Communication Networks. Tech. rep., MASCOTTE - INRIA Sophia Antipolis / Laboratoire I3S, HIPERCOM - INRIA Rocquencourt (2002)
8. Khurana, U., Deshpande, A.: Efficient snapshot retrieval over historical graph data. In: Proc. ICDE 2013, pp. 997–1008 (2013)
9. Labouseur, A.G., Birnbaum, J., Olsen, P.W., Spillane, S.R., Vijayan, J., Hwang, J.H., Han, W.S.: The G* graph database: efficiently managing large distributed dynamic graphs. Distributed and Parallel Databases (2014)
10. Labouseur, A.G., Olsen, J.P.W., Hwang, J.H.: Scalable and Robust Management of Dynamic Graph Data. The VLDB Journal, 1–6 (2013)
11. Nicosia, V., Tang, J., Mascolo, C., Musolesi, M., Russo, G., Latora, V.: Graph metrics for temporal networks. In: Temporal Networks, pp. 15–40. Springer (2013)
12. Raman, R., Raman, V., Rao, S.S.: Succinct indexable dictionaries with applications to encoding k-ary trees and multisets. In: Proc. SODA 2012, pp. 233–242 (2002)

13. Ren, C., Lo, E., Kao, B., Zhu, X., Cheng, R.: On querying historical evolving graph sequences. PVLDB 4(11), 726–737 (2011)
14. Sadakane, K.: New text indexing functionalities of the compressed suffix arrays. Journal of Algorithms 48(2), 294–313 (2003)
15. Zhang, J., Long, X., Suel, T.: Performance of compressed inverted list caching in search engines. In: Proc. WWW 2008, pp. 387–396 (2008)
16. Zobel, J., Moffat, A.: Inverted files for text search engines. ACM Computing Surveys 38(2) (July 2006)
17. Zukowski, M., Héman, S., Nes, N., Boncz, P.A.: Super-scalar ram-cpu cache compression. In: Proc. ICDE 2006, p. 59 (2006)

Succinct Indexes for Reporting Discriminating and Generic Words*

Sudip Biswas, Manish Patil, Rahul Shah, and Sharma V. Thankachan

Louisiana State University, USA
{sbiswas,mpatil,rahul,thanks}@csc.lsu.edu

Abstract. We consider the problem of indexing a collection $\mathcal{D}$ of D strings (documents) of total n characters from an alphabet set of size σ, such that whenever a pattern P (of p characters) and an integer $\tau \in [1, D]$ comes as a query, we can efficiently report all (i) *maximal generic words* and (ii) *minimal discriminating words* as defined below:

  – *maximal generic word* is a maximal extension of P occurring in at least τ documents..
  – *minimal discriminating word* is a minimal extension of P occurring in at most τ documents.

These problems were introduced by Kucherov et al. [8], and they proposed linear space indexes occupying $O(n \log n)$ bits with query times $O(p + output)$ and $O(p + \log \log n + output)$ for Problem (i) and Problem (ii) respectively. In this paper, we describe succinct indexes of $n \log \sigma + o(n \log \sigma) + O(n)$ bits space with near optimal query times i.e., $O(p + \log \log n + output)$ for both these problems.

1 Introduction and Related Work

Let $\mathcal{D} = \{d_1, d_2, d_3, ..., d_D\}$ be a collection of D strings (which we call as documents) of total n characters from an alphabet set Σ of size σ. For simplicity we assume, that every document ends with a special character $ which does not appear any where else in the documents. Our task is to index $\mathcal{D}$ in order to compute all (i) *maximal generic words* and (ii) *minimal discriminating words* corresponding to the given query pattern P (of length p) and threshold τ. The document frequency $df(.)$ of a pattern P is defined as the number of distinct documents in $\mathcal{D}$ containing P. Then, a generic word is an extension $\bar{P}$ of P with $df(\bar{P}) \geq \tau$, and is maximal if $df(P') < \tau$ for all extensions P' of $\bar{P}$. Similarly, a discriminating word is an extension $\bar{P}$ of P with $df(\bar{P}) \leq \tau$, and is called a minimal discriminating word if $df(P') > \tau$ for any proper prefix P' of $\bar{P}$ (i.e., $P' \neq \bar{P}$). These problems were introduced by Kucherov et al. [8], and they proposed indexes of size $O(n \log n)$ bits or $O(n)$ words. The query processing time is optimal $O(p + output)$ for reporting all maximal generic words, and is near optimal $O(p + \log \log n + output)$ for reporting all minimal discriminating words.

---

* This work is supported in part by National Science Foundation (NSF) Grants CCF–1017623 and CCF–1218904.

© Springer International Publishing Switzerland 2014

Later on Gawrychowski et al. [6] gave $O(n)$ words space with optimal query time index for minimal discriminating words problem. In this paper, we describe succinct indexes of $n \log \sigma + o(n \log \sigma) + O(n)$ bits space with $O(p + \log \log n + output)$ query times for both these problems.

These problems are motivated from applications in computational biology. For example, it is an interesting problem to identify words that are exclusive to the genomic sequences of one species or family of species [3]. Such patterns that appear in a small set of biologically related DNA sequences but do not appear in other sequences the collection often carries a biological significance. Discriminating and generic words also find applications in text mining and automated text classification.

2   Preliminaries

2.1   Suffix Trees and Generalized Suffix Trees

For a text $S[1...n]$, a substring $S[i...n]$ with $i \in [1, n]$ is called a suffix of T. The suffix tree [13,9] of S is a lexicographic arrangement of all these n suffixes in a compact trie structure of $O(n)$ words space, where the i-th leftmost leaf represents the i-th lexicographically smallest suffix of S. For a node i (i.e., node with pre-order rank i), $path(i)$ represents the text obtained by concatenating all edge labels on the path from root to node i in a suffix tree. The locus node i_P of a pattern P is the node closest to the root such that the P is a prefix of $path(i_P)$. The suffix range of a pattern P is given by the maximal range $[sp, ep]$ such that for $sp \leq j \leq ep$, P is a prefix of (lexicographically) j-th suffix of S. Therefore, i_P is the lowest common ancestor of sp-th and ep-th leaves. Using suffix tree, the locus node as well as the suffix range of P can be computed in $O(p)$ time, where p denotes the length of P. Let $T = d_1 d_2 ... d_D$ be the text obtained by concatenating all documents in $\mathcal{D}$. Recall that each document is assumed to end with a special character \$. The suffix tree of T is called the generalized suffix tree (GST) of $\mathcal{D}$.

Encoding of GST with the goal of supporting navigation and other tree operations has been extensively studied in the literature. We use the data structure by Sadakane and Navarro [12] with focus on following operations:

- $lca(i, j)$: the lowest common ancestor of two nodes i, j
- $child(i, k)$: k-th child of node i
- $level\text{-}ancestor(i, d)$: ancestor j of i such that depth(j)=depth(i)-d
- $subtree\text{-}size(i)$: number of nodes in the subtree of node i

Lemma 1. *[12] An ordinal tree with m nodes can be encoded by $2m + O(\frac{m}{polylog(m)})$ bits supporting lca, k-th child, level-ancestor, and subtree-size queries in constant time.*

Define $count(i) = df(path(i))$. Using the data structure by Sadakane [11] we can answer $count(i)$ query efficiently for any input node i. Following lemma summarizes the result in [11].

Lemma 2. *[11] Generalized suffix tree (GST) with n leaves can be encoded by $2n + o(n)$ bits, supporting count(i) query in constant time.*

2.2 Marking Scheme in GST

Here we briefly explain the marking scheme introduced by Hon et al. [7] which will be used later in the proposed succinct index. We identify certain nodes in the *GST* as marked nodes and prime nodes with respect to a parameter g called the *grouping factor*. The procedure starts by combining every g consecutive leaves (from left to right) together as a group, and marking the lowest common ancestor (LCA) of first and last leaf in each group. Further, we mark the LCA of all pairs of marked nodes recursively. We also ensure that the root is always marked. At the end of this procedure, the number of marked nodes in *GST* will be not more than $2n/g$. Hon et al. [7] showed that, given any node u with u^* being its highest marked descendent (if exists), number of leaves in $GST(u\backslash u^*)$ i.e., the number of leaves in the subtree of u, but not in the subtree of u^* is at most $2g$.

Prime nodes are the children of marked nodes (illustrated in figure 1). Corresponding to any marked node u^* (except the root node), there is a *unique prime* node u', which is its closest prime ancestor. In case u^*'s parent is marked then $u' = u^*$. For every prime node u', the corresponding closest marked descendant u^* (if it exists) is unique.

2.3 Segment Intersection Problem

In [8] authors have shown how the problem of identifying minimal discriminating words can be reduced to orthogonal segment intersection problem. In this article, we rely on this key insight for both the problems under consideration and use the result summarized in lemma below for segment intersection.

Lemma 3. *[2] A given set $\mathcal{I}$ of n vertical segments of the form $(x_i, [y_i, y'_i])$, where $x_i, y_i, y'_i \in [1, n]$ can be indexed in $O(n)$-word space (in Word RAM model), such that whenever a horizontal segment $s_q = ([x_q, x'_q], y_q)$ comes as a query, all those vertical segments in $\mathcal{I}$ that intersect with s_q can be reported in $O(\log \log n + output)$ time.*

2.4 Range Maximum Query

Let A be an array of length n, a range maximum query (RMQ) asks for the position of the maximum value between two specified array indices $[i, j]$. i.e., the RMQ should return an index k such that $i \leq k \leq j$ and $A[k] \geq A[x]$ for all $i \leq x \leq j$. We use the result captured in following lemma for our purpose.

Lemma 4. *[4,5] By maintaining a $2n + o(n)$ bits structure, range maximum query (RMQ) can be answered in $O(1)$ time (without accessing the array).*

3   Computing Maximal Generic Words

In this section, we first review a linear space index, which is based on the ideas from the previous results [8]. Later we show how to employ sampling techniques to achieve a space efficient solution.

3.1   Linear Space Index

Let i_P be the locus node of the query pattern P. Then, our task is to return all those nodes j in the subtree of i_P such that $count(j) \geq \tau$ and $count(.)$ of every child node of j is less than τ. Note that corresponding to each such output j, $path(j)$ represents a maximal generic word with respect to the query (P, τ). The mentioned task can be performed efficiently by reducing the original problem to a segment intersection problem as follows. Each node i in GST is mapped to a vertical segment $(i, [count(i_{max}) + 1, count(i)])$, where i_{max} is the child of i with the highest $count(.)$ value. If i is a leaf node, then we set $count(i_{max}) = 0$. Moreover, if $count(i) = count(i_{max})$ we do not maintain such a segment as it can not possibly lead to a generic word. The set $\mathcal{I}$ of these segments is then indexed using a linear space structure as described in Section 2.3. Additionally we maintain the GST of $\mathcal{D}$ as well.

The maximal generic words corresponding to a query (P, τ) can be computed by issuing an orthogonal segment intersection query on $\mathcal{I}$ with $s_q = ([i_P, i'_P], \tau)$ as the input, where i_P is the locus node of pattern P and i'_P represents the rightmost leaf in the subtree of i_P. It can be easily verified than $path(j)$ corresponding to each retrieved interval $(j, [count(j_{max}) + 1, count(j)])$ is a maximal generic word. In conclusion, we have a linear space index of $O(n)$ words with $O(\log \log n + output)$ query time. By combining with the space for GST, and the initial $O(p)$ time for pattern search where p is the size of pattern, we have the following lemma.

Lemma 5. *There exists an $O(n)$ word data structure for reporting maximal generic word queries in $O(p + \log \log n + output)$ time.*   □

3.2   Succinct Space Index

Our succinct space index for computing maximal generic words has the following key components:

- For finding maximal generic words corresponding to the marked nodes, we store a data structure similar to the one in [8] described above only for the marked nodes, which takes space linear to the number of marked nodes.
- To capture the outputs falling in the path between two marked nodes, we store a segment intersection index along with encoding of the path between a marked node and its unique lowest prime ancestor.
- The remaining output fall in small subtrees which can be efficiently found using bit encodings of subtrees and table structure.

At first,, we begin by extending the marking scheme (illustrated in figure 1) of Hon et al. [7] described earlier in Section 2.2 and then discuss our succinct space index. We introduce the notion of *orphan* and *maximal orphan* nodes in GST based on the marking scheme of Hon et al. [7] as follows:

1. *Orphan node* is a node with no marked node in its subtree. Note that the number of leaves in the subtree of an orphan node is at most g.
2. *maximal orphan* is an orphan node with *non-orphan* parent. Therefore, every orphan node has a unique maximal orphan ancestor. The number of leaves in the subtree of any maximal orphan node is at most g and the number of maximal orphan nodes can be $\omega(n/g)$.

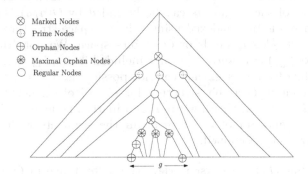

Fig. 1. Marking Scheme

We now describe a structure to solve a variant of computing maximal generic words summarized in lemma below that forms the basis of our final space-efficient index. We choose $g = \lfloor \frac{1}{8} \log n \rfloor$ as a grouping parameter in the marking scheme of Hon et al. [7] and nodes in GST are identified by their pre-order rank.

Lemma 6. *The collection $\mathcal{D}=\{d_1, d_2, d_3, ..., d_D\}$ of strings of total n characters can be indexed in $(n \log \sigma + O(n))$ bits, such that given a query (P, τ), we can identify all marked nodes j^* satisfying either of the following condition in $O(p + \log \log n + output)$ time.*

- *$path(j^*)$ is a maximal generic word for input (P, τ)*
- *there is at least one node in $N(j^*) = GST(j'/j^*)$ that corresponds to desired maximal generic word, where j' is a unique lowest prime ancestor of j^**
- *j^* is a highest marked descendant of locus node for pattern P*

Proof. It can be noted that, $N(j^*)$ is essentially a set of all nodes in the subtree of j' but not in the subtree of j^* ($N(j^*)$ does not include node j^*). To be able to retrieve the required marked nodes in succinct space, we maintain index consisting of following components:

- Instead of GST, we use its space efficient version i.e., a compressed suffix array (CSA) of T. There are many versions of CSA's available in literature, however for our purpose we use the one in [1] that occupies $n \log \sigma + o(n \log \sigma) + O(n)$ bits space and retrieves the suffix range of pattern P in $O(p)$ time.
- A $4n + o(n)$ bits encoding of GST structure (Lemma 1) to support the (required) tree navigational operations in constant time.
- We keep a bit vector $B_{mark}[1...2n]$, where $B_{mark}[i] = 1$ iff node i is a marked node, with constant time rank-select supporting structures over it in $O(n)$ bits space [10]. This bit vector enables us to retrieve the unique highest marked descendant of any given node in GST in constant time.
- We map each marked node i^* to a vertical segment $(i^*, [count(i^*_{max}) + 1, count(i^*)])$, where i^*_{max} is the child of i^* with the highest $count(.)$ value. The number of such segments can be bounded by $O(n/g)$. The set $\mathcal{I}_1$ of these segments is then indexed using a linear space structure as before in $O(n/g) = O(n/\log n)$ words or $O(n)$ bits space. We note that segments with $count(i^*_{max}) = count(i^*)$ are not included in set $\mathcal{I}_1$, as marked nodes corresponding to those segments can not possibly lead to a generic word.
- We also maintain $O(n)$ bit structure for set $\mathcal{I}_2$ of segments of the form $(i', [count(i^*) + 1, count(i')])$, where i^* is a marked node and i' is the unique lowest prime ancestor of i^*. Once again the segments with $count(i^*) = count(i')$ are not maintained.

Given an input (P, τ), we first retrieve the suffix range in $O(p)$ time using CSA and the locus node i_P in another $O(1)$ time using GST structure encoding. Then we issue an orthogonal segment intersection query on $\mathcal{I}_1$ and $\mathcal{I}_2$ with $s_q = ([i_P, i'_P], \tau)$ as the input, i'_P being the rightmost leaf in the subtree of i_P. Any marked node that corresponds to a maximal generic word for query (P, τ) is thus retrieved by querying set $\mathcal{I}_1$. Instead of retrieving non-marked nodes corresponding to the maximal generic word, the structure just described returns their representative marked nodes instead.

For a segment in $\mathcal{I}_2$ that is reported as an answer, corresponding to a marked node j^* with j' being its lowest prime ancestor, we have $count(j') \geq \tau$ and $count(j^*) < \tau$. Therefore, there must exist a parent-child node pair (u, v), on the path from j' to j^* such that $count(u) \geq \tau$ and $count(v) < \tau$. As before, let u_{max} be the child of node u with the highest $count(.)$ value. It can be easily seen that if $(v = u_{max})$ or $(v \neq u_{max}$ with $count(u_{max}) < \tau)$, then $path(u)$ is a maximal generic word with respect to (P, τ). Otherwise, consider the subtree rooted at node u_{max}. For this subtree $count(u_{max}) \geq \tau$ and $count(.) = 1$ for all the leaves and hence it is guaranteed to contain at least one maximal generic word for query (P, τ).

We highlight that the segment intersection query on set $\mathcal{I}_2$ will be able to capture the node pairs (j^*, j') where both j^* and j' are in the subtree of locus node i_P as the segments in $\mathcal{I}_2$ use (pre-order rank of) prime node as their x coordinate. The case when both j^*, j' are outside the subtree of i_P can be ignored as in this case none of the nodes in $N(j^*)$ will be in the subtree of i_P and hence can not lead to a desired output. Further we observe that for the remaining scenario when locus node i_P is on the path from j' to j^* (both exclusive), there can be at-most one such pair (j^*, j'). This is true due to the way nodes are marked in GST and moreover j^* will be the highest marked descendant of i_P (if exists). Such j^* can be obtained in constant time by first obtaining the marked node using query $select_1(rank_1(i_P) + 1)$ on B_{mark} and then evaluating if it is in the subtree of i_P. We note that in this case $N(j^*)$ (j^* being the highest marked descendant of i_P) may or may not result in a maximal generic word for query (P, τ), however we can afford to verify it irrespective of the output due to its uniqueness.   □

If a marked node j^* reported by the data structure just described is retrieved from set $\mathcal{I}_1$ then $path(j^*)$ can be returned as an maximal generic word for query (P, τ) directly. However, every other marked node retrieved needs to be decoded to obtain the actual maximal generic word corresponding to one or more non-marked nodes it represents. Before we describe the additional data structures that enable such decoding, we classify the non-marked answers as orphan and non-orphan outputs based on whether or not it has any marked node in its subtree as defined earlier. Let j^* be a marked node and j' be its lowest prime ancestor. A non-marked node that is an output is termed as orphan if it is not on the path from j' to j^*. Due to Lemma 6, every marked node retrieved from $\mathcal{I}_2$ leads to either a orphan output or non-orphan outputs or both. Below, we describe how to report all orphan and non-orphan outputs for a given query (P, τ) and a marked node j^*. We first append the index in Lemma 6 by data structure by Sadakane (Lemma 2) without affecting its space complexity to answer $count(.)$ query in constant time for any GST node.

Retrieving Non-orphan Outputs: To be able to report a non-orphan output of query (P, τ) for a given marked node (if it exists), we maintain a collection of bit vectors as follows:

- We associate a bit vector to each marked node i^* that encodes $count(.)$ information of the nodes on the (top-down) path from i' to i^*, i' being the (unique) lowest prime ancestor of i^*. Let $x_1, x_2, ..., x_r$ be the nodes on this path inclusive of both i' and i^*. Note that $r \leq g$, $\delta_i = count(x_{i-1}) - count(x_i) \geq 0$, and $count(x_1) - count(x_r) \leq 2g$ from the properties of the marking scheme. Now we maintain a bit vector $B_{i^*} = 10^{\delta_1} 10^{\delta_2} ... 10^{\delta_r}$ along with constant time rank-select supporting structures at marked node i^*. Number of 0 in the bit-vector is $count(x_1) - count(x_2) + count(x_2) - count(x_3) + ... + count(x_{r-1}) - count(x_r) = count(x_1) - count(x_r)$. As length of the bit vector is bounded by $O(2g)$ and number of marked nodes is bounded by $O(n/g)$, total space required for these structures is $O(n)$ bits.
- Given a node i we would like to retrieve the node i_{max} i.e., child of node i with the highest $count(.)$ value in constant time. To enable such a lookup we maintain a bit vector $B = 10^{\delta_1} 10^{\delta_2} ... 10^{\delta_n}$, where $child(i, \delta_i) = i_{max}$. If i is leaf then we assume $\delta_i = 0$. As each node contributes exactly one bit with

value 1 and at most one bit with value 0, length of the bit vector B is also bounded by $2g$ and subsequently occupies $O(n)$ bits.

Given a marked node j^* and a query (P, τ), we need to retrieve a parent-child node pair (u, v) on the path from j' to j^* such that $count(u) \geq \tau$ and $count(v) < \tau$. We can obtain the lowest prime ancestor j' of j^* in constant time to begin with [7]. Then to obtain a node u, we probe bit vector B_{j^*} by issuing a query $rank_1(select_0(count(j') - \tau))$. We note that these rank-select operations only returns the distance of the node u from j^* which can be then used along with level-ancestor query on j^* to obtain u. To verify if $path(u)$ indeed corresponds to a maximal generic word, we need to check if $count(.) \leq \tau$ for all the child nodes of u. To achieve this, we retrieve the $j_{max} = child(u, select_1(u+1) - select_1(u) - 1)$ and obtain its count value $count(j_{max})$ using a data structure by Sadakane (Lemma 2) in constant time. Finally, if $count(j_{max}) < \tau$ then u can be reported as an maximal generic word for (P, τ) query. If the input node j^* is highest marked descendant of locus node i_P then we need to verify if node u is within the subtree of i_P before it can be reported as an maximal generic word. Thus overall time spent per marked node to output the associated maximal generic word (if any) is $O(1)$. We note that unsuccessfully querying the marked node for non-orphan output does not hurt the overall objective of optimal query time since, such a marked node is guaranteed to generate orphan outputs (possibly except the highest marked ancestor of locus node i_P).

Retrieving Orphan Outputs: In this part we take advantage of the smaller size of the subtrees rooted at any maximal orphan node. We use bit encodings for every possible combination of the subtrees rooted at any maximal orphan node and table for storing all the answers for each possible subtree. Query procedure follows efficiently finding the encoding of the subtree and retrieving the answers from the table.

Instead of retrieving orphan outputs of the query based on the marked nodes as we did for non-orphan outputs, we retrieve them based on maximal orphan nodes by following two step query algorithm: (i) identify all maximal orphan nodes i in the subtree of the locus node i_P of P, with $count(i) \geq \tau$ and (ii) explore the subtree of each such i to find out the actual (orphan) outputs. If $count(i) \geq \tau$, then there exists at least on output in the subtree of i, otherwise there will not be any output in the subtree of i.

Since an exhaustive search in the subtree of a maximal orphan node is prohibitively expensive, we rely on the following insight to achieve the optimal query time. For a node i in the GST, let $subtree\text{-}size(i)$, $leaves\text{-}size(i)$ represents the number of nodes and number of leaves in the subtree rooted at node i respectively. The subtree of i can be then encoded (simple balanced parenthesis encoding) in $2subtree\text{-}size(i)$ bits. Also the $count(.)$ values of all nodes in the subtree of i in GST can be encoded in $2leaves\text{-}size(i)$ bits using the encoding scheme by Sadakane [11]. Therefore $2subtree\text{-}size(i) + 2leaves\text{-}size(i)$ bits are sufficient to encode the subtree of any node i along with the $count(.)$ information. Since $subtree\text{-}size(i) < 2leaves\text{-}size(i)$ and there are less than g leaves in the subtree of a maximal orphan node, for a maximal orphan node i we have $2subtree\text{-}size(i) + 2leaves\text{-}size(i) < 6g = \frac{3}{4}\log n$. This implies the number of distinct maximal orphan nodes possible with respect to the above encoding is bounded by $\sum_{k=1}^{\frac{3}{4}\log n} 2^k = \Theta(n^{3/4})$.

To be able to efficiently execute the two step algorithm to retrieve all orphan outputs we maintain following components:

- A bit vector $B_{orph}[1...2n]$, where $B_{orph}[i] = 1$ iff node i is a maximal orphan node, along with constant time rank-select supporting structures over it occupying $O(n)$ bits space.
- Define an array $E[1...]$, where $E[i] = count(select_1(i))$ with $select$ operation applied to a bit vector B_{orph} (i.e., the count of i-th maximal orphan node). Array E is not maintained explicitly, instead a $2n + o(n)$ bits RMQ structure over it is maintained.
- For each distinct encoding of a maximal orphan node out of total $\Theta(n^{3/4})$ of them, we shall maintain the list of top-g answers for $\tau = 1, 2, 3, ..., g$. Note that for each τ, number of answers is bounded by g. Overall space is therefore $O(n^{3/4}g^2) = o(n)$ bits.
- The total $n^{3/4}$ distinct encodings of maximal orphan nodes can be thought to be categorized into groups of size 2^k for $k = 1, ..., \frac{3}{4}\log n$. Encodings for all possible distinct maximal orphan nodes i having $k = 2subtree\text{-}size(i) + 2leaves\text{-}size(i)$ are grouped together and let L_k be this set. Then for a given maximal orphan node i in GST with $k = 2subtree\text{-}size(i) + count(i)$, we maintain a pointer so as to enable the lookup of all answers (at most g) corresponding to the encoding of subtree of i among all the encoding in set L_k. With number of bits required to represent such a pointer being proportional to the number leaves in the subtree of a maximal orphan node i i.e. $2subtree\text{-}size(i) + count(i)$, overall space can be bounded by $O(n)$ bits.

Query processing can now be handled as follows. Begin by identifying the x-th and y-th maximal orphan nodes, which are the first and last maximal orphan nodes in the subtree of the locus node i_P in $O(1)$ time as $x = 1 + rank_1(i_P - 1)$ and $y = rank_1(i'_P)$ using bit vector B_{orph}, where i'_P is the rightmost leaf of the subtree rooted at i_P. Then, all those $z \in [x, y]$ where $E[z] \geq \tau$ can be obtained in constant time per z using recursive range maximum queries on E as follows: obtain $z = RMQ_E(x, y)$, and if $E[z] < \tau$, then stop recursion, else recurse the queries on intervals $[x, z-1]$ and $[z+1, y]$. Recall that even if $E[z]$ is not maintained explicitly, it can be obtained in constant time using B_{orph} as $E[z] = count(select_1(z))$. Further, the maximal orphan node corresponding to each z can be obtained in constant time as $select_1(z)$. In conclusion, step (i) of query algorithm can be performed in optimal time. Finally, for each of these maximal orphan nodes we can find the list of pre-computed answers based on given τ and report them in optimal time. It can be noted that, for a given maximal orphan node i, we first obtain $subtree\text{-}size(i)$ and $count(i)$ in constant time using Lemma 1 and 2 respectively and than use the pointer stored as in index in the set L_k, with $k = subtree\text{-}size(i) + count(i)$.

Combining all pieces together we achieve the result summarized in following theorem.

Theorem 1. *The collection* $\mathcal{D}=\{d_1, d_2, d_3, ..., d_D\}$ *of strings of total n charac- ters can be indexed in* $(n \log \sigma + O(n))$ *bits, such that given a query* (P, τ), *all maximal generic words can be reported in* $O(p + \log \log n + output)$ *time.*

4   Computing Minimal Discriminating Words

In the case of minimal discriminating words, given a query pattern P and a threshold τ, the objective is to find all nodes i in the subtree of locus node i_P such that $count(i) \leq \tau$ and $count(i_{parent}) > \tau$, where i_{parent} is the parent of a node i. Then each of these nodes represent a minimal discriminating word given by $path(i_{parent})$ concatenated with the first leading character on the edge connecting nodes i_{parent} and i. A linear space index with same query bounds as summarized in Lemma 5 can be obtained for minimal discriminating word queries by following the description in Section 3.1, except in this scenario, we map each node i in GST to a vertical segment $(i, [count(i), count(i_{parent}) + 1])$. Similarly, the succinct space solution can be obtained by following the same index framework as that of maximal generic words. Below we briefly describe the changes required in the index and query algorithm described in Section 3.2 so as to retrieve minimal discriminating words instead of maximal generic words.

We need to maintain all the components of index listed in the proof for Lemma 6 with a single modification. The set $\mathcal{I}_1$ consists of seg- ments obtained by mapping each marked node i^* to a vertical segment $(i^*, [count(i^*), count(i^*_{parent}) + 1])$. By following the same arguments as before, we can rewrite the Lemma 6 as follows:

Lemma 7. *The collection* $\mathcal{D}=\{d_1, d_2, d_3, ..., d_D\}$ *of strings of total n characters can be indexed in* $(n \log \sigma + O(n))$ *bits, such that given a query* (P, τ), *we can identify all marked nodes* j^* *satisfying either of the following condition in* $O(p + \log \log n + output)$ *time.*

- *$path(j^*_{parent})$ appended with leading character on edge j^*_{parent}-j^* is a minimal discriminating word for input (P, τ)*
- *there is at least one node in $N(j^*) = GST(j'/j^*)$ that corresponds to desired minimal discriminating word, j' being the unique lowest prime ancestor of marked node j^**
- *j^* is a highest marked descendant of i_P*

We append the components required in the above lemma by data structure by Sadakane (Lemma 2) to answer $count(.)$ query in constant time for any GST node and retrieve the non-orphan, orphan outputs separately as before.

Though we maintain same collection of bit vectors as required for maximal generic words to retrieve non-orphan outputs, query processing differs slightly in this case. Let i^* be the input marked node and i' be its lowest prime ancestor.

Then we can obtain parent-child node pair (u, v) on the path from i' to i^* such that $count(u) > \tau$ and $count(v) \leq \tau$ in constant time. Node v can now be returned as an answer since concatenation of $path(u)$ with first character on edge u-v will correspond to a minimal discriminating word. Thus, every marked node obtained by segment intersection query on set $\mathcal{I}_2$ produces a non-orphan output in this case as opposed to the case of maximal generating words where it may or may not produce a non-orphan output. Also if the input node i^* is highest marked descendant of locus node i_P then we need to verify if node v is within the subtree of i_P before it can be reported as an output.

Data structures and query algorithm to retrieve the orphan outputs remain the same described earlier in Section 3.2. It is to be noted that top-g answers to be stored for $\tau = 1, 2, 3, ..., g$ corresponding to each of the distinct maximal orphan node encoding now corresponds to the minimal discriminating word.

Based on the above description, following theorem can be easily obtained.

Theorem 2. *The collection* $\mathcal{D} = \{d_1, d_2, d_3, ..., d_D\}$ *of strings of total n characters can be indexed in $(n \log \sigma + O(n))$ bits, such that given a query (P, τ), all minimal discriminating words can be reported in $O(p + \log \log n + output)$ time.*

5   Concluding Remarks

In this paper, we revisited the maximal generic word and minimal discriminating word problem and proposed a first succinct index for both the problems. It would be interesting to see if succinct index can be obtained for these problems achieving optimum query time.

References

1. Belazzougui, D., Navarro, G.: Alphabet-independent compressed text indexing. In: Demetrescu, C., Halldórsson, M.M. (eds.) ESA 2011. LNCS, vol. 6942, pp. 748–759. Springer, Heidelberg (2011)
2. Chan, T.M.: Persistent predecessor search and orthogonal point location on the word ram. In: SODA, pp. 1131–1145 (2011)
3. Fadiel, A., Lithwick, S., Ganji, G., Scherer, S.W.: Remarkable sequence signatures in archaeal genomes. Archaea 1(3), 185–190 (2003)
4. Fischer, J., Heun, V.: A New Succinct Representation of RMQ-Information and Improvements in the Enhanced Suffix Array. In: Chen, B., Paterson, M., Zhang, G. (eds.) ESCAPE 2007. LNCS, vol. 4614, pp. 459–470. Springer, Heidelberg (2007)
5. Fischer, J., Heun, V., Stühler, H.M.: Practical Entropy-Bounded Schemes for O(1)-Range Minimum Queries. In: IEEE DCC, pp. 272–281 (2008)
6. Gawrychowski, P., Kucherov, G., Nekrich, Y., Starikovskaya, T.: Minimal discriminating words problem revisited. In: Kurland, O., Lewenstein, M., Porat, E. (eds.) SPIRE 2013. LNCS, vol. 8214, pp. 129–140. Springer, Heidelberg (2013)
7. Hon, W.-K., Shah, R., Vitter, J.S.: Space-efficient framework for top-k string retrieval problems. In: FOCS, pp. 713–722 (2009)

8. Kucherov, G., Nekrich, Y., Starikovskaya, T.: Computing discriminating and generic words. In: Calderón-Benavides, L., González-Caro, C., Chávez, E., Ziviani, N. (eds.) SPIRE 2012. LNCS, vol. 7608, pp. 307–317. Springer, Heidelberg (2012)
9. McCreight, E.M.: A space-economical suffix tree construction algorithm. J. ACM 23(2), 262–272 (1976)
10. Raman, R., Raman, V., Rao, S.S.: Succinct Indexable Dictionaries with Applications to Encoding k-ary Trees and Multisets. In: ACM-SIAM SODA, pp. 233–242 (2002)
11. Sadakane, K.: Succinct data structures for flexible text retrieval systems. J. Discrete Algorithms 5(1), 12–22 (2007)
12. Sadakane, K., Navarro, G.: Fully-functional succinct trees. In: SODA, pp. 134–149 (2010)
13. Weiner, P.: Linear pattern matching algorithms. In: SWAT (FOCS), pp. 1–11 (1973)

Fast Construction of Wavelet Trees

J. Ian Munro[1], Yakov Nekrich[1], and Jeffrey S. Vitter[2]

[1] David R. Cheriton School of Computer Science, University of Waterloo
[2] Department of Electrical Engineering & Computer Science, University of Kansas

Abstract. In this paper we describe a fast algorithm that creates a wavelet tree for a sequence of symbols. We show that a wavelet tree can be constructed in $O(n\lceil \frac{\log \sigma}{\sqrt{\log n}} \rceil)$ time where n is the number of symbols and σ is the alphabet size.

1 Introduction

Wavelet tree, introduced in [5], is one of the most extensively studied succinct data structures. It is used in succinct representations of graphs, strings, points and other geometric objects on a grid, indexes, data structures for document retrieval, XML documents, and binary relations. We refer to an extensive survey of Navarro [7] for a description of these and other applications of wavelet trees. In this paper we describe the first algorithm that constructs a wavelet tree in $o(n \log \sigma)$ time.

Let X be a sequence of length n over an alphabet of size σ. We can assume w. l. o. g. that the i-th element $X[i]$ of X is an integer in the range $[1, \sigma]$. Essentially constructing a wavelet tree for a sequence X requires re-grouping the bits of X into a bit sequence of total length $n \log \sigma$. Since different bits of an element $X[i]$ are stored in different parts of the bit sequence, it appears that we need $\Omega(n \log \sigma)$ time to construct a wavelet tree. In this paper we show that the cost of the straightforward solution can be reduced by an $O(\sqrt{\log n})$ factor. The main idea of our method is usage of bit parallelism, i.e. we use bit operations to keep $\Omega(1)$ elements of X in one word and perform certain operations on elements packed into one word in constant time. Suppose that we can pack L symbols of a sequence X into one machine word. Then we can generate the wavelet tree for the resulting sequence of symbols in $O(n(\log \sigma/L))$ time by processing $O(L)$ symbols in constant time.

Previous and Related Work. Since wavelet trees were introduced in 2003 [5], a large number of papers that use this data structure appeared in the literature. We refrain from listing previous works due to space reasons; an interested reader is referred to e.g., the recent survey [7]. In spite of a significant number of previous papers, no results for constructing a wavelet tree in $o(n \log \sigma)$ time were previously described. Algorithms that generate a wavelet tree and use little additional workspace were considered by Claude et al. [4] and Tischler [9].

Chazelle [3] described a linear space ($O(n \log n)$-bit) geometric data structure that answers certain kinds of two-dimensional range searching queries.

E. Moura and M. Crochemore (Eds.): SPIRE 2014, LNCS 8799, pp. 101–110, 2014.
© Springer International Publishing Switzerland 2014

Data organization in [3] is quite similar to the approach of wavelet trees. We remark, however, that the general concept and the usage of the wavelet tree are different; in particular, the data structure of Chazelle [3] is not succinct. Some other linear-space geometric data structures also use similar ways of structuring data. By the same argument, we need $O(n \log n)$ time to construct these data structures. Chan and Pătraşcu [2] showed that bit parallelism can be used to obtain linear-space data structures with faster construction time. In [2] they describe data structures that use linear space and can be constructed in $O(n\sqrt{\log n})$ time. Their approach is based on recursively reducing the original problem to several problems of smaller size. When point coordinates are sufficiently small, we can pack L points into one machine word and process data associated to L points in constant time.

In this paper we show how bit parallelism can be applied to speed-up the construction of the standard wavelet tree data structure. Our simple two-stage approach improves the construction time of the wavelet tree by $O(\sqrt{\log n})$. After recalling the basic concepts in Section 2, we describe the main algorithm and its variants in Section 3. In Section 4 we show how we can construct secondary data structures stored in the wavelet tree nodes.

2   Wavelet Tree

Let X denote a sequence over alphabet $\Sigma = \{1, \ldots, \sigma\}$. The standard wavelet tree for X is a balanced binary tree with bit sequences stored in each internal node. These bit sequences can be obtained as follows: we start by dividing the alphabet symbols into two subsets Σ_0 and Σ_1 of equal size, $\Sigma_0 = \{1, \ldots, \sigma/2\}$ and $\Sigma_1 = \{\sigma/2+1, \ldots, \sigma\}$. Let X_0 and X_2 denote the subsequences of X induced by symbols from Σ_0 and Σ_1 respectively. The bit sequence $X(v_R)$ stored in the root v_R of the wavelet tree indicates for each symbol $X[i]$ whether it belongs to X_0 or X_1: $X(v_R)[i] = 0$ if $X[i]$ is in X_0 and $X(v_R)[i] = 1$ if $X[i]$ is in X_1. The left child of v_R is the wavelet tree for X_0 and the right child of v_R is the wavelet tree for X_1.

A symbol from an alphabet Σ can be represented as a bit sequence of length $\lfloor \log \sigma \rfloor$ or $\lceil \log \sigma \rceil$. Bit sequences $X(u)$ in the nodes of the wavelet tree consist of the same bits as the symbols in X, but the bits are ordered in a different way. The sequence $X(v_R)$ contains the first bit from each symbol $X[i]$ in the same order as symbols appear in X. Let v_l and v_r be the left and the right children of v_R. The sequence $X(v_l)$ contains the second bit of every symbol in X_0. That is, $X(v_l)$ contains the second bit of every symbol $X[i]$, such that the first bit of $X[i]$ is 0. $X(v_r)$ contains the second bit of every $X[i]$ such that the first bit of $X[i]$ is 1, etc.

Some generalizations of the wavelet tree often lead to improved results. We can consider t-ary wavelet tree for $t = \log^\varepsilon n$ and a small constant $\varepsilon > 0$. In this case the original alphabet Σ is divided into t parts Σ_0, ..., Σ_{t-1}. The sequence $X(v_R)$ in the root node is a sequence over an alphabet $\{0, \ldots, t-1\}$ such that $X(v_R)[i] = j$ iff $X[i]$ is a symbol from Σ_j for $1 \leq j \leq t$. Let X_j be the subsequence of X induced by symbols from Σ_j. The j-th child v_j of v_R is the root of the wavelet tree for X_j. The advantage of the t-ary wavelet tree is that the tree height is reduced from $O(\log \sigma)$ to $O(\log \sigma / \log \log n)$. Another useful improvement is to modify the shape of the tree so that the average leaf depth is (almost) minimized. Finally we can also keep the binary or t-ary sequences $X(u)$, stored in the nodes, in compressed form. Two latter improvements enable us to store a sequence X in asymptotically optimal space.

3   Constructing a Wavelet Tree

In this section we describe our algorithm for constructing a wavelet tree. Our method uses bit parallelism in a way that is similar to [2]. However a recursive algorithm employed in [2] to reduce the problem size is not necessary. Our algorithm consists of two stages. During the first stage we construct an L-ary wavelet tree $\mathcal{T}^g$ for $L = 2^{\sqrt{\log n}}$. That is, each internal node $u \in \mathcal{T}^g$ has L children. To avoid tedious details, we assume that L is an integer that divides σ. An L-ary wavelet tree can be defined in the same way as in Section 2. We partition the alphabet $\Sigma = \{ 1, \ldots, \sigma \}$ into L parts $\Sigma_1, \Sigma_2, \ldots, \Sigma_L$. Each Σ_i for $1 \leq i \leq L-1$ contains σ/L alphabet symbols; the last part Σ_L contains at most σ/L symbols. The root node u_R of $\mathcal{T}^g$ contains a sequence $X^g(u_R)$. Every element of $X^g(u_R)$ is a positive integer that does not exceed L. $X^g(u_R)[i] = j$ if $X[i]$ is a symbol from Σ_j. The child u_i of u is the root node of the wavelet tree for the subsequence X_i, where X_i is the subsequence of X induced by symbols from Σ_i. An L-ary tree can be constructed in $O(\log \sigma / L)$ time. During the second stage, we transform an L-ary tree into a binary tree. We replace each internal node u of $\mathcal{T}^g$ with a subtree $T(u)$ of height $\ell = \log L$. $T(u)$ has at most $L-1$ internal nodes; leaves of $T(u)$ correspond to children of u in $\mathcal{T}^g$. If the sequence $X^g(u)$ contains m elements, then all binary sequences $X(v)$ in the nodes $v \in T(u)$ contain $m\ell$ nits. Since we can pack ℓ elements of $X^g(u)$ into one word, $T(u)$ can be constructed in $O(m)$ time. A more technical description is provided below. We start by showing in Lemma 1 how the wavelet tree can be constructed in linear time when elements are bounded by L. Then we show in Theorem 1 how a binary wavelet tree for any sequence X can be constructed following the method outlined above. The result for a balanced binary wavelet tree can be easily extended to a t-ary tree of an arbitrary shape. Finally we can also obtain the original sequence X from its wavelet tree by reversing the algorithm that constructs the wavelet tree.

Lemma 1. *Let X be a sequence of L positive integers such that $L \leq 2^{\sqrt{\log n}}$ and $X[i] \leq 2^{\sqrt{\log n}}$ for all i, $1 \leq i \leq L$. A binary balanced wavelet tree for X can be constructed in $O(L)$ time using workspace $O(L)$. The algorithm employs a universal look-up table of $o(n)$ bits.*

Proof: We start by constructing a packed sequence $\overline{X}$; $\overline{X}$ consists of $\lceil L/\ell \rceil$ words and every word contains $\ell = \sqrt{\log n}$ elements of X. We initialize $\overline{X}(u_R) = \overline{X}$ for the root node u_R and visit all nodes in the depth-first order. When a node u is visited, we traverse $\overline{X}(u)$ and construct the bit sequence $X(u)$ that must be stored in the root u of the wavelet tree. We extract the first bit from each $\overline{X}[i]$ and append it to the end of $X(u)$. We also produce two sequences $\overline{X}(u_l)$ and $\overline{X}(u_r)$ unless u is a leaf node. If the first bit in $\overline{X}(u)[i]$ is 0, we append the value v to the end of $\overline{X}(u_l)$, where v is $\overline{X}(u)[i]$ without the first bit; if the first bit in $\overline{X}(u)[i]$ is 1, we append v to the end of $\overline{X}(u_r)$. Sequences $\overline{X}(u_l)$ and $\overline{X}(u_r)$ are also stored in packed form. When $X(u)$, $\overline{X}(u_l)$ and $\overline{X}(u_r)$ are generated, we can discard $\overline{X}(u)$.

The key observation is that each $\overline{X}(u)$ can be processed in $O(\lceil |\overline{X}(u)|/\ell \rceil)$ time using universal look-up tables $\mathbb{T}$ and $\mathbb{T}_1$. For any sequence of $\ell/4$ elements $\overline{Y}[1]\ldots\overline{Y}[\ell/4]$ of $p \leq \ell$ bits each, $\mathbb{T}$ can output (i) the bit sequence $Y[1]\ldots Y[\ell/4]$, where $Y[i]$ is the first bit of $\overline{Y}[i]$ (ii) sequences $\overline{Y}_l$ and $\overline{Y}_r$ defined below. The sequence $\overline{Y}_l$ contains elements $\overline{Y}[i]$ whose first bit is 0 in the same order as in $\overline{Y}$; the sequence $\overline{Y}_r$ contains elements $\overline{Y}[i]$ whose first bit is 1 in the same order as in $\overline{Y}$. Another look-up table, $\mathbb{T}_1$ can produce for any sequence $\overline{Y}[1]\ldots\overline{Y}[\ell/4]$ a sequence $\overline{Z}[1]\ldots\overline{Z}[\ell/4]$, where $\overline{Z}[i]$ equals to $\overline{Y}[i]$ without the first bit. Using these two look-up tables, we can read $\ell/4$ elements of $\overline{X}(u)$ and produce the next $\ell/4$ elements of $X(u)$, $\overline{X}(u_l)$, and $\overline{X}(u_r)$ in $O(1)$ time. $\mathbb{T}$ and $\mathbb{T}_1$ contain one entry for each p, $1 \leq p \leq \ell$, and for each sequence of $\ell/4$ integers of p bits each. Hence both tables have $O(n^{1/4})$ entries and use $o(n^{1/2})$ bits. Since we spend $O(\lceil |X(u)|/\ell \rceil)$ time in each node u and the total length of all $X(u)$ is $O(L\ell)$, we can construct the binary wavelet tree for X in $O(L)$ time.   □

Theorem 1. *Let X be a sequence of n positive integers such that $1 \leq X[i] \leq \sigma$ for $1 \leq i \leq n$. A binary balanced wavelet tree for X can be constructed in $O(n\lceil \frac{\log \sigma}{\sqrt{\log n}} \rceil)$ time.*

Proof: We employ the two-stage procedure described at the beginning of this section. During the first stage we construct a wavelet tree with node degree $L = 2^{\sqrt{\log n}}$. We will consider elements of X as binary sequences of length σ. For an integer v, we denote by $v.\mathrm{bits}(a..b)$ the bit sequence obtained by extracting bits at positions a, $a+1$, ..., b from v (bit positions are in the left-to-right order so that the most significant bit is at position 1). The process of recursive alphabet division can be re-formulated as recursive division of symbols according to their prefixes. That is, elements of X are distributed among 2^ℓ subsequences according to their prefixes of length $\ell = \sqrt{\log n}$. Each subsequence is further divided into 2^ℓ subsequences, etc. Let $\ell = \sqrt{\log n}$. We process the sequence X and generate sequences X_α. Initially all X_α are empty. For every $j = 0, 1, \ldots, \lceil \log \sigma/\sqrt{\log n} \rceil - 1$ we append $X[i].\mathrm{bits}(j\ell+1..j(\ell+1))$ to the sequence X_α for $\alpha = X[i].\mathrm{bits}(1..j\ell)$. Sequences X_α are stored in an L-ary wavelet tree $\mathcal{T}^g$. First ℓ bits of each $X[i]$ are kept in a sequence X_ϵ for an empty string ϵ. X_ϵ is stored in the root node that has 2^ℓ children labeled with bit sequences of length ℓ. The child that is labeled with α contains the sequence X_α. Every internal node also has 2^ℓ children that are labeled by bit sequences of length ℓ. Thus there are ℓ^i nodes of depth i. A node u of depth i contains the sequence X_α where α is the concatenation of node labels on the path from the root to u. We spend $O(n\lceil \frac{\log \sigma}{\sqrt{\log n}} \rceil)$ time to produce $\mathcal{T}^g$.

It remains to show how to construct a binary wavelet tree for each X_α. We divide X_α into subsequences $X_{\alpha,i}$ for $i = 1, \ldots, \lceil |X_\alpha|/2^{\sqrt{\log n}} \rceil$, where $|X_{\alpha,i}| = 2^{\sqrt{\log n}}$ for $1 \leq i \leq \lfloor |X_\alpha|/2^{\sqrt{\log n}} \rfloor$ and $|X_{\alpha,i}| \leq 2^{\sqrt{\log n}}$ for $i = \lceil |X_\alpha|/2^{\sqrt{\log n}} \rceil$. Then we apply Lemma 1 to each $X_{\alpha,i}$. □

The result of Theorem 1 can be easily extended to the case when the wavelet tree has an arbitrary shape.

Theorem 2. *Let X be a sequence of n positive integers such that $1 \leq X[i] \leq \sigma$ for $1 \leq i \leq n$. Any binary wavelet tree for X can be constructed in time $O(n \lceil \frac{h}{\sqrt{\log n}} \rceil + \sigma)$ where h is the average leaf depth.*

Proof: We assume in this theorem that the shape of the wavelet tree is already known. Let the *codeword* for a symbol $a \in \Sigma$ denote the bit string α obtained by following the path from the root to the leaf that contains a; we start with an empty string α and append 0 (1) to α every time when the left (respectively, the right) edge is taken. We start by replacing each element $X[i]$ with its codeword. Then we proceed exactly as in Theorem 1. If the codeword for a symbol $X[i]$ is of length $l[i]$, then the bits of $X[i]$ will be stored at $\lceil l[i]/\sqrt{\log n} \rceil$ nodes of the L-ary wavelet tree. Hence the first stage takes $O(n \lceil \frac{h}{\sqrt{\log n}} \rceil)$ time and the total number of symbols in all sequences X_α is also $O(n \lceil \frac{h}{\sqrt{\log n}} \rceil)$. We showed in Theorem 1 that a wavelet tree for each X_α is constructed in linear time. Hence node u of an L-ary wavelet tree is transformed into a binary tree in $O(|X(u)|)$ time, where $|X(u)|$ denotes the number of symbols in $X(u)$. Hence the total time to construct the wavelet tree is $O(n \lceil \frac{h}{\sqrt{\log n}} \rceil)$.

□

Besides our algorithm can be also modified for the case when the wavelet tree has arity $\log^\alpha n$ for a small constant α.

Theorem 3. *Let X be a sequence of n positive integers such that $1 \leq X[i] \leq \sigma$ for $1 \leq i \leq n$. A wavelet tree with node degree $\log^\alpha n$ for the sequence X can be constructed in time $O(n \lceil \frac{h\alpha \log \log n}{\sqrt{\log n}} \rceil + \sigma)$ where h is the average leaf depth.*

Proof: Let $\mathcal{T}$ denote the wavelet tree to be constructed. We extend $\mathcal{T}$ to a binary tree $\mathcal{T}^E$ by inserting some dummy nodes. Each node $u \in \mathcal{T}$ with descendants $u_1, \ldots, u_d$ is extended to a full binary tree[1] of height $\alpha \log \log n$ with root u and leaves $u_1, \ldots, u_d$. The nodes of the original tree will be called *data nodes*; all other nodes will be called *auxiliary nodes*. Our procedure constructs wavelet tree $\mathcal{T}^E$ in the same way as in Theorem 2, but we generate sequences $X(u)$ only for the data nodes u. Suppose that we visit a node and generate sequences $\overline{X}(u_l), \overline{X}(u_r)$. If u_l and u_r are auxiliary nodes, then $\overline{X}(u_l)$ and $\overline{X}(u_r)$ contain the elements of $\overline{X}(u)$; unlike Theorem 2, the leftmost bits of $\overline{X}(u)$ are not removed. We simply assign the elements of $\overline{X}(u)$ to $\overline{X}(u_l)$ and $\overline{X}(u_r)$ according to the t-th bit of $\overline{X}(u)$ where t is the distance from u_l to its lowest ancestor that is a data node. If

---

[1] To simplify the description we assume that $\log^\alpha n$ is a power of 2.

u_l and u_r are data nodes, we generate $X(u_l)$ and $X(u_r)$ according to the value of the d-th bit in $\overline{X}(u)$ for $d = \alpha \log\log n$; depending on the value of the d-th bit in $\overline{X}(u)$ we append lshift($X(u)[i]$) to $X(u_l)$ or $X(u_r)$, where lshift(v) denotes the value of v with $\alpha \log\log n$ leftmost bits removed. If a data node u is visited, we also generate a sequence $X(u)$ such that $X(u)[i] = \overline{X}(u)[i].\mathrm{bits}(1..\alpha \log\log n)$. That is, we retrieve the $\alpha \log\log n$ leftmost bits from each $\overline{X}(u)[i]$ and store them in $X(u)$; we note that $\overline{X}(u)[i]$ do not change when u is visited. ☐

Finally we can also restore the original sequence X from its wavelet tree.

Theorem 4. *We can obtain a sequence X from its binary wavelet tree $\mathcal{T}$ in $O(n\frac{h}{\sqrt{\log n}}+\sigma)$ time, where h is the average leaf depth. We can obtain a sequence X from its wavelet tree $\mathcal{T}$ with node degree $\log^\alpha n$ in $O(n\frac{h\alpha \log\log n}{\sqrt{\log n}} + \sigma)$ time.*

Proof: We say that a node $u \in \mathcal{T}$ is special, if its depth is divisible by $\ell = \sqrt{\log n}$. Our algorithm consists of two stages. First, we create sequences $\overline{X}(u)$ stored in special nodes u, so that each $\overline{X}(u)[i]$ is an integer of at most ℓ bits. That is, we turn $\mathcal{T}$ into a wavelet tree $\mathcal{T}_b$ such that each internal node of $\mathcal{T}_b$ has up to 2^ℓ children. The total bit length of all sequences stored in the nodes of $\mathcal{T}_b$ equals the total bit length of all sequences stored in the nodes of $\mathcal{T}$. Thus the total space usage does not increase. The procedure for converting $\mathcal{T}$ into $\mathcal{T}_b$ works as follows. For every node u, such that both its children are special nodes, we assume that $\overline{X}(u) = X(u)$. Then we work up the tree and produce $\overline{X}(v)$ for ancestors v of u until a special node is reached. Suppose that sequences $\overline{X}(u_l)$ and $\overline{X}(u_r)$ for children of a node w are already produced. We generate $\overline{X}(w)$ according to the following rule: if $X(w)[i] = 0$, then $\overline{X}(w)[i] = 0\overline{X}(u_l)[i]$; if $X(w)[i] = 1$, then $\overline{X}(w)[i] = 1\overline{X}(u_r)[i]$. The total time to construct $\mathcal{T}_b$ is $O(n\frac{h}{\sqrt{\log n}} + \sigma)$.

Finally we collect the values of $\overline{X}(u)[i]$ in special nodes u and obtain the sequence of integers X by concatenating those values. The procedure starts in the root node u_R of $\mathcal{T}_b$. Depending on the value of $\overline{X}(u_R)[i]$ we visit the corresponding child u of u_R in $\mathcal{T}_b$ (we observe that u_R can have up to $2^{\sqrt{\log n}}$ children) and retrieve the next element e_u in $\overline{X}(u)$. Then we replace $\overline{X}(u_R)[i]$ with the concatenation of $\overline{X}(u_R)[i]$ and e_u. Proceeding in the same way for nodes of $\mathcal{T}_b$ on all levels, we obtain the values of the original sequence X in $\overline{X}(u_R)$. ☐

4   Rank and Select Queries in Wavelet Trees

In this section we consider the problem of storing a sequence $S = s_1 s_2 \ldots s_n$ over an alphabet σ that supports the following queries:
-access(i, S) returns $S[i]$
-rank$_a(i, S)$ computes the number of times a occurs in $S[1..i] = s_1 \ldots s_i$
-select$_a(i, S)$ finds the position j of the i-th occurrence of a, i.e., select$_a(i, S) = j$ such that access(j, S) = a and rank$_a(j, S) = i$.

Wavelet trees support rank, select, and access queries in $O(\log \sigma)$ time. This is achieved by augmenting sequences $X(u)$, stored in the nodes, with data structures that answer rank and select queries. If queries on sequences $X(u)$ are answered in constant time, then queries on the original sequence X are answered in $O(\log \sigma)$ time. The sequence $X(u)$ stored in a node of a binary wavelet tree is a sequence of bits. In the case of a t-ary wavelet tree, sequences $X(u)$ are over an alphabet of size t. Thus a wavelet tree reduces rank, select, access queries on a sequence X to $O(\log \sigma)$ queries on binary sequences or $O(\log \sigma / \log t)$ queries on t-ary sequences. For details we refer to e.g., [7]. It remains to show how data structures for $X(u)$ can be constructed quickly.

Rank and Select on Binary and t-ary Sequences. We will describe below several results on constructing rank-select data structures for sequences over a small alphabet. We remark that the data structures are not new and are based on standard techniques. However, we show that these data structures can be constructed in less than linear time, provided that the original sequence is available in packed form.

We show in the following Theorem that the data structure of Jacobson [6] can be constructed in $O(m/\log n)$ time.

Theorem 5. *A bit sequence B of length m can be stored in data structure that answers* rank, select, *and* access *queries in constant time. This data structure uses $m + O(m \log \log n/\log n)$ bits and can be constructed in $O(m/\log n)$ time. The construction algorithm relies on a universal table of size $o(n)$.*

Proof: B is divided into blocks of $d_1 = \log^2 n$ bits. We compute and store the number of 0's and the number of 1's in the first i blocks for $i = 1, \ldots, m/\log^2 n$. Since the number of blocks is $O(m/\log^2 n)$, this information takes $O(m/\log n)$ bits. We assume that B is kept in packed form, so that this information can be computed in $O(m/\log n)$ time. Each block is divided into sub-blocks of size $d_2 = \log n/2$. We compute and store the number of 0's and the number of 1's in the first j sub-blocks of a block for each block and for $j = 1, \ldots, 2\log n$. The number of 0's or 1's in a block is at most $\log^2 n$. Hence, we can keep information about 0 and 1's in the first j sub-blocks of a block in $O(\log \log n)$ bits. The total number of sub-blocks is $O(m/\log n)$. Hence, all sub-block counts take $O(m \log \log n/\log n)$ bits.

Using a pre-computed universal table of size $O(n^{1/2} \log n)$ we can find the number of 0's and the number of 1's within the first t positions of a sub-block for $t = 1, 2, \ldots, \log n/2$ in $O(1)$ time. A rank query $\mathrm{rank}_0(i, B)$ is answered by finding the block j_1 that contains the i-th bit and the sub-block j_2 within the j_1-th block that contains the i-th bit. We also find the position j_3 of the i-th bit within that sub-block. Now $\mathrm{rank}_0(i, B) = c_1 + c_2 + c_3$ where c_1 is the number of 0's in the first $j_1 - 1$ blocks, c_2 is the number of 0's in the first $j_2 - 1$ sub-blocks of the j_1-th block, and c_3 is the number of 0's among the first j_3 bits in the j_2-nd sub-block of the j_1-th block. Queries $\mathrm{rank}_1(i, B)$ are answered in the same way.

The data structure for rank queries can be created in $O(m/\log n)$ time. Using the same universal table, we can compute the number of 0's and the number of 1's in each sub-block in $O(1)$ time. Using this information, we count the number of 0's in the first t sub-blocks of each block for $t = 1, \ldots, \log^2 n$. Then we count the number of 0's in the first i blocks for $i = 1, 2, \ldots, m/\log^2 n$. The total number of sub-blocks in all blocks is $O(m/\log n)$ and the total number of blocks is $O(m/\log^2 n)$. Since we spend $O(1)$ time in every sub-block, auxiliary data structures for rank queries can be computed in $O(m/\log n)$ time.

The data structure for select queries is based on a similar approach. Suppose that we want to answer queries $\text{select}_0(i, B)$. We divide B into chunks, so that each chunk contains $\log^2 n$ 0-bits. If the size of a chunk exceeds, $\log^4 n$, we say that this chunk is sparse; otherwise a chunk is dense. We keep left boundaries of each chunk in an array. If a chunk is sparse, we also keep the position of the t-th 0-bit in that chunk for $t = 1, \ldots, \log n$. A dense chunk is divided into words, so that each word consists of $\log n/2$ bits. We keep the number of 0-bits in every word in a data structure M. The number of 0-bits in every word is at most $\log n/2$, the number of words in a chunk is $O(\log^3 n)$. We can implement M so that it uses $O(\log \log n)$ bits per word; moreover we can find the word that contains the t-th 0-bit in a block and the number of 0-bits in the first d words for any t, d in time $O(1)$.

A query $\text{select}_0(i, B)$ is answered by finding the starting position of the i_0-th chunk for $i_0 = \lfloor i/\log n \rfloor$. Let $i_1 = i - i_0 \cdot \log n$. Clearly $\text{select}_0(i, S) = j$ where j is the position of the i_1-th 0 in the i_0-th chunk. If this chunk is sparse, then the position of the i_1-th 0 is stored. Otherwise we find the word W_j that contains the i_1-th 0-bit using M. We can find the position of the i_1-th bit in W_j using a universal look-up table.

There are $O(n/\log^2 n)$ chunks and $O(n/\log^4 n)$ sparse chunks. The number of 0-bits in sparse chunks is $O(n/\log^2 n)$; hence, we can store positions of all 0-bits in $O(n/\log n)$ bits. We can create the array that contains left boundaries of all chunks and positions of 0-bits in all chunks in $O(m/\log n)$ time. The time needed to create data structures M for all dense chunks is proportional to the number of words in all dense chunks. Hence all M are created in $O(m/\log n)$ time.

Data structures that support rank_1 and select_1 on B are implemented in the same way. $\square$

Theorem 6. *A bit sequence B of length m can be stored in data structure that answers* rank, select, *and* access *queries in constant time. This data structure uses $mH_0(B) + O(m \log \log n/\log n)$ bits and can be constructed in $O(m/\log n)$ time, where $H_0(B)$ is the zero-order entropy of B. The construction algorithm relies on a universal table of size $o(n)$.*

Proof: The only difference is that the bit sequence B itself is stored in compressed form. We employ the method of Raman et al. [8] that splits the sequence into pieces of size $\Theta(\log n)$ and keeps all pieces in $mH_0(B) + O(m \log \log n/\log n)$ bits. $\square$

Theorem 7. *A sequence B of length m over an alphabet $\{1, 2, \log^\varepsilon n\}$, where $\varepsilon > 0$ is a constant, can be stored in data structure that answers* rank, select, *and* access *queries in constant time. This data structure uses $mH_0(B) + O(m \log^\varepsilon n \log \log n/\log n)$ bits and can be constructed in $O(m/\log^{1-\varepsilon} n)$ time, where $H_0(B)$ is the zero-order entropy of B. The construction algorithm relies on a universal table of size $o(n)$.*

Proof: We keep auxiliary structures that answer rank_a and select_a for every a such that $1 \leq a \leq \log^\varepsilon n$. These data structures are implemented as in Theorems 5 and 6 and need $O(m(\log \log n/ \log n))$ bits. All auxiliary data structures use $O(m(\log \log n/ \log^{1-\varepsilon} n))$ bits. Since data structures for a fixed symbol a can be constructed in $O(m/ \log n)$ time, all auxiliary structures are constructed in $O(m/ \log^{1-\varepsilon} n)$ time. $\qquad\square$

Wavelet Trees. In Section 3 we showed how bit sequences $X(u)$ stored in the nodes of the wavelet tree of a sequence X can be obtained. Using Theorems 5, 6, and 7, we can augment $X(u)$ with secondary data structures that enable us to answer rank, access, and select queries on X.

Corollary 1. *Let X be a sequence of n positive integers such that $1 \leq X[i] \leq \sigma$ for $1 \leq i \leq n$. We can construct a binary balanced wavelet tree T for a sequence X, such that T uses $nH_0(X) + o(n \log \sigma)$ bits and answers queries rank, select, and access in $O(\log \sigma)$ time. The wavelet tree T can be constructed in $O(n\lceil \frac{\log \sigma}{\sqrt{\log n}} \rceil)$ time.*

Proof: We construct a balanced binary wavelet tree as in Theorem 1. Sequences $X(u)$ are stored in compressed form. It can be shown that the total space usage of all $X(u)$ is $n \log \sigma + o(n \log \sigma)$ bits; see e.g., [7], Section 3.1. $\qquad\square$

We can further improve the construction time if a wavelet tree of special shape is used.

Corollary 2. *Let X be a sequence of n positive integers such that $1 \leq X[i] \leq \sigma$ for $1 \leq i \leq n$. We can construct a wavelet tree T for a sequence X, such that T uses $n(H_0(X) + 2) + o(n \log \sigma)$ bits and answers queries rank, select, and access in $O(\log \sigma)$ time. The wavelet tree T can be constructed in $O(n\lceil \frac{H_0(X)}{\sqrt{\log n}} \rceil)$ time.*

Proof: Barbay and Navarro [1] describe a wavelet tree T, such that the average leaf depth in T is $O(H_0(X))$ and the maximum leaf depth is $O(\log \sigma)$. Furthermore the total number of bits in all sequences $X(u)$ stored in nodes of T is bounded by $n(H_0(X) + 2)$. We construct T using Theorem 2. Data structures for sequences $X(u)$ are constructed as in Theorem 5. $\qquad\square$

We also remark that we can construct wavelet trees for X with node degree $\log^\varepsilon n$ where ε is a small positive constant. In this case the space usage and construction times are the same as in Corollaries 1 and 2, but queries are supported in $O(\log \sigma/ \log \log n)$ time.

5   Conclusions

In this paper we described fast algorithms for constructing a wavelet tree. We showed that this important data structure can be constructed in $O(n\lceil \frac{\log \sigma}{\sqrt{\log n}} \rceil)$ time. If the wavelet tree with a special shape is used, then construction cost can be further reduced.

The problem of designing faster algorithms (e.g., algorithms that work in $O(n)$ or $O(n \log \log n)$ time for an alphabet $\{1, \ldots, n\}$) remains open.

References

1. Barbay, J., Navarro, G.: Compressed representations of permutations, and applicatiqons. In: Proc. 26th International Symposium on Theoretical Aspects of Computer Science (STACS 2009), pp. 111–122 (2009)
2. Chan, T.M., Patrascu, M.: Counting inversions, offline orthogonal range counting, and related problems. In: Proc. 21st Annual ACM-SIAM Symposium on Discrete Algorithms (SODA 2010), pp. 161–173 (2010)
3. Chazelle, B.: A functional approach to data structures and its use in multidimensional searching. SIAM Journal on Computing 17(3), 427–462 (1988)
4. Claude, F., Nicholson, P.K., Seco, D.: Space efficient wavelet tree construction. In: Grossi, R., Sebastiani, F., Silvestri, F. (eds.) SPIRE 2011. LNCS, vol. 7024, pp. 185–196. Springer, Heidelberg (2011)
5. Grossi, R., Gupta, A., Vitter, J.S.: High-order entropy-compressed text indexes. In: Proc. 14th Annual ACM-SIAM Symposium on Discrete Algorithms (SODA 2003), pp. 841–850 (2003)
6. Jacobson, G.: Space-efficient static trees and graphs. In: Proc. 30th Annual Symposium on Foundations of Computer Science (FOCS 1989), pp. 549–554 (1989)
7. Navarro, G.: Wavelet trees for all. J. Discrete Algorithms 25, 2–20 (2014)
8. Raman, R., Raman, V., Rao, S.S.: Succinct indexable dictionaries with applications to encoding k-ary trees and multisets. In: Proc. 13th Annual ACM-SIAM Symposium on Discrete Algorithms (SODA 2002), pp. 233–242 (2002)
9. Tischler, G.: On wavelet tree construction. In: Proc. 22nd Annual Symposium on Combinatorial Pattern Matching (CPM 2011), pp. 208–218 (2011)

Order Preserving Prefix Tables

Md. Mahbubul Hasan[1,*], A.S.M. Sohidull Islam[2,4],
Mohammad Saifur Rahman[1,5], and M. Sohel Rahman[1,5]

[1] AℓEDA Group, Department of CSE, BUET, Dhaka 1000, Bangladesh
[2] Department of Computational Engineering and Science
McMaster University, Hamilton, Ontario, Canada
[3] shanto86@gmail.com,
[4] sohanas@mcmaster.ca,
[5] {mrahman,msrahman}@cse.buet.ac.bd

Abstract. In the Order Preserving Pattern Matching (OPPM) problem, we have a text T and a pattern P on an integer alphabet as input. And the goal is to locate a fragment which is order-isomorphic with the pattern. Two sequences over integer alphabet are order-isomorphic if the relative order between any two elements at the same positions in both sequences is the same. In this paper we present an efficient algorithm to construct an interesting and useful data structure, namely, prefix table, from the order preserving point of view.

1 Introduction

In the Order Preserving Pattern Matching (OPPM) problem, we have a text T and a pattern P on an integer alphabet as input. And, instead of looking for a substring of the text which is identical to the given pattern, we are interested in locating a fragment which is order-isomorphic with the pattern. Two sequences over integer alphabet are order-isomorphic if the relative order between any two elements at the same positions in both sequences is the same. Very recently OPPM has received much attention [6,7,4,3,2,5].

In this paper, our main focus is an interesting data structure known as the prefix table (also *prefix array*)[1] [8,9]. Briefly speaking, at each position i, the prefix table π of a string $S[1..n]$ keeps track of the length of the longest substring of S that starts at position i and matches a prefix of S. Here, we present the first efficient algorithm for constructing a prefix table from Order Preserving Point of view. In what follows we will conveniently refer to this as the *order preserving prefix table*.

2 Preliminaries

Let Σ denote the set of numbers such that a comparison of two numbers can be done in constant time, and let Σ^* denote the set of strings over the alphabet Σ.

---

[*] Currently working at Google Zürich.
[1] We prefer "table" because of the possible confusion with "suffix array", a completely different data structure.

E. Moura and M. Crochemore (Eds.): SPIRE 2014, LNCS 8799, pp. 111–116, 2014.
© Springer International Publishing Switzerland 2014

Let $|S| = n$ denote the length of a string S, which is described by a sequence of numbers as $(S[1], S[2], ..., S[n])$. For $n = 0$, $S = \varepsilon$, the empty string. If $S = UVW$, then U is said to be a prefix, V a substring (also called a factor) and W a suffix of S; if $VW \neq \varepsilon$, $UW \neq \varepsilon$, $UV \neq \varepsilon$, respectively, then U, V, W is, respectively, a proper prefix, proper substring, proper suffix of S. A substring $(S[i], S[i+1], ..., S[j])$ of S is denoted by $S[i, j]$, the ith prefix $S[1, i]$ by $\text{prefix}_i(S)$ and the ith suffix $S[i, |S|]$ by $\text{suffix}_i(S)$.

The rank of a number c in string S is defined as $rank_S(c) = 1 + |\{i : S[i] < c \text{ for } 1 \leq i \leq |S|\}|$. For simplicity, we assume that all the numbers in a string are distinct. However, this does not lose generality because when a number occurs more than once in a string, we can extend our number definition to a pair of number and index in the string so that the numbers in the string become distinct. For comparing two such pairs, we first compare the numbers in the pair and if they are equal, we compare the two indexes. The order preserving representation (OPR) $\sigma(S)$ of a string S can be defined as $\sigma(S) = rank_S(S[1]), rank_S(S[2]), ..., rank_S(S[|S|])$. In what follows, if two strings S_1 and S_2 have identical OPR, i.e., $\sigma(S_1) = \sigma(S_2)$, then we say that the two strings are OPR-equal.

3   Order Preserving Prefix Tables

We start with the following formal definition of the order preserving prefix table.

Definition 1. *Given a string S, the order preserving prefix table π, $1 \leq i \leq |S|$ is the length of the longest prefix P of $\text{suffix}_i(S)$ such that $\sigma(P) = \sigma(S[1, |P|])$.*

In other words, $\pi_S[i]$ is the length of the longest substring of S that starts at position i and its OPR is same as the OPR of some prefix of S. An example of an order preserving prefix table is given below:

$$S = (\ 11,\ 18,\ 24,\ 20,\ 25,\ 29\ )$$
$$\pi = \quad 6 \quad 2 \quad 1 \quad 3 \quad 2 \quad 1$$

In Algorithm 1, we formally present the construction procedure. To discuss the correctness of algorithm, we start with the following lemma which will be useful shortly.

Lemma 1. *Let S and T be two strings such that $\sigma(S) \neq \sigma(T)$. Now assume that s and t are any numbers. Then we must also have $\sigma(Ss) \neq \sigma(Tt)$.*

Proof. We prove it by contradiction. Assume for the sake of contradiction that $\sigma(Ss) = \sigma(Tt)$. So the rank of s in S is same as the rank of t in T. If we remove s and t from the corresponding strings, the ranks of the numbers greater than s and t will reduce by 1; ranks of the numbers less than those will remain the same. Hence we must have $\sigma(S) = \sigma(T)$, a contradiction.   □

It is very easy to extend Lemma 1 to get the following lemma.

Algorithm 1. High Level overview of order preserving prefix table Construction

```
 1: procedure OPPTAB(S)
 2:     π[1] := |S|
 3:     L := R := 1
 4:     for i := 2 to |S| do
 5:         π[i] := 0
 6:         if i ≤ R then
 7:             π[i] := min(R − i + 1, π[i − L + 1])
 8:         end if
 9:         while i + π[i] ≤ |S| and σ(S[i, i + π[i]]) = σ(S[1, π[i] + 1]) do
10:             π[i] := π[i] + 1
11:         end while
12:         if i + π[i] − 1 > R then
13:             L := i
14:             R := i + π[i] − 1
15:         end if
16:     end for
17:     return π
18: end procedure
```

Lemma 2. *Let S and T are two strings such that $|S| = |T|$ and $\sigma(S) \neq \sigma(T)$. Now let U and V are any two strings. Then $\sigma(SU) \neq \sigma(TV)$.*

Now we discuss the correctness of our algorithm. In this algorithm we maintain two pointers L and R such that $\sigma(S[L, R]) = \sigma(S[1, R − L + 1])$ and R is as large as possible for any particular L. The algorithm always maintain S the invariant that $L \leq i$. This is ensured because the value of L is updated only in Line 13 to i under some condition. Now as we proceed from $i = 2$ to $|S|$, for each i we check whether $i \leq R$ and if so we can safely say that $\pi[i]$ is at least the minimum of $R − i + 1$ and $\pi[i − L + 1]$; otherwise $\pi[i]$ is set to 0. The reason for such bound of $\pi[i]$ is the invariant that we always have $\sigma(S[L, R]) = \sigma(S[1, R − L + 1])$. Additionally, we in fact have a stronger invariant: OPR of any substring of $S[L, R]$ is same as the OPR for the corresponding substring of $S[1, R − L + 1]$. So, OPR of $S[i, R]$ will be same as the OPR of $S[i − L + 1, R − L + 1]$. Since i is increasing, we have already calculated the length of the largest prefix of suffix$_{i−L+1}$ which has the same OPR as some prefix of the main string S and that length is $\pi[i − L + 1]$. Therefore $S[i, R]$ has the same OPR as $S[i − L + 1, R − L + 1]$ and we also know $S[i − L + 1, i − L + \pi(i − L + 1)]$ has the same OPR as $S[1, \pi[i − L + 1]]$. Hence we can say that, $\sigma(S[i, i + M − 1]) = \sigma(S[1, M])$ for $M = min(R − i + 1, \pi[i − L + 1])$. So M is set as the lower bound in Line 7. But as we said, this is a lower bound; the real $\pi[i]$ may be larger. So in the **while** loop at Line 9, we check whether $\pi[i]$ can be incremented by 1. When we reach the string boundary or $\pi[i]$ can not be incremented more maintaining the invariant, we break the loop.

Now, according to lemma 1, if $\sigma(S[i, i + \pi[i]]) \neq \sigma(S[1, \pi[i] + 1])$ then for all the values $j > \pi[i]$ we can say that $\sigma(S[i, i + j − 1]) \neq \sigma(S[1, j])$. So we will have the correct value in $\pi[i]$ as soon as execution of the *while* loop of Line 9 is complete. For the efficiency of the algorithm we try to update the value of L and R in the **if** block of Line 12 which will be discussed shortly.

3.1 OPR Matching

In the condition of the **while** loop at Line 9 of Algorithm 1, we compare OPR of two strings $S[i, i+\pi(i)]$ and $S[1, \pi(i)+1]$. In a naive approach, it takes $O(n \log n)$ time to compute OPR of a string S of length n. For this we can use a balanced binary search tree $\mathcal{T}$ that would support the first two operations in Table 1. Note that, all the operations in the table 1 can be implemented in logarithmic time complexity. Using first two operations, OPR of a string S can be computed in $O(n \log n)$ time (Algorithm 2). So it will take $O(n \log n)$ time in the worst case to compute the OPRs in Line 9 of Algorithm 1 and an additional $O(n)$ time to check the equality.

Table 1. Operations of tree $\mathcal{T}$

Function	Description
Insert(x, i)	Inserts a (key, value) pair (x, i) in the tree $\mathcal{T}$
Rank(x)	Calculates the rank of the number x among the keys present in $\mathcal{T}$. In other words, it returns number of the keys in $\mathcal{T}$ that are at least x
PreviousIndex(x)	Finds the value of the pair having largest key less than x.
NextIndex(x)	Finds the value of the pair having smallest key greater than x.

Algorithm 2. Calculation of OPR ofastring S

```
 1: procedure COMPUTEROPR(S)
 2:     T := Empty Tree
 3:     for i := 1 to |S| do
 4:         T.Insert(S[i], i)
 5:     end for
 6:     for i := 1 to |S| do
 7:         OPR(i) := T.Rank(S[i])
 8:     end for
 9:     return OPR
10: end procedure
```

However, we can implement this checking in constant time with an additional $O(n \log n)$ preprocessing. Later, we will see that it will significantly improve the time complexity of our algorithm. However to achieve this improvement, we have to use tree $\mathcal{T}$ which can perform all the operations of Table 1. The preprocessing phase processes the initial string S and computes two arrays named **Prev** and **Next**. Here, Prev(i) is equal to some j such that $j < i$ and $S[j]$ is the largest value that is less than $S[i]$. Similarly, Next(i) is equal to some j such that $j < i$ and $S[j]$ is the smallest value that is greater than $S[i]$. This preprocessing algorithm is formally presented in Algorithm 3.

Algorithm 3. Preprocessing Phase

```
1: procedure PREPROCESS(S)
2:     T := Empty Tree
3:     T.Insert(−∞, −infty)
4:     T.Insert(∞, infty)
5:     for i := 1 to |S| do
6:         T.Insert(S[i], i)
7:         Prev(i) := T.PreviousIndex(S[i])
8:         Next(i) := T.NextIndex(S[i])
9:     end for
10:     return (Prev, Next)
11: end procedure
```

With the help of **Prev** and **Next** arrays we now can check the equality of two OPRs in constant time. In Line 9 of Algorithm 1, we know that $\sigma(S[i, i + \pi(i) - 1]) = \sigma(S[1, \pi(i)])$ and we would like to increase the value of $\pi(i)$ by one. Instead of appending the next number to the strings and evaluating the entire OPRs, we proceed as follows. We simply find the position of the next number in the already calculated OPRs of the strings $S[i, i + \pi(i) - 1]$ and $S[1, \pi(i)]$. This is where our preprocessing comes handy. From the precalculated **Prev** and **Next** arrays we know that immediate smaller and larger values of $S[\pi(i) + 1]$ in $S[1, \pi(i)]$ are at $Prev[\pi(i) + 1]$ and $Next[\pi(i) + 1]$ respectively. So if the immediate smaller and larger values of $S[i + \pi(i)]$ are at the corresponding places of $S[i, i + \pi(i)]$ that is at $Prev[\pi(i) + 1] + i - 1$ and $Next[\pi(i) + 1] + i - 1$ respectively, we can say that the OPRs of the new strings will also be same. Since $\sigma(S[i, i + \pi(i) - 1]) = \sigma(S[1, \pi(i)])$, it would be enough to check if $S[Prev[\pi(i) + 1] + i - 1] < S[i + \pi(i)] < S[Next[\pi(i) + 1] + i - 1]$. This follows readily following a similar line of arguments as discussed in the proof of Lemma 1. Our improved algorithm is presented in Algorithm 4.

Algorithm 4. Improved algorithm for order preserving prefix table Construction

```
1: procedure OPPTAB(S)
2:     (Prev, Next) = Preprocess(S)
3:     π(1) := |S|
4:     L := R := 1
5:     for i := 2 to |S| do
6:         π(i) := 0
7:         if i ≤ R then
8:             π(i) := min(R − i + 1, π(i − L + 1))
9:         end if
10:         while i + π(i) ≤ |S| and S[Prev[π(i) + 1] + i − 1] < S[i + π(i)] < S[Next[π(i) + 1] + i − 1]
    do
11:             π(i) := π(i) + 1
12:         end while
13:         if i + π(i) − 1 > R then
14:             L := i
15:             R := i + π(i) − 1
16:         end if
17:     end for
18:     return π
19: end procedure
```

The time complexity analysis of Algorithm 4 is a bit tricky. Firstly, the pre-processing phase takes $O(n \log n)$ time. Now observe carefully that, if $\pi(i)$ is set to the value of $R - i + 1$ in Line 8, then, there is a chance that the value of $\pi(i)$ may increase in the following **while** loop and that results in the same amount of increase in the value of R in Line 15. However, if $\pi(i)$ is not set to $R - i + 1$ then $\pi(i)$ will not increase. Condition checking inside the **if** blocks or **while** takes constant time. We can say that the number of iterations of the **while** loop is equal to $R + O(n)$. But $R \leq |S|$. Hence the total running time of Algorithm 4 is $O(n \log n)$. Space complexity is quite straight forward to analyze as follows. The space complexity of the tree $\mathcal{T}$ is $O(n)$ and also for the arrays π, Prev and Next we need $O(n)$ memory. So the space complexity of the algorithm is $O(n)$.

References

1. Bland, W., Kucherov, G., Smyth, W.F.: Prefix table construction and conversion. In: Lecroq, T., Mouchard, L. (eds.) IWOCA 2013. LNCS, vol. 8288, pp. 41–53. Springer, Heidelberg (2013)
2. Cho, S., Na, J.C., Park, K., Sim, J.S.: Fast order-preserving pattern matching. In: Widmayer, P., Xu, Y., Zhu, B. (eds.) COCOA 2013. LNCS, vol. 8287, pp. 295–305. Springer, Heidelberg (2013)
3. Crochemore, M., Iliopoulos, C.S., Kociumaka, T., Kubica, M., Langiu, A., Pissis, S.P., Radoszewski, J., Rytter, W., Waleń, T.: Order-preserving incomplete suffix trees and order-preserving indexes. In: Kurland, O., Lewenstein, M., Porat, E. (eds.) SPIRE 2013. LNCS, vol. 8214, pp. 84–95. Springer, Heidelberg (2013)
4. Crochemore, M., Iliopoulos, C.S., Kociumaka, T., Kubica, M., Langiu, A., Pissis, S.P., Radoszewski, J., Rytter, W., Walen, T.: Order-preserving suffix trees and their algorithmic applications. CoRR, abs/1303.6872 (2013)
5. Gawrychowski, P., Uznanski, P.: Order-preserving pattern matching with k mismatches. CoRR, abs/1309.6453 (2013)
6. Kim, J., Eades, P., Fleischer, R., Hong, S.-H., Iliopoulos, C.S., Park, K., Puglisi, S.J., Tokuyama, T.: Order preserving matching. CoRR, abs/1302.4064 (2013)
7. Kubica, M., Kulczynski, T., Radoszewski, J., Rytter, W., Walen, T.: A linear time algorithm for consecutive permutation pattern matching. Inf. Process. Lett. 113(12), 430–433 (2013)
8. Main, M.G., Lorentz, R.J.: An o(n log n) algorithm for finding all repetitions in a string. J. Algorithms 5(3), 422–432 (1984)
9. Smyth, W.F.: Computing patterns in strings. Pearson Addison-Wesley (2003)

Alphabet-Independent Algorithms for Finding Context-Sensitive Repeats in Linear Time

Enno Ohlebusch and Timo Beller

Institute of Theoretical Computer Science, University of Ulm, D-89069 Ulm
{Enno.Ohlebusch,Timo.Beller}@uni-ulm.de

Abstract. The identification of repetitive sequences (repeats) is an essential component of genome sequence analysis, and there are dozens of algorithms that search for exact or approximate repeats. The notions of maximal and supermaximal (exact) repeats have received special attention, and it is possible to simultaneously compute them on index data structures like the suffix tree or the enhanced suffix array. Very recently, this research has been extended in two directions. Gallé and Tealdi devised an alphabet-independent linear-time algorithm that finds all context-diverse repeats (which subsume maximal and supermaximal repeats as special cases), while Taillefer and Miller gave a quadratic-time algorithm that simultaneously computes and classifies maximal, near-supermaximal, and supermaximal repeats. In this paper, we provide new alphabet-independent linear-time algorithms for both tasks.

1 Introduction

In the analysis of a genome, a basic task is to locate and characterize the repetitive sequences (repeats). While bacterial genomes usually do not contain large amounts of repetitive sequences, a considerable portion of the genomes of higher organisms is composed of repeats. For example, more than half of the 3 billion basepairs of the haploid human genome consists of repeats.

In order to avoid redundant output, the search for exact repeats is usually restricted to so-called maximal and supermaximal repeats. Let us briefly recall these notions. A substring ω of a (long) string S is an exact repeat if it occurs at least twice in S. A repeat ω of S is a maximal repeat if any extension of ω occurs fewer times in S than ω. A supermaximal repeat is a maximal repeat that is not a proper substring of another repeat. Section 3 discusses articles that describe how maximal and supermaximal repeats can be computed efficiently.

This paper is inspired by two papers published recently. In the first paper, Gallé and Tealdi [10] introduced context-diverse repeats, which subsume maximal and supermaximal repeats as special cases. They provided three algorithms for finding all context-diverse repeats:

1. An $O(n\sigma)$ time algorithm based on the enhanced suffix array of S (using a variant of the bottom-up traversal of the lcp-interval tree [1], a method of simulating a bottom-up traversal of the suffix tree), where σ is the size of the underlying alphabet Σ.

E. Moura and M. Crochemore (Eds.): SPIRE 2014, LNCS 8799, pp. 117–128, 2014.
© Springer International Publishing Switzerland 2014

2. An $O(n \log n)$ time algorithm based on Fenwick trees (a Fenwick tree permits to calculate a prefix sum of an array of values in $O(\log n)$ time; it is dynamic because the table can be modified in $O(\log n)$ time).

3. An $O(n)$ time algorithm that (a) computes right-diverse repeats based on the enhanced suffix array of S, (b) computes left-diverse repeats based on the enhanced suffix array of the reverse string of S, and (c) merges them.

Here, we provide a simpler $O(n)$ time algorithm. In essence, we replace the Fenwick tree with the correction terms devised by Hui [13].

The second paper that inspired our work is [24]. In that paper, Taillefer and Miller study length distributions of repeats in genome sequences. To that end, they used a simple algorithm that simultaneously computes and classifies maximal, near-supermaximal, and supermaximal repeats. The worst-case time complexity of the simple algorithm is $O(n^2)$, but it improves to $O(n\sigma)$ if, in the bottom-up traversal of the lcp-interval tree, information at child intervals is propagated to their parent interval. The real challenge is to devise an $O(n)$ time algorithm for this task. Note that the result on context-diverse repeats does not apply because context-diverse repeats do not comprise near-supermaximal repeats. In this paper, we give the first alphabet-independent linear-time algorithm that simultaneously finds and classifies maximal, near-supermaximal, and supermaximal repeats.

Experimental results show that our algorithms are not only of theoretical interest: both have implementations that are faster in practice than all known algorithms. For space reasons, proofs and experimental results had to be omitted.

2   Preliminaries

Let Σ be an ordered alphabet of size σ whose smallest element is the so-called sentinel character \$. In the following, S is a string of length n on Σ having the sentinel character at the end (and nowhere else). For $1 \leq i \leq n$, $S[i]$ denotes the *character at position* i in S. For $i \leq j$, $S[i..j]$ denotes the *substring* of S starting with the character at position i and ending with the character at position j. Furthermore, S_i denotes the i-th suffix $S[i..n]$ of S. The *suffix array* SA of the string S is an array of integers in the range 1 to n specifying the lexicographic ordering of the n suffixes of S, that is, it satisfies $S_{\mathsf{SA}[1]} < S_{\mathsf{SA}[2]} < \cdots < S_{\mathsf{SA}[n]}$; see Fig. 1 for an example. We refer to the overview article [21] for suffix array construction algorithms (some of which have linear runtime).

The Burrows and Wheeler transform [6] converts a string S into the string $\mathsf{BWT}[1..n]$ defined by $\mathsf{BWT}[i] = S[\mathsf{SA}[i]-1]$ for all i with $\mathsf{SA}[i] \neq 1$ and $\mathsf{BWT}[i] = \$$ otherwise; see Fig. 1.

The suffix array SA is often enhanced with the so-called LCP-array containing the lengths of longest common prefixes between consecutive suffixes in SA; see Fig. 1. Formally, the LCP-array is an array so that $\mathsf{LCP}[1] = -1 = \mathsf{LCP}[n+1]$ and $\mathsf{LCP}[i] = |\mathsf{lcp}(S_{\mathsf{SA}[i-1]}, S_{\mathsf{SA}[i]})|$ for $2 \leq i \leq n$, where $\mathsf{lcp}(u, v)$ denotes the longest common prefix between two strings u and v. Kasai et al. [15] showed that the LCP-array can be computed in linear time from the suffix array and its

i	SA	LCP	BWT	$S_{SA[i]}$	lcp-intervals
1	12	-1	i	$	
2	11	0	p	i	
3	8	1	s	$ippi$	
4	5	1	s	$issippi$	
5	2	4	m	$ississippi$	
6	1	0	$	$mississippi$	
7	10	0	p	pi	
8	9	1	i	ppi	
9	7	0	s	$sippi$	
10	4	2	s	$sissippi$	
11	6	1	i	$ssippi$	
12	3	3	i	$ssissippi$	

Fig. 1. Suffix array, LCP-array, BWT and lcp-intervals of the string $S = mississippi$. The entry $\mathsf{LCP}[13] = -1$ is not shown in the table.

inverse. Abouelhoda et al. [1] introduced the concept of lcp-intervals; see Fig. 1. An interval $[i..j]$, where $1 \leq i < j \leq n$, in the LCP-array is called an *lcp-interval of lcp-value* ℓ (denoted by $\ell\text{-}[i..j]$) if

1. $\mathsf{LCP}[i] < \ell$,
2. $\mathsf{LCP}[k] \geq \ell$ for all k with $i + 1 \leq k \leq j$,
3. $\mathsf{LCP}[k] = \ell$ for at least one k with $i + 1 \leq k \leq j$,
4. $\mathsf{LCP}[j + 1] < \ell$.

Every index k, $i + 1 \leq k \leq j$, with $\mathsf{LCP}[k] = \ell$ is called ℓ-*index* or *lcp-index*. Note that each lcp-interval has at least one and at most $\sigma - 1$ many ℓ-indices. Abouelhoda et al. [1] showed that there is a one-to-one correspondence between the set of all lcp-intervals and the set of all internal nodes of the suffix tree of S (we assume a basic knowledge of suffix trees). Consequently, there are at most $n - 1$ lcp-intervals for a string of length n.

A substring ω of S is a *repeat* if it occurs at least twice in S. Let ω be a repeat of length ℓ and let $[i..j]$ be the ω-interval in the suffix array SA of S (i.e., ω is a prefix of $S_{SA[k]}$ for all $i \leq k \leq j$, but ω is not a prefix of any other suffix of S). Clearly, the number $occ(\omega)$ of occurrences of ω in S is $j - i + 1$. The left and right contexts of ω in S are defined by

$$lc(\omega) = \{S[SA[k] - 1] \mid i \leq k \leq j\} = \{BWT[k] \mid i \leq k \leq j\}$$
$$rc(\omega) = \{S[SA[k] + \ell] \mid i \leq k \leq j\}$$

So the left (right) context is the set of characters that appear to the left (right) of the occurrences of ω in S; see Fig. 2 for an example.

A repeat ω of S is *left-maximal* (*right-maximal*, respectively) if and only if $|lc(\omega)| \geq 2$ ($|rc(\omega)| \geq 2$, respectively). A left and right maximal repeat is said

ω	$lc(\omega)$	$rc(\omega)$	maximal	near supermaximal	supermaximal
i	$\{p,s,m\}$	$\{\$,p,s\}$	✓	✓	–
$issi$	$\{s,m\}$	$\{p,s\}$	✓	✓	✓
p	$\{p,i\}$	$\{i,p\}$	✓	✓	✓
s	$\{s,i\}$	$\{i,s\}$	✓	–	–
si	$\{s\}$	$\{p,s\}$	–	–	–
ssi	$\{i\}$	$\{p,s\}$	–	–	–

Fig. 2. Classification of the repeats of the string $S = mississippi\$$

to be a *maximal repeat*. A repeat ω of S is a *supermaximal repeat* if and only if $|lc(\omega)| = |rc(\omega)| = occ(\omega)$.

We call the occurrence of a maximal repeat ω that starts at position $SA[k]$ in S the *occurrence at index k* (in the suffix array). The occurrence at index k, where $i \le k \le j$, has a *unique left context* if and only if $BWT[k] = a$ is the only occurrence of the character a in $BWT[i..j]$, where $[i..j]$ is the ω-interval. It has a *unique right context* if and only if $S[SA[k] + \ell] = b$ is the only occurrence of the character b in the list $[S[SA[k] + \ell] \mid i \le k \le j]$. We say that an occurrence has a *unique context* if it has a unique left and a unique right context.

A maximal repeat ω of S is a *near-supermaximal repeat* if and only if it has an occurrence with a unique context. Such an occurrence of ω is said to witness the near-supermaximality of ω. With this terminology, a supermaximal repeat ω is a maximal repeat in which every occurrence of ω is a witness to its near-supermaximality.

Given two natural numbers p and q, a repeat ω of S is said to be $\langle p, q \rangle$-*context-diverse* if and only if $|lc(\omega)| \ge p$ and $|rc(\omega)| \ge q$. With this terminology, ω is a maximal repeat if and only if it is $\langle 2, 2 \rangle$-context-diverse and it is a supermaximal repeat if and only if it is $\langle occ(\omega), occ(\omega) \rangle$-context-diverse. Note that near-supermaximal repeats cannot be characterized in terms of $\langle p, q \rangle$-context-diversity. In the following, we tacitly assume that the threshold values p and q are strictly greater than 1 because we are not interested in non-maximal repeats.

3 Related Work

Gusfield [11, 7.12.1] describes an $O(n)$ time algorithm to find all maximal repeats in S, using the suffix tree of S. Subsequently, several other authors provided algorithms for the same task, using different data structures: Raffinot [23] uses a compact suffix automaton of S, Franek et al. [9] use the suffix arrays of both S and its reversed string, Narisawa et al. [18] use the suffix array, the inverse suffix array, and the LCP-array of S, Prieur and Lecroq [20] use a compact suffix vector of S, and Puglisi et al. [22] use the suffix array and the LCP-array of S. Many of these algorithms (implicitly or explicitly) use the fact that there is a one-to-one correspondence between the set of lcp-intervals and the set of all right maximal repeats. More recently, three software tools have been developed with the purpose to find maximal repeats in whole genomes. Becher et al. [2] presented an algorithm that accesses LCP-values in increasing order to identify lcp-intervals in increasing order of their lcp-values by using a dynamic data structure: a balanced binary tree of height $\log n$ that is queried and updated in $O(\log n)$ time. Consequently, their repeat finding algorithm has a worst-case time complexity of $O(n \log n)$. The test whether a candidate repeat is left-maximal is done with the aid of the suffix array and the inverse suffix array of S. The algorithm of Külekci et al. [16] shares the same high level idea and the $O(n \log n)$ time complexity with that of Becher et al., but their implementation uses succinct data structures. By further developing the techniques introduced in [4], Beller et al. [3] showed that lcp-intervals can be computed space-efficiently on the wavelet tree of the Burrows-Wheeler transform of S. Their algorithm to find all maximal repeats runs in $O(n \log \sigma)$ time.

Gusfield [11, 7.12.2] also presented an $O(n\sigma)$ time algorithm to find all near-supermaximal and supermaximal repeats in S, again using the suffix tree of S. Abouelhoda et al. [1] sketched an algorithm for finding supermaximal repeats that is based on the suffix array and the LCP-array of S, and Puglisi et al. [22] improved that algorithm. Lian et al. [17] derived an auxiliary data structure from the suffix tree of S and computed supermaximal repeats based on that data structure. The classification of repeats is not only interesting in bioinformatics [19] but it is also important in other areas: e.g. it was used to cluster process instances in the field of process mining [5].

4   An Algorithm for Finding $\langle p, q \rangle$-Context-Diverse Repeats in Linear Time

We start with the following characterization of $\langle p, q \rangle$-context-diverse repeats.[1]

Lemma 1. *A substring ω of S is a $\langle p, q \rangle$-context-diverse repeat if and only if (a) the ω-interval $[i..j]$ is an lcp-interval of lcp-value $\ell = |\omega|$, (b) $|\{\text{BWT}[i], \text{BWT}[i + 1], \ldots, \text{BWT}[j]\}| \geq p$, and (c) the number of ℓ-indices in the lcp-interval $[i..j]$ is greater than or equal to $q - 1$.*

Abouelhoda et al. [1] have shown how the lcp-interval tree can be traversed in a bottom-up fashion. Algorithm 1 is a slight variation of their algorithm that simply enumerates all lcp-intervals (i.e., it neglects the parent-child relationship between lcp-intervals). For later purposes, Algorithm 1 also computes all lcp-indices of an lcp-interval. According to Lemma 1, Algorithm 1 can be used as the basis of the computation of all $\langle p, q \rangle$-context-diverse repeats: For each lcp-interval $[i..j]$ encountered, test in constant time whether the number of lcp-indices in $[i..j]$ is greater than or equal to $q - 1$.[2] If so, one must check whether condition (b) is satisfied. Because there are at most $n - 1$ lcp-intervals, we will obtain

---

[1] Recall that we assume $p > 1$ and $q > 1$.

[2] Therefore, in this context it suffices to compute the *number* of lcp-indices of an lcp-interval; it is not necessary to compute the lcp-indices themselves.

Algorithm 1. Enumeration of lcp-intervals (in a bottom-up fashion)

$push(\langle 0, 1, [\ ]\rangle)$
for $k \leftarrow 2$ **to** $n + 1$ **do**
$\quad lb \leftarrow k - 1$
$\quad$ **while** $\mathsf{LCP}[k] < top().lcp$ **do**
$\quad\quad \langle \ell, lb, lcpIndices \rangle \leftarrow pop()$
$\quad\quad rb \leftarrow k - 1$
$\quad\quad process(\langle \ell, lb, rb, lcpIndices \rangle)$
$\quad$ **if** $\mathsf{LCP}[k] > top().lcp$ **then**
$\quad\quad push(\langle \mathsf{LCP}[k], lb, [k]\rangle)$
$\quad$ **else** $\quad\quad$ /* $\mathsf{LCP}[k] = top().lcp$ */
$\quad\quad top().lcpIndices \leftarrow add(top().lcpIndices, k)$

---

an alphabet-independent linear-time algorithm if this test can be performed in constant time. The latter is indeed possible by solving the *color set size* problem, defined as follows: Given a rooted tree with n leaves colored from 1 to σ, for each node v find the number of different leaf colors in the subtree rooted at v. In our problem, the tree is the suffix tree of S and each character in Σ is a color. The number of distinct colors in a subtree of an internal node v can be obtained by subtracting the number of duplicate colors, called the *correction term* $CT(v)$, from the number of leaves in that subtree. The key idea of Hui's algorithm is to use constant-time *lowest common ancestor* (LCA) queries [12]. His algorithm does a depth-first traversal of the tree. For each color c, it keeps track of the last leaf (seen so far) colored c. To this end, it uses an array $last[1..\sigma]$, whose entries are initially undefined. If the algorithm encounters a leaf i with color c and $last[c] = j$, then it computes the LCA v of i and j in the tree and increments the counter $count(v)$ by one because the color c is duplicated once in the subtree rooted at v (initially $count(v) = 0$). Moreover, it updates $last[c]$ by $last[c] \leftarrow i$. After the depth-first traversal, the correction terms are calculated by a bottom-up traversal of the tree as follows:

$$CT(v) = \sum_{u \in subtree(v)} count(u) = count(v) + \sum_{w \in children(v)} CT(w)$$

where $subtree(v)$ is the subtree rooted at v and $children(v)$ is the set of child nodes of v; see [13] and [11, 9.7.1] for more details.

Since we do not want to work with an explicit tree structure, we replace constant-time LCA queries with constant-time *range minimum queries* (RMQs) on the LCP-array; see [8,7]. Moreover, we use an array of counters $counter[1..n]$, whose entries are initially set to zero. We scan the BWT from left to right and, for each "color" $c = \mathsf{BWT}[i]$, we keep track of the last index $last[c]$ (seen so far) colored c. If we find an index i with $c = \mathsf{BWT}[i]$ and $last[c] = j$, then—in terms of the suffix tree—this means that we encountered two leaves (with suffix numbers $\mathsf{SA}[j]$ and $\mathsf{SA}[i]$) for which the counter $count(v)$ of their LCA v in the suffix tree must be incremented by one. The longest common prefix ω of the

---

Algorithm 2. Computation of CT

---

prepare the LCP-array for constant-time range minimum queries
for $c \leftarrow 1$ **to** σ **do**   /* initialize the array $last[1..\sigma]$ */
   $last[c] \leftarrow 0$
for $i \leftarrow 1$ **to** n **do**   /* initialize the counter array $CT[1..n]$ */
   $CT[i] \leftarrow 0$
for $i \leftarrow 1$ **to** n **do**   /* increment counters */
   $c \leftarrow \mathsf{BWT}[i]$
   if $last[c] \neq 0$ **then**
      $k \leftarrow \mathsf{RMQ_{LCP}}(last[c] + 1, i)$
      $CT[k] \leftarrow CT[k] + 1$
   $last[c] \leftarrow i$
for $i \leftarrow 2$ **to** n **do**   /* compute prefix sums of counter values */
   $CT[i] \leftarrow CT[i-1] + CT[i]$

---

suffixes $S_{\mathsf{SA}[j]}$ and $S_{\mathsf{SA}[i]}$ can be obtained by concatenating the edge labels on the path from the root of the suffix tree to node v. As mentioned in Section 2, there is a one-to-one correspondence between the set of all lcp-intervals and the set of all internal nodes of the suffix tree of S; see [1]. Let $[lb..rb]$ be the lcp-interval that corresponds to v. Note that $[lb..rb]$ is the ω-interval and the length ℓ of ω is the lcp-value of $[lb..rb]$. To simulate Hui's method, we must increment a counter at an index k with $lb \leq k \leq rb$. This index k should belong to the lcp-interval $[lb..rb]$ but not to one of its child intervals. In other words, it should be an ℓ-index of $[lb..rb]$. Recall that $\mathsf{LCP}[m] \geq \ell$ for all m with $lb + 1 \leq m \leq rb$ by the definition of lcp-intervals. Moreover, there must exist an index k with $j + 1 \leq k \leq i$ and $\mathsf{LCP}[k] = \ell$ because ω is the longest common prefix of the suffixes $S_{\mathsf{SA}[j]}$ and $S_{\mathsf{SA}[i]}$. Thus, the ℓ-index we are searching for can be found by the range minimum query $\mathsf{RMQ_{LCP}}(j + 1, i)$.

Algorithm 2 gives pseudo-code for the computation of the correction terms. In the penultimate for-loop the counters are incremented as explained above. Then we can compute the correction term of an lcp-interval $[lb..rb]$ as follows (by the definition of lcp-intervals we have $\mathsf{LCP}[lb] < \ell$, so the counter of lb must not be taken into accout):

$$\sum_{k=lb+1}^{rb} count(k) = \sum_{k=1}^{rb} count(k) - \sum_{k=1}^{lb} count(k)$$

In other words, we need the prefix sums of the counter values. These are calculated in the last for-loop of Algorithm 2. Now we have all the ingredients to calculate the number of distinct colors in an lcp-interval $[lb..rb]$. We simply have to subtract the correction term

$$CT[rb] - CT[lb] = \sum_{k=1}^{rb} count(k) - \sum_{k=1}^{lb} count(k)$$

from the size $rb - lb + 1$ of the interval $[lb..rb]$.

Algorithm 3. This implementation of the procedure *process* decides whether an lcp-interval $\langle \ell, lb, rb, lcpIndices \rangle$ induces a repeat $\omega = S[\text{SA}[lb]..\text{SA}[lb] + \ell - 1]$ of length $\geq \ell_{min}$ with $lc(\omega) \geq p$ and $rc(\omega) \geq q$

$process(\langle \ell, lb, rb, lcpIndices \rangle)$

   if $\ell \geq \ell_{min}$ **then**   /* $\ell_{min} \geq 1$ */

      if $(rb - lb + 1) - (CT[rb] - CT[lb]) \geq p$ **and** $|lcpIndices| \geq q - 1$ **then**

         output $\langle \ell, lb, rb \rangle$

In combination with Algorithm 1, Algorithm 3 computes all $\langle p, q \rangle$-context-diverse repeats of length $\geq \ell_{min}$ in $O(n)$ time. That is, the worst-case time complexity does not dependent on the alphabet size.

Our algorithm has the following advantage over the $O(n)$ time algorithm of Gallé and Tealdi [10]. Suppose one wants to calculate $\langle p, q \rangle$-context-diverse repeats for different values of the parameters p and q. In our method, once the correction terms are precomputed, they can be stored along with the enhanced suffix array. So in subsequent computations of context-diverse repeats, a renewed computation of the correction terms is unnecessary. By contrast, all three steps in the linear time algorithm presented in [10] (these steps are decribed in Section 1) must be performed for every choice of p and q.

5 Finding Maximal, Near-SuperMaximal, and Supermaximal Repeats in Linear Time

As already mentioned, context-diverse repeats subsume maximal and supermaximal repeats, but they do not subsume near-supermaximal repeats. Our next goal is to simultaneously compute and classify all three kinds of repeats. As we have seen in the previous section, maximal and supermaximal repeats can be detected with the help of correction terms. The linear-time precomputation of the correction terms, however, needs constant-time RMQs, which are slow in practice. As we shall see next, we can solve the problem without them.

Recall that every supermaximal repeat is a near-supermaximal repeat and every near-supermaximal repeat is a maximal repeat. Thus, the first task is to find all maximal repeats. We use Algorithm 1 to enumerate all lcp-intervals and test for each lcp-interval $[lb..rb]$ whether $|\{\text{BWT}[lb], \text{BWT}[lb + 1], \ldots, \text{BWT}[rb]\}| \geq 2$. This test can be done by keeping track of the largest index $lastdiff < k$ at which $\text{BWT}[lastdiff - 1]$ and $\text{BWT}[lastdiff]$ differ (initially $lastdiff = 1$). Since $lastdiff \leq rb$, the characters $\text{BWT}[lb], \text{BWT}[lb + 1], \ldots, \text{BWT}[rb]$ are not all the same if and only if $lastdiff > lb$. Pseudo-code for the computation of maximal repeats by this approach can be found in Algorithms 4 and 5 (apart from the first five and the last five lines of code, Algorithm 4 is identical with Algorithm 1). In Algorithm 5, the procedure *process* first tests whether an lcp-interval ℓ-$[lb..rb]$ induces a maximal repeat $\omega = S[\text{SA}[lb]..\text{SA}[lb] + \ell - 1]$ of length $\ell \geq \ell_{min}$. If so, it further checks whether this maximal repeat ω is near-supermaximal or

Algorithm 4. $O(n)$ time computation

$lastdiff \leftarrow 1$
for $c \leftarrow 1$ **to** σ **do**
  $penultimate[c] \leftarrow 0$
  $last[c] \leftarrow 0$
$last[\mathsf{BWT}[1]] \leftarrow 1$
$push(\langle 0, 1, [\ ]\rangle)$
for $k \leftarrow 2$ **to** $n + 1$ **do**
  $lb \leftarrow k - 1$
  while $\mathsf{LCP}[k] < top().lcp$ **do**
    $\langle \ell, lb, lcpIndices \rangle \leftarrow pop()$
    $rb \leftarrow k - 1$
    $process(\langle \ell, lb, rb, lcpIndices, lastdiff, penultimate, last \rangle)$
  if $\mathsf{LCP}[k] > top().lcp$ **then**
    $push(\langle \mathsf{LCP}[k], lb, [k]\rangle)$
  else   /* $\mathsf{LCP}[k] = top().lcp$ */
    $top().lcpIndices \leftarrow add(top().lcpIndices, k)$
  $c \leftarrow \mathsf{BWT}[k]$
  $penultimate[c] \leftarrow last[c]$
  $last[c] \leftarrow k$
  if $\mathsf{BWT}[k - 1] \neq c$ **then**
    $lastdiff \leftarrow k$

---

even supermaximal. In the following, we explain how this is done. By definition, ω is near-supermaximal if it has an occurrence with a unique context.

Lemma 2. *Let ℓ-$[lb..rb]$ be an lcp-interval and consider the induced repeat $\omega = S[\mathsf{SA}[lb]..\mathsf{SA}[lb] + \ell - 1]$. The occurrence of ω at index*

- *j, where $lb < j < rb$, has a unique right context if and only if both j and $j + 1$ are ℓ-indices,*
- *lb has a unique right context if and only if $lb + 1$ is an ℓ-index,*
- *rb has a unique right context if and only if rb is an ℓ-index.*

Based on Lemma 2 and the ℓ-indices of the lcp-interval ℓ-$[lb..rb]$, Algorithm 6 computes all indices at which the occurrence of the repeat ω has a unique right context.[3] The final task is to find out which of these occurrences also have a unique left context. To this end, we use two arrays *penultimate* and *last* of size σ. Algorithm 4 maintains the following invariant: Before the body of its second for-loop is executed for value k, $last[c]$ is the index at which the last occurrence of $c \in \Sigma$ in $\mathsf{BWT}[1..k - 1]$ can be found and *penultimate*$[c]$ points to the penultimate occurrence of c in $\mathsf{BWT}[1..k - 1]$.

---

[3] In the terminology of Abouelhoda et al. [1], each of these indices corresponds to a singleton child interval of $[lb..rb]$. That is why the set of these indices is called *singletons*.

Algorithm 5. This implementation of the procedure *process* decides whether an lcp-interval $\langle \ell, lb, rb, lcpIndices \rangle$ induces a repeat $\omega = S[SA[lb]..SA[lb] + \ell - 1]$ of length $\geq \ell_{min}$ that is maximal, near-supermaximal or supermaximal

---

$process(\langle \ell, lb, rb, lcpIndices, lastdiff, penultimate, last \rangle)$

   if $\ell \geq \ell_{min}$ **and** $lastdiff > lb$ **then**

      /* ℓ-$[lb..rb]$ induces a maximal repeat of length $\ell \geq \ell_{min}$ */

      $witnesses \leftarrow \emptyset$

      $singletons \leftarrow computeSingletons(lb, rb, lcpIndices)$

      for each $j \in singletons$ **do**

         $c \leftarrow \mathsf{BWT}[j]$

         if $j = last[c]$ **and** $penultimate[c] < lb$ **then**

            $witnesses \leftarrow witnesses \cup \{j\}$

      if $|witnesses| = rb - lb + 1$ **then**

         output ℓ-$[lb..rb]$ induces a supermaximal repeat

      else if $witnesses \neq \emptyset$ **then**

         output ℓ-$[lb..rb]$ induces a near-supermaximal repeat and the

         occurrences starting at $SA[j]$, $j \in witnesses$, have a unique context

      else

         output ℓ-$[lb..rb]$ induces a maximal repeat

Lemma 3. *Let ℓ-$[lb..rb]$ be an lcp-interval and consider the induced repeat $\omega = S[SA[lb]..SA[lb] + \ell - 1]$. The occurrence of ω at index j, $lb \leq j \leq rb$, has a unique left context if and only if $j = last[c]$ and $penultimate[c] < lb$.*

Based on the preceding lemma, we can now show that Algorithms 4, 5, and 6 correctly compute and classify maximal repeats in $O(n)$ time. As already mentioned, there is a one-to-one correspondence between the set of lcp-intervals and the set of all repeats ω with $rc(\omega) \geq 2$. Algorithm 4 finds all these repeats in $O(n)$ time because it enumerates all lcp-intervals in $O(n)$ time. We next discuss what happens when the procedure *process* (Algorithm 5) is applied to the lcp-interval ℓ-$[lb..rb]$ and the parameters $lcpIndices$, $lastdiff$, $penultimate$, and $last$. As already mentioned, the lcp-interval $[lb..rb]$ induces a maximal repeat if and only if $lastdiff > lb$. If this test fails, Algorithm 5 terminates. Otherwise, a maximal repeat $\omega = S[SA[lb]..SA[lb] + \ell - 1]$ is detected. Algorithm 5 calls the procedure *computeSingletons* with the parameters lb, rb, and the set $lcpIndices$ of all ℓ-indices of the lcp-interval ℓ-$[lb..rb]$. The procedure *computeSingletons* returns the set $singletons$, the set of all indices at which the occurrence of the repeat ω has a unique right context. Note that the overall time spent for *all* calls to the procedure *computeSingletons* is $O(n)$ because each index is an lcp-index of exactly one lcp-interval. For each $j \in singletons$, Algorithm 5 checks whether the occurrence of the maximal repeat ω at index j has a unique left context; cf. Lemma 3. If so, it adds j to the set $witnesses$. In total, the for-loop in Algorithm 5 is executed at most n times because the overall number of indices at which an occurrence of a maximal repeat has a unique right context is bounded by n. When the for-loop in Algorithm 5 terminates, the set $witnesses$ contains all

Algorithm 6. Given the list $[i_1, \ldots, i_m]$ of all ℓ-indices of an lcp-interval ℓ-$[lb..rb]$, this procedure returns the set of all indices j so that the occurrence of $\omega = S[\text{SA}[lb]..\text{SA}[lb] + \ell - 1]$ at index j has a unique right context

---

$computeSingletons(lb, rb, [i_1, \ldots, i_m])$
   $singletons \leftarrow \emptyset$
   if $lb = i_1 - 1$ **then**
      $singletons \leftarrow singletons \cup \{lb\}$
   for $k \leftarrow 2$ **to** m **do**
      if $i_{k-1} = i_k - 1$ **then**
         $singletons \leftarrow singletons \cup \{i_{k-1}\}$
   if $i_m = rb$ **then**
      $singletons \leftarrow singletons \cup \{i_m\}$
   return $singletons$

---

indices at which the maximal repeat ω has a unique context. Now the size of this set is used to classify the maximal repeat ω. If $|witnesses| = rb - lb + 1$, then every occurrence of ω has a unique context and ω is reported as a supermaximal repeat. Otherwise, Algorithm 5 tests whether $witnesses \neq \emptyset$. If this is the case, there is at least one index at which the maximal repeat ω has a unique context and ω is reported as a near-supermaximal repeat. If this is not the case, ω is reported as a maximal repeat. In summary, Algorithm 5 correctly classifies maximal repeats. The worst-case time complexity of Algorithms 4, 5, and 6 is $O(n)$ as argued above.

References

1. Abouelhoda, M.I., Kurtz, S., Ohlebusch, E.: Replacing suffix trees with enhanced suffix arrays. Journal of Discrete Algorithms 2, 53–86 (2004)
2. Becher, V., Deymonnaz, A., Heiber, P.: Efficient computation of all perfect repeats in genomic sequences of up to half a gigabyte, with a case study on the human genome. Bioinformatics 25(14), 1746–1753 (2009)
3. Beller, T., Berger, K., Ohlebusch, E.: Space-efficient computation of maximal and supermaximal repeats in genome sequences. In: Calderón-Benavides, L., González-Caro, C., Chávez, E., Ziviani, N. (eds.) SPIRE 2012. LNCS, vol. 7608, pp. 99–110. Springer, Heidelberg (2012)
4. Beller, T., Gog, S., Ohlebusch, E., Schnattinger, T.: Computing the longest common prefix array based on the Burrows-Wheeler transform. In: Grossi, R., Sebastiani, F., Silvestri, F. (eds.) SPIRE 2011. LNCS, vol. 7024, pp. 197–208. Springer, Heidelberg (2011)
5. Bose, R.P.J.C., van der Aalst, W.M.P.: Trace clustering based on conserved patterns: Towards achieving better process models. In: Rinderle-Ma, S., Sadiq, S., Leymann, F. (eds.) BPM 2009. LNBIP, vol. 43, pp. 170–181. Springer, Heidelberg (2010)
6. Burrows, M., Wheeler, D.J.: A block-sorting lossless data compression algorithm. Research Report 124, Digital Systems Research Center (1994)

7. Fischer, J., Heun, V.: Space-efficient preprocessing schemes for range minimum queries on static arrays. SIAM Journal on Computing 40(2), 465–492 (2011)
8. Fischer, J., Heun, V., Kramer, S.: Optimal string mining under frequency constraints. In: Fürnkranz, J., Scheffer, T., Spiliopoulou, M. (eds.) PKDD 2006. LNCS (LNAI), vol. 4213, pp. 139–150. Springer, Heidelberg (2006)
9. Franěk, F., Smyth, W.F., Tang, Y.: Computing all repeats using suffix arrays. Journal of Automata, Languages and Combinatorics 8(4), 579–591 (2003)
10. Gallé, M., Tealdi, M.: On context-diverse repeats and their incremental computation. In: Dediu, A.-H., Martín-Vide, C., Sierra-Rodríguez, J.-L., Truthe, B. (eds.) LATA 2014. LNCS, vol. 8370, pp. 384–395. Springer, Heidelberg (2014)
11. Gusfield, D.: Algorithms on Strings, Trees, and Sequences. Cambridge University Press (1997)
12. Harel, D., Tarjan, R.E.: Fast algorithms for finding nearest common ancestors. SIAM Journal on Computing 13, 338–355 (1984)
13. Hui, L.C.K.: Color set size problem with applications to string matching. In: Apostolico, A., Galil, Z., Manber, U., Crochemore, M. (eds.) CPM 1992. LNCS, vol. 644, pp. 230–243. Springer, Heidelberg (1992)
14. Kärkkäinen, J., Manzini, G., Puglisi, S.J.: Permuted longest-common-prefix array. In: Kucherov, G., Ukkonen, E. (eds.) CPM 2009 Lille. LNCS, vol. 5577, pp. 181–192. Springer, Heidelberg (2009)
15. Kasai, T., Lee, G.H., Arimura, H., Arikawa, S., Park, K.: Linear-time longest-common-prefix computation in suffix arrays and its applications. In: Amir, A., Landau, G.M. (eds.) CPM 2001. LNCS, vol. 2089, pp. 181–192. Springer, Heidelberg (2001)
16. Külekci, M.O., Vitter, J.S., Xu, B.: Efficient maximal repeat finding using the Burrows-Wheeler transform and wavelet tree. IEEE/ACM Transactions on Computational Biology and Bioinformatics 9(2), 421–429 (2012)
17. Lian, C.N., Halachev, M., Shiri, N.: Searching for supermaximal repeats in large DNA sequences. In: Elloumi, M., Küng, J., Linial, M., Murphy, R.F., Schneider, K., Toma, C. (eds.) BIRD 2008. CCIS, vol. 13, pp. 87–101. Springer, Heidelberg (2008)
18. Narisawa, K., Inenaga, S., Bannai, H., Takeda, M.: Efficient computation of substring equivalence classes with suffix arrays. In: Ma, B., Zhang, K. (eds.) CPM 2007. LNCS, vol. 4580, pp. 340–351. Springer, Heidelberg (2007)
19. Ohlebusch, E.: Bioinformatics Algorithms: Sequence Analysis, Genome Rearrangements, and Phylogenetic Reconstruction. Oldenbusch-Verlag (2013)
20. Prieur, E., Lecroq, T.: On-line construction of compact suffix vectors and maximal repeats. Theoretical Computer Science 407(1-3), 290–301 (2008)
21. Puglisi, S.J., Smyth, W.F., Turpin, A.: A taxonomy of suffix array construction algorithms. ACM Computing Surveys 39(2), article 4 (2007)
22. Puglisi, S.J., Smyth, W.F., Yusufu, M.: Fast, practical algorithms for computing all the repeats in a string. Mathematics in Computer Science 3(4), 373–389 (2010)
23. Raffinot, M.: On maximal repeats in strings. Information Processing Letters 80(3), 165–169 (2001)
24. Taillefer, E., Miller, J.: Exhaustive computation of exact duplications via *super* and *non-nested local* maximal repeats. Journal of Bioinformatics and Computational Biology 12(1), article 1350018 (2014)

A 3-Approximation Algorithm for the Multiple Spliced Alignment Problem and Its Application to the Gene Prediction Task

Regina Beretta Mazaro, Leandro Ishi Soares de Lima, and Said Sadique Adi

Universidade Federal de Mato Grosso do Sul, Faculdade de Computação
Av. Costa e Silva, s/n, CP 549,
79070-900, Campo Grande, MS, Brazil
regina.beretta@ufms.br, leandro.ishi.lima@gmail.com, said@facom.ufms.br

Abstract. The *Spliced Alignment Problem* is a well-known problem in Bioinformatics with application to the gene prediction task. This problem consists in finding an ordered subset of non-overlapping substrings of a subject sequence g that best fits a target sequence t. In this work we present an approximation algorithm for a variant of the Spliced Alignment Problem, called *Multiple Spliced Alignment Problem*, that involves more than one target sequence. Under a metric, this algorithm is proved to be a 3-approximation for the problem and its good practical results compare to those obtained by four heuristics already developed for the Multiple Spliced Alignment Problem.

Keywords: Approximation algorithm, gene prediction, multiple spliced alignment problem.

1 Introduction

The term *Bioinformatics* has been used since 1970, when Hogeweg and Hesper defined it as "the study of informatic process in biotic systems" [4]. Since then, Biology and its branches have been a valuable source of new and interesting computational tasks involving long strings (genomic DNAs, cDNAs, RNAs, proteins, etc). As such, they require robust and efficient algorithms that work well in both theory and practice. A well-known task in this scenario is that of identifying the genes encoded in a genomic DNA of interest.

Given the practical importance and the difficulties associated with the gene prediction task, a number of computational methods has been developed to deal with it. By considering sequence conservation and the large quantity of entire genomes from many species already annotated, *similarity based* approaches are promising techniques that allow the identification of genes by comparing genomic sequences with related transcript sequences. In this context, Gelfand *et al.* proposed in [3] a theoretical/computational problem, called *Spliced Alignment Problem*, that models the gene prediction task as a combinatorial optimization problem involving (substrings of) a subject sequence (genomic DNA) and a target sequence (cDNA).

E. Moura and M. Crochemore (Eds.): SPIRE 2014, LNCS 8799, pp. 129–138, 2014.
© Springer International Publishing Switzerland 2014

In this work we propose an approximation algorithm for a variant of the Spliced Alignment Problem, called *Multiple Spliced Alignment Problem*, where more than one target sequence is involved. This problem was proved to be NP-complete by Kishi and Adi in [5], where they also proposed some heuristics for it. To the best of our knowledge, there are no approximation algorithms for the Multiple Spliced Alignment Problem in the literature, and it is exactly this gap that the present work wants to narrow.

This paper is organized as follows. In the next section we introduce the Spliced and Multiple Spliced Alignment Problem, and relate both with the gene prediction task. A 3−approximation algorithm for the Multiple Spliced Alignment Problem, that constitutes the main result of this work, is shown in Section 3. In Section 4 we give the details about the experimental results obtained by our approach over real-world instances of the gene prediction task. Finally, in the last section we summarize this work and consider future research directions.

2   The Multiple Spliced Alignment Problem

Among the several regions that comprise a genomic DNA, the protein coding regions, or *genes*, are of main interest for biologists. In eukaryotes, these regions are separated by long stretches of intergenic DNA and their coding fragments, called *exons*, are interrupted by non-coding ones, called *introns*. Given a genomic DNA, the gene prediction task consists in finding the correct exon-intron structure of its genes. In computational terms, this task has as input a genomic DNA sequence g and as output the start and end positions of each exon that constitutes the genes of g.

Given the undeniable practical importance of the gene prediction task, since 1980 many different methods have been proposed to address it. These methods can be roughly classified into *extrinsic* methods, that make use of information concerning fully annotated transcript sequences related to the target gene, and *intrinsic* methods, that rely basically on statistical information about the gene being searched for (see [7,8] for surveys on this topic).

Among the different extrinsic approaches suggested for the gene prediction task, the one proposed by Gelfand *et al.* in [3] is of particular interest to the string processing field since it lies on a combinatorial optimization problem involving sequences, namely the *Spliced Alignment Problem*. To a better understanding of this problem consider the following definitions.

Let $s = s_1 s_2 \ldots s_n$ be a finite string over an alphabet Σ. We denote the *length* of s by $|s|$. A *substring* $b = s_i \ldots s_j$ of s is an ordered sequence of consecutive symbols of s. We denote by $first(b) = i$ the position of the first symbol of b in s and by $last(b) = j$ the position of the last symbol of b in s. Let $\mathcal{B} = \{b_1, b_2, \ldots, b_k\}$ be a set of k substrings of s. We say that $\mathcal{B}$ is an *ordered set of substrings* if: 1) $first(b_i) < first(b_{i+1})$ or 2) $first(b_i) = first(b_{i+1})$ and $last(b_i) < last(b_{i+1})$, for $1 \leq i \leq k-1$. We also say that a substring $b' = s_o \ldots s_p$ of s *overlaps* another substring $b'' = s_q \ldots s_r$ of s if $q \leq o \leq r$, or $q \leq p \leq r$, or $o \leq q \leq p$, or $o \leq r \leq p$. Moreover, we say that a substring $b' = s_o \ldots s_p$ of s

precedes another substring $b'' = s_q \ldots s_r$ of s if $p < q$, and we denote this relation by $b' \prec b''$. A subset $\Gamma = \{b_i, b_j, \ldots, b_p\}$ of $\mathcal{B}$ is a *chain* if $b_i \prec b_j \prec \ldots \prec b_p$ and we denote the string resulting of the concatenation of the elements of a chain Γ by $\Gamma^\bullet$. That is, $\Gamma^\bullet = b_i \bullet b_j \bullet \ldots \bullet b_p$, where $\bullet$ is the string concatenation operator. Finally, given two strings s and t, we denote by $\mathrm{sim}_\omega(s, t)$ the *similarity* (or the score of an *optimal alignment*) between s and t under a scoring function $\omega : \Sigma \times \Sigma \to \mathcal{R}$ [9].

With the previous definitions in mind, the SAP is defined as follows [3]:

Spliced Alignment Problem (SAP): Given a subject sequence g, a target sequence t and an ordered set of substrings $\mathcal{B} = \{b_1, b_2, \ldots, b_k\}$ of g, find a chain Γ of $\mathcal{B}$ such that $\mathrm{sim}_\omega(\Gamma^\bullet, t)$ is maximum among all chains of $\mathcal{B}$.

An instance of the SAP and its solution can be seen in Figure 1.

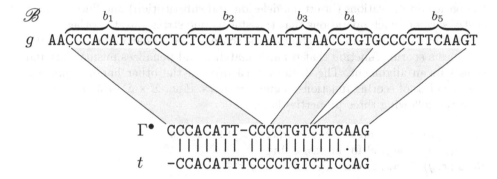

Fig. 1. An instance of the SAP and its solution. The symbols of g that compose each substring $b \in \mathcal{B}$ are disposed below its corresponding horizontal brace. The scoring function used in this instance is $\omega(a, b) = \{1, \text{ if } a = b; -1, \text{ if } a \neq b; -2 \text{ if } a = -\text{ or } b = -\}$ and its solution is $\Gamma = \{b_1, b_4, b_5\}$. Figure adapted from [5].

Looking at the SAP in the context of the gene prediction task, g could be interpreted as a (fragment of a) genomic sequence encoding a gene of interest, t as a transcript sequence related to this gene and $\mathcal{B} = \{b_1, b_2, \ldots, b_k\}$ as a set of potential exons of g. With these relations in mind, and given the fact that the coding regions of a gene are less susceptible to mutations than the non-coding ones, it is very likely that a solution for the SAP will include the exons of the gene being searched for.

In [3], Gelfand *et al.* propose a polynomial time dynamic programming algorithm for the SAP. To understand the main recurrence of this algorithm, consider the following definitions taken from [3]. Let $b_k = g_l \ldots g_i \ldots g_m$ be a substring of g containing a position i. The i-prefix of b_k is defined as $b_k(i) = g_l \ldots g_i$. Let $\Gamma = \{b_1, b_2, \ldots, b_k, \ldots, b_t\}$ be a chain such that some substring b_k contains position i and let $\Gamma^\bullet(i) = b_1 \bullet b_2 \bullet \ldots \bullet b_k(i)$. The algorithm presented in [3] efficiently calculates a three-dimensional matrix S such that

$$S[i][j][k] = \max_{\text{all chains } \Gamma \text{ containing substring } b_k} \text{sim}_\omega(\Gamma^\bullet(i), t[1..j]).$$

After computing S, the value of the optimal solution can be found as

$$\max_{1 \le k' \le k} S[last(b_{k'})][|t|][k'].$$

Finally, it is possible to build the optimal solution itself considering the choices the algorithm made to compute its value. Using the dynamic programming technique, this algorithm for the SAP runs in time $\mathcal{O}(mnc+mk^2)$ and space $\mathcal{O}(mnc)$, with $m = |t|$, $n = |g|$, $k = |\mathcal{B}|$ and $c = \frac{1}{n}\sum_{b_i \in \mathcal{B}} |b_i|$.

As we can see, the SAP was originally proposed as a maximization problem. However, we can address it as a minimization problem as well. For this matter, we need to make use of the concept of distance between two strings, instead of similarity. To calculate the distance between two strings, we assign costs to the basic edit operations (insertion, deletion and substitution) and find the least costly series of such operations that transforms one string into the other.

The similarity, as being by definition the score of an optimal alignment, usually assumes a scoring function that rewards matches and penalizes mismatches and spaces in an alignment. The distance measure, on the other hand, requires a specific class of scoring functions, namely *metrics*. If $\omega : \Sigma \times \Sigma \to \mathcal{R}$ is a metric, then the following three properties hold:

1. $\omega(x, x) = 0$ for all $x \in \Sigma$ and $\omega(x, y) > 0$ for $x \ne y$;
2. $\omega(x, y) = \omega(y, x)$ for all $x, y \in \Sigma$;
3. $\omega(x, y) \le \omega(x, z) + \omega(z, y)$ for all $x, y, z \in \Sigma$.

In summary, the first property assures that the costs of the basic edit operations are positive. The second property establishes that ω is symmetric. The last and most important property is called *triangle inequality*. It assures that the cost of transforming a symbol x into another symbol y is not greater than the cost of transforming x into z and then z into y. This property can be extended to sequences as a whole.

Given a metric ω and two sequences s and t, we denote by $\text{dist}_\omega(s, t)$ the cost, regarding ω, of the least expensive series of edit operations that transforms s into t. We can now reformulate the SAP as follows, noticing that we will refer to this version from now on:

Spliced Alignment Problem (SAP): Given a subject sequence g, a target sequence t, an ordered set of substrings $\mathcal{B} = \{b_1, b_2, ..., b_k\}$ of g and a metric ω, find a chain Γ of $\mathcal{B}$ such that $\text{dist}_\omega(\Gamma^\bullet, t)$ is minimum among all chains of $\mathcal{B}$.

Kishi and Adi started exploring in [5] a variant of the SAP called *Multiple Spliced Alignment Problem*. In this variant, instead of only one target sequence t, we have a set of target sequences $\mathcal{T} = \{t_1, t_2, ..., t_u\}$ and the objective is to find a chain Γ of B such that $\sum_{i=1}^{u} \text{dist}_\omega(\Gamma^\bullet, t_i)$ is minimum among all chains of $\mathcal{B}$. Back to the gene prediction task, the Multiple Spliced Alignment Problem is also of practical interest since now the prediction is obtained by taking more evidences into consideration, which tends to give better practical results. A formal definition of the Multiple Spliced Alignment Problem can be found below:

Multiple Spliced Alignment Problem (MSAP): Given a subject sequence g, a set of target sequences $\mathcal{T} = \{t_1, t_2, ..., t_u\}$, an ordered set of substrings $\mathcal{B} = \{b_1, b_2, ..., b_k\}$ of g and a metric ω, find a chain Γ of $\mathcal{B}$ such that $\sum_{i=1}^{u} \text{dist}_\omega(\Gamma^\bullet, t_i)$ is minimum among all chains of $\mathcal{B}$.

An instance of the MSAP and its solution can be seen in Figure 2.

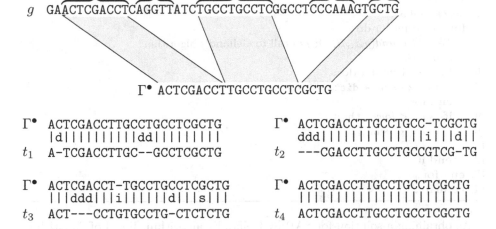

Fig. 2. An instance of the MSAP and its solution. The symbols of g that compose each substring $b \in \mathcal{B}$ are disposed below its corresponding horizontal brace and the set $\mathcal{T}$ is composed by the sequences t_1, t_2, t_3 and t_4. The metric used in this instance is the Levenshtein distance, where d indicates a delete operation, i indicates an insert operation and s indicates a substitution operation. The solution of this instance is $\Gamma = \{b_1, b_3, b_5\}$. Figure adapted from [5].

The MSAP was proved to be NP-complete even for binary sequences by Kishi and Adi in [5]. As a direct result of this fact, two approaches come to mind to deal with such hard problem: heuristics and approximation algorithms. As some heuristics for the MSAP were already developed in [5,6], we present in this work an approximation algorithm for it that deals in a satisfactory way with both theoretical and practical aspects of the problem.

3   A 3-Approximation Algorithm for the MSAP

The approximation algorithm developed in this work is a natural extension of the solution proposed by Gelfand *et al.* in [3] for the Spliced Alignment Problem.

It consists in finding u solutions for the SAP, one for each target sequence t_i, for $1 \leq i \leq u$, and choosing as final solution for the MSAP that chain less distant to all sequences in $\mathcal{T}$.

Algorithm 1, called MSAP-3-APP, details the idea of our approximation. In this algorithm, Γ_i is a chain of $\mathcal{B}$ returned by Gelfand's algorithm taking t_i as target sequence, and Γ corresponds to a Γ_i such that $\sum_{j=1}^{u} \text{dist}_\omega(\Gamma_i^\bullet, t_j)$ is minimum among all Γ_i, for $1 \leq i \leq u$.

Algorithm 1. MSAP-3-APP$(g, \mathcal{T}, \mathcal{B}, \omega)$

Require: Subject sequence g, a set of target sequences $\mathcal{T} = \{t_1, t_2, ..., t_u\}$, a set $\mathcal{B} = \{b_1, b_2, ..., b_k\}$ of ordered substrings of g and a metric ω.
Ensure: A chain Γ of $\mathcal{B}$.
1. $\Gamma \leftarrow \emptyset$;
2. $lower \leftarrow +\infty$;
3. **for** $i \leftarrow 1$ until u **do**
4. $\Gamma_i \leftarrow Gelfand(g, t_i, \mathcal{B}, \omega)$; //a call to Gelfand's algorithm
5. $sum \leftarrow 0$;
6.   **for** $j \leftarrow 1$ until u **do**
7. $sum \leftarrow sum + \text{dist}_\omega(\Gamma_i^\bullet, t_j)$;
8.   **end for**
9.   **if** $sum < lower$ **then**
10. $lower \leftarrow sum$;
11. $\Gamma \leftarrow \Gamma_i$;
12.   **end if**
13. **end for**
14. **return** Γ;

As obtaining a solution for SAP by Gelfand's algorithm (line 4 of Algorithm 1) and calculating the distance between two sequences under some metric ω (line 7 of Algorithm 1) are known tasks that can be done in polynomial time, it is easy to see that Algorithm 1 also has polynomial time complexity. More specifically, algorithm MSAP-3-APP runs in time $\mathcal{O}(um(nc + k^2 + un))$, with u, n, c and k as previously defined, and $m = \max_{1 \leq j \leq u}\{|t_j|\}$.

Now, we will show that Algorithm 1 is a 3−approximation for the MSAP. To this end, let Γ^* be an optimal solution for an instance $I = (g, \mathcal{T}, \mathcal{B}, \omega)$ of the problem, i.e. $\sum_{i=1}^{u} \text{dist}_\omega(\Gamma^{*\bullet}, t_i) = \text{opt}$ is minimum, and consider the following lemma:

Lemma 1. $\sum_{i=1}^{u} dist_\omega(\Gamma_i^\bullet, t_i) \leq \sum_{i=1}^{u} dist_\omega(\Gamma^{*\bullet}, t_i)$

Proof. Suppose, by contradiction, that $\sum_{i=1}^{u} \text{dist}_\omega(\Gamma_i^\bullet, t_i) > \sum_{i=1}^{u} \text{dist}_\omega(\Gamma^{*\bullet}, t_i)$. Then, there is some i such that $\text{dist}_\omega(\Gamma_i^\bullet, t_i) > \text{dist}_\omega(\Gamma^{*\bullet}, t_i)$. But this fact contradicts our hypothesis that $\text{dist}_\omega(\Gamma_i^\bullet, t_i)$ is minimum as assured by Gelfand's algorithm. ∎

The relation between the value of a solution Γ computed by our algorithm, equals to $\sum_{i=1}^{u} \text{dist}_\omega(\Gamma^\bullet, t_i)$, and the value of an optimal solution Γ^* for MSAP, equals to $\sum_{i=1}^{u} \text{dist}_\omega(\Gamma^{*\bullet}, t_i)$, is given by Theorem 1.

Theorem 1. MSAP-3-APP *is a 3-approximation for MSAP.*

Proof. Firstly, consider the following inequality, that can be verified by the definitions of Γ and Γ_i:

$$\sum_{j=1}^{u}\sum_{i=1}^{u} \text{dist}_\omega(\Gamma^\bullet, t_i) \leq \sum_{j=1}^{u}\sum_{i=1}^{u} \text{dist}_\omega(\Gamma_j^\bullet, t_i) \tag{1}$$

Given the triangular inequality property of ω, we have that $\text{dist}_\omega(\Gamma_j^\bullet, t_i) \leq \text{dist}_\omega(\Gamma_j^\bullet, \Gamma^{**}) + \text{dist}_\omega(\Gamma^{**}, t_i)$. Replacing the right side of Inequation 1 with this inequality, we get:

$$\sum_{j=1}^{u}\sum_{i=1}^{u} \text{dist}_\omega(\Gamma^\bullet, t_i) \leq \sum_{j=1}^{u}\sum_{i=1}^{u}(\text{dist}_\omega(\Gamma_j^\bullet, \Gamma^{**}) + \text{dist}_\omega(\Gamma^{**}, t_i))$$

$$u\sum_{i=1}^{u} \text{dist}_\omega(\Gamma^\bullet, t_i) \leq u\sum_{j=1}^{u} \text{dist}_\omega(\Gamma_j^\bullet, \Gamma^{**}) + u\sum_{i=1}^{u} \text{dist}_\omega(\Gamma^{**}, t_i) \tag{2}$$

Replacing j by i in Inequation 2, dividing its both sides by u, and making use of the equality $\sum_{i=1}^{u} \text{dist}_\omega(\Gamma^{**}, t_i) = \text{opt}$, we get:

$$\sum_{i=1}^{u} \text{dist}_\omega(\Gamma^\bullet, t_i) \leq \sum_{i=1}^{u} \text{dist}_\omega(\Gamma_i^\bullet, \Gamma^{**}) + \text{opt} \tag{3}$$

Using again the triangular inequality property of ω, we have that $\text{dist}_\omega (\Gamma_i^\bullet, \Gamma^{**}) \leq \text{dist}_\omega(\Gamma_i^\bullet, t_i) + \text{dist}_\omega(t_i, \Gamma^{**})$. So, we can expand the term $\sum_{i=1}^{u} \text{dist}_\omega(\Gamma_i^\bullet, \Gamma^{**})$ in Inequation 3 as shown below:

$$\sum_{i=1}^{u} \text{dist}_\omega(\Gamma^\bullet, t_i) \leq \sum_{i=1}^{u} \text{dist}_\omega(\Gamma_i^\bullet, t_i) + \sum_{i=1}^{u} \text{dist}_\omega(t_i, \Gamma^{**}) + \text{opt}$$

Now the equality $\sum_{i=1}^{u} \text{dist}_\omega(\Gamma^{**}, t_i) = \text{opt}$ can be applied again:

$$\sum_{i=1}^{u} \text{dist}_\omega(\Gamma^\bullet, t_i) \leq \sum_{i=1}^{u} \text{dist}_\omega(\Gamma_i^\bullet, t_i) + \text{opt} + \text{opt} \tag{4}$$

By Lemma 1, we can replace $\sum_{i=1}^{u} \text{dist}_\omega(\Gamma_i^\bullet, t_i)$ by $\sum_{i=1}^{u} \text{dist}_\omega(\Gamma^{**}, t_i)$ in Inequation 4:

$$\sum_{i=1}^{u} \text{dist}_\omega(\Gamma^\bullet, t_i) \leq \sum_{i=1}^{u} \text{dist}_\omega(\Gamma^{**}, t_i) + \text{opt} + \text{opt}$$

Finally, applying again the equality $\sum_{i=1}^{u} \text{dist}_\omega(\Gamma^{**}, t_i) = \text{opt}$, we get:

$$\sum_{i=1}^{u} \text{dist}_\omega(\Gamma^\bullet, t_i) \leq 3 * \text{opt} \tag{5}$$

Therefore, the value of the solution computed by algorithm MSAP-3-APP is no worse than 3 times the value of an optimal solution for the MSAP.   ∎

4   Experimental Results

In order to assess the practical accuracy of our approximation, algorithm MSAP-3-APP was implemented in ANSI C++ and tested on real-world instances of the gene prediction task.

The benchmark taken to evaluate our program was the same one used by Kishi and Adi in [5], so we could compare our approach with the heuristics proposed by them. This benchmark consists of 240 fragment sequences of human DNA, obtained from the chromosomes analyzed by the ENCODE project [10]. All these fragments include only one gene and the corresponding targets were obtained by a search in the HOMOLOGENE [11] database for cDNAs sequences evolutionarily related to the genes being searched for. Finally, the ordered set of substrings for each instance was obtained by means of a HMM-based algorithm implemented by a gene prediction tool called GENSCAN [1].

To assess the accuracy of the programs, we made use of the following measures, introduced by Burset and Guigó in [2] and commonly used in the evaluation of gene prediction tools:

(1) Specificity at the nucleotide level ($Sp_n = \frac{TP}{TP+FP}$): proportion of nucleotides predicted as coding that are really coding;
(2) Sensitivity at the nucleotide level ($Sn_n = \frac{TP}{TP+FN}$): proportion of really coding nucleotides correctly predicted as coding;
(3) Specificity at the exon level ($Sp_e = \frac{NCE}{NPE}$): proportion of predicted exons that match an annotated exon;
(4) Sensitivity at the exon level ($Sn_e = \frac{NCE}{NAE}$): proportion of annotated exons that were correctly predicted.

The approximate correlation, AC, defined as

$$AC = \frac{1}{2}(\frac{TP}{TP+FN} + \frac{TP}{TP+FP} + \frac{TN}{TN+FP} + \frac{TN}{TN+FN}) - 1$$

has been introduced to summarize sensitivity and specificity in a single measure. At the exon level, the average $Av_e = (Sp_e + Sn_e)/2$ is used instead.

In the previous definitions, TP (true positives) is the number of really coding nucleotides correctly predicted as coding, TN (true negatives) represents the number of really non-coding nucleotides correctly predicted as non-coding, FP (false positives) is the number of really non-coding nucleotides incorrectly predicted as coding and FN (false negatives) is the number of really coding nucleotides incorrectly predicted as non-coding. On the level of complete exons, NCE is defined as the number of correctly predicted exons, NPE as the number of predicted exons and NAE as the number of annotated exons. Here, a predict exon is considered as correctly predicted when its start and end positions match the start and end positions of an annotated exon of the input sequence.

Table 1 summarizes the results obtained by our approach and by the heuristics proposed in [5,6] on the detailed benchmark. In this table, each column stores the average values of Sn, Sp, AC and Av_e.

Table 1. Results obtained on 240 real-world instances of the gene prediction task

Approach	Nucleotide			Exon		
	Sn_n	Sp_n	AC	Sn_e	Sp_e	Av_e
MSAP-3-APP	0.95	0.96	0.95	0.85	0.81	0.83
Heuristic H1	0.96	0.96	0.95	0.86	0.81	0.83
Heuristic H2	0.96	0.91	0.93	0.83	0.73	0.78
Heuristic H3	0.93	0.96	0.94	0.86	0.84	0.85
Heuristic H4	0.77	0.80	0.77	0.54	0.51	0.53

The values in Table 1 show that our approach presented a good level of sensitivity and specificity on both nucleotide and exon levels. From all the nucleotides predicted as coding by our approximation, 96% are in fact coding. Furthermore, our approach correctly identified 95% of the coding nucleotides. At the exon level, 81% of the predicted exons match an annotated exon, and 85% of the annotated exons were correctly identified by our program.

Obviously, the accuracy of our approach in identifying the correct exon-intron structure of a gene is strongly dependent on the input set $\mathcal{B}$. If this set includes all the annotated exons of the target gene, it is very likely that all of them will be included in the chain returned by our approximation. From a total of 1677 annotated exons, 1550 were included in the sets of candidate exons and only 67 of them were missed by our approach. On the other hand, if an annotated exon is not included in the input set $\mathcal{B}$, it will be missed by our approach. From a total of 1677 annotated exons, 127 were missed by our approach since they could not be found in $\mathcal{B}$.

In comparison with the four heuristics developed so far for the MSAP, our 3-approximation algorithm achieved results comparable to all of them. It outperformed heuristic H4 in all measures, and performed very close to the other three heuristics. At the nucleotide level, for example, our approximation was slightly less sensitive than heuristics H1 and H2, but its value of specificity compares with that obtained by H1 and H3. In summary, looking at the AC column, our algorithm and Heuristic H1 were the approaches with the best values. At the exon level, our approximation outperformed H2, achieved results comparable to H1 and was overwhelmed only by H3. Anyway, in this last case, H3 outperformed our approach with only 1% and 3% of improvement in sensitivity and specificity, respectively.

5   Discussion

In this work we presented a 3-approximation algorithm for the Multiple Spliced Alignment Problem, a combinatorial optimization problem directly related with to gene prediction task. We also compared our approach with 4 previously proposed heuristics for the MSAP, achieving results comparable to the best one. This fact is very encouraging since it shows that our approach can perform as good as previously proposed heuristics for the MSAP when applied to the gene prediction task, beside ensuring its results are not worse than 3 times the optimal solution, no matter which instance is considered.

In a more detailed observation, and taking into account the measures AC and Av_e that summarize the experimental results at the nucleotide and exon levels respectively, our algorithm showed the same accuracy of Heuristic H1, being the best on nucleotide level and the second best in exon level. As Heuristic H1 is based on the idea of choosing a central sequence of $\mathcal{T}$, applying it to obtain a SAP solution with Gelfand's algorithm and extending it to the MSAP in question, it becomes clear that both approaches share similar aspects and therefore such close results are expected.

In further studies, we intend to handle the MSAP by proposing a linear programming model in order to attack it from a third perspective. We already have a preliminary integer linear programming formulation, and experimental tests with it are in course.

References

1. Burge, C., Karlin, S.: Prediction of Complete Gene Structures in Human Genomic DNA. Journal of Molecular Biology 268(1), 78–94 (1997)
2. Burset, M., Guigo, R.: Evaluation of Gene Structure Prediction Programs. Genomics 34(298), 353–367 (1996)
3. Gelfand, M.S., Mironov, A.A., Pevzner, P.A.: Gene Recognition Via Spliced Sequence Alignment. Proceedings of the National Academy of Sciences of the United States of America 93, 9061–9066 (1996)
4. Hogeweg, P.: The Roots of Bioinformatics in Theoretical Biology. PLoS Computational Biology 7(3), 1–5 (2011)
5. Kishi, R.M., dos Santos, R.F., Adi, S.S.: Gene Prediction by Multiple Spliced Alignment. In: Norberto de Souza, O., Telles, G.P., Palakal, M. (eds.) BSB 2011. LNCS, vol. 6832, pp. 26–33. Springer, Heidelberg (2011)
6. Kishi, R.M., dos Santos, R.F., Montera, L., Adi, S.S.: A Similarity-based Genetic Algorithm for the Gene Prediction Problem. In: BSB & EBB Digital Proceedings, Campo Grande, pp. 84–89 (2012)
7. Majoros, W.H.: Methods for Computational Gene Prediction, 1st edn. Cambridge University Press (2007)
8. Mathé, C., Sagot, M.-F., Schiex, T., Rouzé, P.: Current Methods of Gene Prediction, Their Strengths and Weaknesses. Nucleic Acids Research 30(19), 4103–4117 (2002)
9. Needleman, S.B., Wunsch, C.D.: A General Method Applicable to the Search for Similarities in the Amino Acid Sequence of Two Proteins. Journal of Molecular Biology 48, 443–453 (1970)
10. The ENCODE Project Consortium: The ENCODE (Encyclopedia of DNA Elements) Project. Science 306(5696), 636–640 (2004)
11. Sayers, E.W., Barrett, T., Benson, D.A., Bolton, E., Bryant, S.H., Canese, K., Chetvernin, V., Church, D.M., DiCuccio, M., Federhen, S., Feolo, M., Fingerman, I.M., Geer, L.Y., Helmberg, W., Kapustin, Y., Krasnov, S., Landsman, D., Lipman, D.J., Lu, Z., Madden, T.L., Madej, T., Maglott, D.R., Marchler-Bauer, A., Miller, V., Karsch-Mizrachi, I., Ostell, J., Panchenko, A., Phan, L., Pruitt, K.D., Schuler, G.D., Sequeira, E., Sherry, S.T., Shumway, M., Sirotkin, K., Slotta, D., Souvorov, A., Starchenko, G., Tatusova, T.A., Wagner, L., Wang, Y., Wilbur, W.J., Yaschenko, E., Ye, J.: Database resources of the National Center for Biotechnology Information. Nucleic Acids Research 40 (D1), D13–D25 (2012)

Improved Filters for the Approximate
Suffix-Prefix Overlap Problem

Gregory Kucherov[12] and Dekel Tsur[2]

[1] CNRS/LIGM, Université Paris-Est Marne-la-Vallée, France
[2] Department of Computer Science, Ben-Gurion University of the Negev, Israel

Abstract. Computing suffix-prefix overlaps for a large collection of
strings is a fundamental building block for the analysis of genomic next-
generation sequencing data. The approximate suffix-prefix overlap problem is to find all pairs of strings from a given set such that a prefix of
one string is similar to a suffix of the other. Välimäki et al. (Information
and Computation, 2012) gave a solution to this problem based on suffix filters. In this work, we propose two improvements to the method of
Välimäki et al. that reduce the running time of the computation.

1 Introduction

Genomic sequences are deciphered by reading short overlapping fragments. Modern *next-generation* sequencing technologies, can produce, in a single run, tens
or hundreds of millions of such fragments, called *reads*, each of the order of a
hundred of letters. Dealing with these gigabytes of sequence data raises a number
of algorithmic challenges.

A basic operation on a collection of genomic reads is the computation of overlaps: we need to be able to quickly retrieve reads which have a significant overlap
with a given read. This operation is a prerequisite for many algorithms dealing
with reads, and most prominently for *genome assembly* algorithms which follow
the so-called *overlap-layout-consensus* paradigm [6]. These algorithms are based
on *overlap graphs* (also called *string graphs* or *assembly graphs*) that represent
all significant overlaps between reads. A recent example is provided by SGA
assembler [12]. Earlier, this approach was taken by several "first-generation"
methods of genome assembly, such as Celera assembler [8] used to assemble one
of the first versions of the human genome.

The goal of this work is to propose an efficient way of computing significant
approximate suffix-prefix overlaps of a set of strings. Previously, several solutions have been proposed to compute all *exact* suffix-prefix overlaps [2, 11, 12].
However, in practice, we are interested in the *approximate* case when strings can
overlap within a certain number of errors.

Most practical methods for computing approximate string similarities are
based on the filtering approach, when the search is done in two steps: at the
first step, *candidate* regions are identified that *potentially* correspond to sought
matches, and at the second step, those candidates are checked to *actually* verify

E. Moura and M. Crochemore (Eds.): SPIRE 2014, LNCS 8799, pp. 139–148, 2014.
© Springer International Publishing Switzerland 2014

the desired matching condition. Filtering algorithms usually do not yield interesting theoretical time bounds but are often very efficient in practice. As an example, *spaced seeds* [1, 7] constitute one of the filtering techniques that has been successfully used for DNA sequence comparison, e.g. [7, 10].

To compute approximate suffix-prefix overlap, the above-mentioned SGA assembler [12] uses a basic *substring filtering*. Välimäki et al. [13] proposed to apply a modified version of *suffix filters* earlier proposed by Kärkkäinen and Na [3]. Suffix filter provide a more selective filtering criterion and therefore a more efficient algorithm.

In this paper, we show how the method of [13] can be further improved. We propose two improvements that reduce the search space and therefore the running time of the algorithm: a new family of suffix filters and a new partitioning scheme. We report on estimations on random datasets that support the superiority of our schemes.

Throughout the paper, we present our method for the Hamming distance between strings, although it can be generalized to the edit distance, similar to [3, 13]. This, however, would entail additional technical details and a more involved presentation that we wanted to avoid.

2   Preliminaries

2.1   Notation

For a sequence of integers A, let $\mathsf{PrefixSum}(A)$ be a sequence of integers of the same length as A in which $\mathsf{PrefixSum}(A)[i] = \sum_{j=1}^{i} A[j]$. For two sequences of integers A and B of the same length, we define $A \leq B$ iff $A[i] \leq B[i]$ for all i.

The Hamming distance between strings A and B of the same length, denoted $\mathsf{Ham}(A, B)$, is the number of indices i for which $A[i] \neq B[i]$. If $\mathsf{Ham}(A, B) \leq k$, we say that A and B k-*match*.

Let A be a string that has a partition $A = A_1 A_2 \cdots A_k$ into k disjoint parts. Let B be a string with the same length as A. The partition of A induces a partition $B = B_1 B_2 \cdots B_k$ of B in which $|B_i| = |A_i|$ for all i. The *partition distance* between A and B, denoted $\mathsf{pd}(A, B)$, is a sequence of integers of length k in which $\mathsf{pd}(A, B)[i] = \mathsf{Ham}(A_i, B_i)$. The *accumulated partition distance* between A and B, denoted $\mathsf{apd}(A, B)$, is the sequence $\mathsf{PrefixSum}(\mathsf{pd}(A, B))$. That is, $\mathsf{apd}(A, B)[i]$ is the total number of mismatches between the first i parts of A and B.

2.2   Suffix Filters

For the problem of approximate pattern matching with q errors, the most basic filtering method, called PEX in [9], consists in splitting the pattern into $k = q + 1$ parts and searching for those parts independently. Once one of the parts is found, it provides a candidate location for the whole pattern. Kärkkäinen and Na [3] proposed an interesting extension of this principle called *suffix filters*. Suffix filters have been shown to generate many fewer candidates than substring filters, and therefore to be much more efficient. Since the present work builds on this idea, we briefly explain it here.

We still split pattern P into $k = q + 1$ parts $P = P_1 P_2 \cdots P_k$, but instead of searching for substrings P_i, we will be searching for *suffixes* $P_i P_{i+1} \cdots P_k$ for $i = 1, \ldots, k$ allowing errors distributed according to a specific pattern

$$\mathsf{filter}_{k,i} = 01 \cdots (k - i).$$

This means that for every i, we are searching for strings B such that $\mathsf{apd}(P_i \cdots P_k, B)$ $\leq \mathsf{filter}_{k,i}$. The key observation of [3] is that this scheme detects all possible occurrences of P within q errors.

For example, let $q = 2$, namely, we want to find the substrings of the text that 2-match to $P = P_1 P_2 P_3$. All such substrings are detected using three filters, denoted by sequences 012, 01, and 0. Filter 012 detects substrings B such that $\mathsf{apd}(P_1 P_2 P_3, B) \leq 012$. That is, $B = B_1 B_2 B_3$, such that $|B_i| = |P_i|$ for all i, $B_1 = P_1$, $\mathsf{Ham}(B_2, P_2) \leq 1$, and $\mathsf{Ham}(B_2 B_3, P_2 P_3) \leq 2$. By a slight abuse of language, the set of all such strings B will be said to be *enumerated* by the filter. Similarly, filter 01 detects substrings $B = B_2 B_3$ such that $B_2 = P_2$ and B_3 is within one error from P_3. Each such string is a suffix of a candidate approximate occurrence of P. Finally, filter 0 detects substrings B_3 with $B_3 = P_3$, which provides again a suffix of a candidate occurrence of P.

Observe that there are 9 cases for the partition distance between P and B — 011, 101, 110, 002, 020, 200, 100, 010, and 001 — which are all covered by suffix filters 012, 01 and 0. Indeed, filter 012 covers cases 011, 002, 010, and 001, filter 01 covers cases 101, 200 and 100, and filter 0 covers cases 020 and 110.

The set of strings enumerated by a filter can be naturally represented by a *trie* (see Figure 1), where branching nodes correspond to positions where the filter allows a possible mismatch to occur. The number of nodes in the tries of all the filters is a crucial parameter for the efficiency of a filtering scheme.

2.3   Suffix Filters and Full-Text Indexes

One of the advantages of suffix filters (as opposed e.g. to spaced seeds) is that they can be naturally implemented using full-text indexes that support incremental string matching. Those indexes include classical indexes such as suffix trees, but also succinct indexes such as FM-index and its variants [14]. These indexes allow reading a pattern left-to-right[1] and quickly updating the index point after each letter, so that all occurrences of the prefix read so far can be retrieved efficiently.

Implementing suffix filters on a full-text index can be done simply by *enumerating* all "approximate suffixes" of the pattern detected by a filter, and reporting the occurrences of all these strings in the text, thus generating the candidate set.

---

[1] For some indexes, such as FM-index, matching is performed right-to-left, in which case we can just assume that the indexed sequences are reversed. On the other hand, FM-index can also be modified to perform matching left-to-right, see [5].

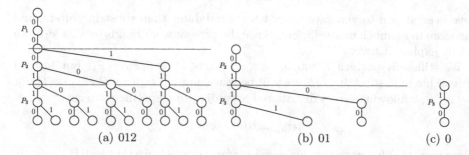

Fig. 1. The tries of the filters 012, 01, and 0 applied on the pattern $P = 000110$, for binary alphabet and $q = 2$ errors. The pattern P is partitioned into 3 parts $P = P_1 P_2 P_3$ where $P_1 = 00$, $P_2 = 01$, and $P_3 = 10$. The filter 012 enumerates the strings corresponding to the leaves of the tries in Figure (a), namely 000110, 000111, ..., 001100. The filter 01 enumerates the strings 0110, 0111, and 0100, and the filter 0 enumerates the string 10.

More precisely, for each of the tries that correspond to the filters, the algorithm traverses the trie in depth-first search order, and updates the index point after each descend in the trie. When reaching a node in the trie whose corresponding string does not appear in the index, the search does not continue to the descendants of this node in the trie. When reaching a leaf of the trie, all occurrences of the string corresponding to the leaf are retrieved from the index, and are added to the set of candidates.

2.4 Suffix Filters Applied to Suffix-Prefix Overlap Problem

Given a set S of strings, the *approximate suffix-prefix overlap* problem is to compute all significant approximate overlaps between pairs of strings of S. "Significance" is defined by a lower threshold l_{min} on the overlap size. Since the overlap size is variable, imposing a fixed number of errors is not reasonable, and a relative error rate is specified instead. Formally, given an integer l_{min}, and $\epsilon > 0$, we have to find all pairs of strings $S, S' \in S$ such that there is an integer $l \geq l_{min}$ for which the prefix of S of length l and the suffix of S' of length l $\lceil \epsilon l \rceil$-match.

In this Section, we explain how suffix filters can be used to solve the approximate suffix-prefix overlap problem. We proceed by enumerating all strings $S \in S$. For each $S \in S$ and for each $l \geq l_{min}$, we want to identify all strings $S' \in S$ whose suffix of length l $\lceil \epsilon l \rceil$-match the prefix $S[1..l]$. We want to do it by applying suffix filters designed for patterns of length l and $\lceil \epsilon l \rceil$ mismatches. If such a filter applies, then a candidate overlap is generated, which is a triplet (S, S', l). At the verification stage, the actual Hamming distance between the prefix of S of length l and the suffix of S' of length l is computed, and the overlap is reported if this distance is no more than $\lceil \epsilon l \rceil$.

Let us now explain how the filtering algorithm works with the filters of Section 2.2. Fix S and a partition $S = S_1 S_2 \cdots$ of S into disjoint parts. Fix $l \geq l_{min}$

and let $P = S[1..l]$. The partition of S induces a partition $P = P_1 \cdots P_k$ of P, where $P_i = S_i$ for $i < k$ and P_k is a prefix of S_k. Suppose that the partition of S was chosen a way to ensure that $k \geq \lceil \epsilon l \rceil + 1$. This allows us to apply suffix filters of Section 2.2.

Consider the above filters $\text{filter}_{k,1}, \ldots, \text{filter}_{k,k}$, where $\text{filter}_{k,i} = 01 \cdots (k - i)$. Each filter $\text{filter}_{k,i}$ enumerates all strings B such that $\text{apd}(P_i \cdots P_k, B) \leq F$, and B appears as a substring of some string in $\mathcal{S}$. For each such string B, the algorithm further selects all the strings $S' \in \mathcal{S}$ that *end with* B and adds the triplets (S, S', l) to the list of candidates.

The main difficulty in applying suffix filters to the approximate suffix-prefix overlap problem is that the length l and, consequently, the number of errors are not fixed. Once the partition of S is defined, our goal is to deal with all values of l in one left-to-right traversal of S. For each $l \geq l_{\min}$, we should be able to efficiently identify appropriate filters that apply to the corresponding partition of $S[1..l]$. We now describe how it is done for the filters of Section 2.2.

A key point here is that the enumeration processes for different values of l are connected. Let k_l denote the number of parts in the partition of $S[1..l]$ that is induced by the partition of S. Let $\mathcal{B}_{S,i,l}$ be the set of strings enumerated by the filter $\text{filter}_{k_l,i}$ when it is applied to $S[1..l]$. That is, $\mathcal{B}_{S,i,l}$ are the strings enumerated by $\text{filter}_{k_l,i}$ when considering the filtering scheme for the fixed value of l. Let $\text{trie}_{S,i}$ be the trie representing the set $\mathcal{B}_{S,i,|S|}$. We have the following property: For every l, where $l_{\min} \leq l < |S|$, and every $i \leq k_l$, the prefix of $\text{filter}_{k_{|S|},i}$ of length $k_l + 1 - i$ is equal to $\text{filter}_{k_l,i}$. It follows that the set $\mathcal{B}_{S,i,l}$ is equal to the set of strings that correspond to the nodes of $\text{trie}_{S,i}$ at depth $l - \sum_{j<i} |S_j|$. Therefore, generation of candidates can be done for all values of l by the following algorithm. For $i = 1, 2, \ldots, k_S$, traverse $\text{trie}_{S,i}$. When the traversal is at a node that corresponds to a string B, if $B \in \mathcal{B}_{S,i,l}$ for some $l \geq l_{\min}$, find all the strings $S' \in \mathcal{S}$ that ends with B, and for each such S' add the triplet $(S, S', |B| + \sum_{j<i} |S_j|)$ to the list of candidates.

Checking whether $B \in \mathcal{B}_{S,i,l}$ is done with the following *candidate generation condition*.

Condition 1. $B \in \mathcal{B}_{S,i,l}$ *if and only if*

$$|B| + \sum_{j<i} |S_j| \geq l_{\min} \tag{C1}$$

2.5   The Filtering Scheme of Välimäki et al. [13]

Välimäki et al. [13] observed that the filtering procedure of Section 2.4 is very inefficient. The inefficiency is caused by the filters $\text{filter}_{k,k} = 0$ that has to be applied, during the search, to short strings including those consisting of a single letter. Formally, when traversing the trie $\text{trie}_{S,i}$, a node of the trie at depth 1, whose corresponding single-letter string is $B = S[1 + \sum_{j<i} |S_j|]$, can generate candidates if Condition (C1) is satisfied. Since B has length 1, there can be many strings $S' \in \mathcal{S}$ that ends with B, generating many spurious candidates.

More generally, assuming the strings of S are sampled randomly from an i.i.d. source, the expected number of candidates generated by a string B corresponding to some node in a trie is $m/\sigma^{|B|}$, where m is the number of strings in S, and σ is the size of the alphabet. Therefore, the total number of candidates is dominated by the number of candidates generated by nodes of small depth.

The solution of Välimäki et al. to this problem is to drop the filters $\text{filter}_{k,k}$ from the filtering scheme. In order to handle the cases of partition distances that were covered by this filter, filters $\text{filter}_{k,1} = 01 \cdots k$ are modified to $\text{filter}_{k,1} = 12 \cdots (k+1)$. It is shown [13] that with this modification, all combinations of partition distance are still covered.

Due to the dropping of filters $\text{filter}_{k,k}$, Condition 1 should be modified for the new filtering scheme. The modified candidate generating condition will now be as follows.

Condition 2. $B \in \mathcal{B}_{S,i,l}$ *if and only if Condition (C1) is satisfied and*

$$|B| > |S_i| \tag{C2}$$

Clearly, the additional condition (C2) reduces the number of generated candidates.

3   New Filtering Scheme

In the filtering scheme of Välimäki et al., the filters $\text{filter}_{k,1}$ start with 1 whereas the other filters start with 0. This has the consequence that the number of nodes in the trie $\text{trie}_{S,1}$ is much larger than the number of nodes in the tries $\text{trie}_{S,i}$ for $i > 1$. We now present a new filtering scheme without this property. Our filtering scheme consists of filters $\text{filter}_{k,1}, \ldots, \text{filter}_{k,k-1}$, where $\text{filter}_{k,i}$ is a sequence of length $k - i + 1$ whose first $k - i$ elements are $0, 1, \ldots, (k - i - 1)$, and the last element is $k - i - 1$. For example, for $k = 4$ the filters are $\text{filter}_{4,1} = 0122$, $\text{filter}_{4,2} = 011$, and $\text{filter}_{4,3} = 00$. Our filtering scheme requires a difference of at least 2 between the number of parts in the partition of $S[1..l]$ and $\lceil \epsilon l \rceil$ (recall that in the scheme of Välimäki et al., the required difference is at least 1). We show the correctness of this scheme in the following lemma.

Lemma 1. *For every $k \geq 2$ and every sequence of integers M of length k whose sum is at most $k-2$, there is an integer i such that $\text{PrefixSum}(M[i..k]) \leq \text{filter}_{k,i}$.*

Proof. The proof is by induction of k. The base of the induction $k = 2$ is trivial since in this case $M = 00$ and therefore $\text{PrefixSum}(M) \leq \text{filter}_{k,k-1}$. Now consider $k > 2$. Let M be a sequence of length k whose sum is at most $k - 2$. If $\text{PrefixSum}(M) \leq \text{filter}_{k,1}$ we are done. Otherwise, there is an index i such that $\text{PrefixSum}(M)[i] > \text{filter}_{k,1}[i]$. Since $\text{PrefixSum}(M)[i] \leq k - 2$ and $\text{filter}_{k,1}[j] = k - 2$ for $j > k - 2$, it follows that $i \leq k - 2$. Therefore $\text{filter}_{k,1}[i] = i - 1$ and $\text{PrefixSum}(M)[i] \geq i$. Let $M' = M[i+1..k]$. The length of M' is $k - i$ and the sum of M' is at most $k - 2 - \text{PrefixSum}(M)[i] \leq k - 2 - i$. By the induction hypothesis, there is an index i' such that $\text{PrefixSum}(M'[i'..k - i]) \leq \text{filter}_{k-i,i'}$. The lemma follows since $M'[i'..k - i] = M[i + i'..k]$ and $\text{filter}_{k-i,i'} = \text{filter}_{k,i+i'}$.   □

To illustrate Lemma 1, observe that the three above-mentioned filters $\text{filter}_{4,i}$, $i = 1, 2, 3$ cover all possible partition distances for the case of 4 parts and 2 errors. Indeed, filter 0122 cover cases 0020, 0002, 0101 and 0110, filter 011 cover cases 0011, 1001 and 1010, and finally filter 00 covers cases 2000, 0200 and 1100.

In addition to reducing the number of nodes in the tries, our filtering scheme also reduces the number of generated candidates. In the filtering scheme of Välimäki et al., $\text{filter}_{k,i}$ is a prefix of $\text{filter}_{k',i}$ for all $k < k'$ and i. Our filtering scheme does not have this property. Therefore, we need a new condition for checking whether $B \in \mathcal{B}_{S,i,l}$.

Condition 3. $B \in \mathcal{B}_{S,i,l}$ *if and only if conditions (C1) and (C2) are satisfied and*

$$\text{apd}(S', B) \leq \text{filter}_{k,i} \qquad \qquad (C3)$$

where S' is the prefix of $S_i S_{i+1} \cdots$ of length $|B|$ and k is the number of parts in the partition of S' induced by the partition of S.

Observe that in the filtering scheme of Välimäki et al., every node of $\text{trie}_{S,i}$ with depth $|S_i| + 1$ generated candidates. However, in our new scheme, only the node whose corresponding string B is equal to S' generates candidates among the nodes of depth $|S_i| + 1$ (for every other node, $\text{apd}(S', B) = 01$ and therefore Condition (C3) is not satisfied). The same is true for depths $|S_i| + 2, \ldots, |S_i| + |S_{i+1}|$.

Our scheme can be generalized by introducing a parameter $s \geq 2$. The filters for a given value of s are $\text{filter}_{k,i} = 01 \cdots (k - i - s)(k - i - s + 1)^s$ for $i = 1, \ldots, k - s + 1$. This scheme requires a difference of at least s between the number of parts in the partition of $S[1..l]$ and $\lceil \epsilon l \rceil$.

4   Partition Schemes

The efficiency of the algorithm of the previous section depends on the sizes of the parts in the partition of S: having larger parts reduces the number of trie nodes and the number of candidates. For correctness of the algorithm, the partitioning of a string S must satisfy the following property.

(P1) For every $l \geq l_{\min}$, the number of parts in the partition of $S[1..l]$ induced by the partition of S is at least $\lceil \epsilon l \rceil + s$, where $s = 1$ for the filtering scheme of Välimäki et al. and $s = 2$ for our new scheme (or some fixed value of s for our extended scheme).

Välimäki et al. used a partition of S into equal sized parts of size p, except for the last part whose size is at most p. The value of p is chosen to be the maximum integer for which Property (P1) is satisfied.

We propose a partitioning scheme in which most parts are larger than those of the equal sized parts partitioning. Since the efficiency of the filtering approach depends on the sizes of the parts, the new partitioning scheme gives better performance.

Let S be a string to be partitioned. Let $l_0 < l_1 < \cdots < l_q$ be all the indices in the range $l_{\min}, \ldots, |S|$ for which $\lceil \epsilon(l-1) \rceil < \lceil \epsilon l \rceil$, and let $l_{q+1} = |S| + 1$. Let $k = \lceil \epsilon l_0 \rceil + s - 1$. We partition S as follows. The sizes of the first k parts are chosen in order to satisfy the following properties.

1. The total length of the first k parts is $l_0 - 1$.
2. The length of the k-th part is at least $l_0 - l_{\min}$.

We can set for example the length of the k-th part to be $L = \max(\lceil (l_0 - 1)/k \rceil, l_0 - l_{\min})$, and set the lengths of the first $k - 1$ parts to be p or $p + 1$, where $p = \lfloor (l_0 - 1 - L)/(k-1) \rfloor$. The lengths of the remaining parts in the partition are $l_1 - l_0, l_2 - l_1, \ldots, l_{q+1} - l_q$. We now show that this partition satisfies Property (P1). Moreover, the partitioning is optimal in the sense that the inequalities of Property (P1) are satisfied with equality.

Lemma 2. *For every $l \geq l_{\min}$, the number of parts in the partition of $S[1..l]$ induced by the partition of S is $\lceil \epsilon l \rceil + s$.*

Proof. For $l_{\min} \leq l < l_0$ (assuming $l_0 > l_{\min}$) we have by construction that the number of parts in the partition of $S[1..l]$ is $k = \lceil \epsilon l_0 \rceil + s - 1$. By the definition of l_0, $\lceil \epsilon l \rceil = \lceil \epsilon l_0 \rceil - 1$, so the equality of the lemma is satisfied. Similarly, if $l_i \leq l < l_{i+1}$ then the number of parts in the partition of $S[1..l]$ is $k + 1 + i$. Moreover, $\lceil \epsilon l \rceil = \lceil \epsilon l_0 \rceil + i$. Therefore, the lemma follows.   □

As an example, consider partitioning a string of length 200 with the parameters $l_{\min} = 40$ and $\epsilon = 0.1$. If $s = 1$, the equal sized partition uses parts of size 8. In our new partitioning scheme, the first 5 parts have size 8, and the remaining parts have size 10.

5   Experimental Results

In this Section, we compare the performance of 4 filtering schemes:

(1) the filtering scheme of Välimäki et al. [13] using partitioning into equal parts,
(2) the filtering scheme of Välimäki et al. combined with our new partitioning scheme (Section 4),
(3) our new filtering scheme (Sections 3-4), and
(4) our extended filtering scheme (see end of Section 3), with $s = 3$.

For our filtering scheme (3), string partitioning is done with our new partitioning scheme. The comparisons have been done using the technique described in Kucherov et al. [4]. In this analysis, we assume that characters of the strings of S are randomly chosen uniformly and independently from the alphabet. Under this assumption, we analytically estimate the expected number of nodes in the tries $\text{trie}_{S,i}$ and the expected number of generated candidates, following the method developed in [4]. The results are summarized in Table 1. Columns 2 to 5 of the Table correspond to schemes (1) to (4) above, respectively.

Table 1. Expected performance of the filtering scheme of Välimäki et al. [13] and our filtering schemes. It is assumed that S contains m random strings of length 300 over an alphabet of size 4. For each scheme, the first column shows the expected number of nodes in the tries $trie_{S,i}$ for all i (for a single $S \in \mathcal{S}$), and the second column is the expected number of candidates generated for S.

m	l_{min}	ϵ	method of [13]		[13] with un-equal parts		our scheme		our scheme extended for $s = 3$	
10^6	20	0.1	31257	18950	11782	144	8582	341	20026	95
10^7	20	0.1	64201	189506	22161	1449	13219	3412	46662	952
10^6	40	0.1	10416	839	8260	65	6912	28	8868	0.5
10^7	40	0.1	14391	8391	11138	651	8921	280	12916	4
10^7	40	0.15	207504	857318	71271	82559	40671	18842	82164	116

Note that for different parameters, the bottleneck of the computation can be either the size of traversed tries, or the number of generated candidates. In all cases, we observe a significant decrease of both these measures compared to the original method of Välimäki et al. [13]. When the threshold l_{min} is small (in our experiments, 20 for the sequence length 300), the filters of [13] combined with our partitioning scheme presents a trade-off with our filtering scheme: our scheme yields a smaller number of traversed trie nodes but a larger number of generated candidates. However, when l_{min} is large enough (in our experiments, 40 for the sequence length 300), our scheme outperforms the one of Välimäki et al. in both the number of nodes in the tries and the number of generated candidates. The extended scheme with $s = 3$ yields a smaller number of candidates, but at the cost of increased number of traversed trie nodes.

6 Conclusions

In this paper, we proposed an improved filtering scheme for the approximate suffix-prefix overlap problem directly raised by bioinformatics applications. Two improvements are proposed: we provide a more efficient filtering scheme as well as new way of partitioning the query string. We show, through analytical estimations, the superiority of our scheme in terms of the size of the search space (size of traversed tries) as well as the selectivity (number of generated candidates).

Several directions for future work can be envisaged. We did not compare the actual performance of the different filtering schemes on real data. However, previous work [4] provides strong grounds to assume that the better performance will be supported by real data too. This, however, remains to be verified experimentally. Another direction, already mentioned in Introduction, concerns the generalization of our results to the case of edit distance. While we don't expect significant obstacles in this generalization, it does bring an additional technical difficulty.

Acknowledgements. GK has been supported by the ABS2NGS grant of the French government (program *Investissement d'Avenir*) as well as by a EU Marie-Curie Intra-European Fellowship for Carrier Development. DT has been supported by ISF grant 981/11.

References

1. Burkhardt, S., Kärkkäinen, J.: Better filtering with gapped q-grams. Fundamenta Informaticae 56(1,2), 51–70 (2003)
2. Gusfield, D., Landau, G., Schieber, B.: An efficient algorithm for the all pairs suffix-prefix problem. Inf. Process. Lett. 41(4), 181–185 (1992)
3. Kärkkäinen, J., Na, J.C.: Faster filters for approximate string matching. In: Proc. 9th Workshop on Algorithm Engineering and Experiments (ALENEX), pp. 84–90 (2007)
4. Kucherov, G., Salikhov, K., Tsur, D.: Approximate string matching using a bidirectional index. In: Kulikov, A.S., Kuznetsov, S.O., Pevzner, P. (eds.) CPM 2014. LNCS, vol. 8486, pp. 222–231. Springer, Heidelberg (2014), Full version at http://arxiv.org/abs/1310.1440
5. Lam, T.W., Li, R., Tam, A., Wong, S.C.K., Wu, E., Yiu, S.-M.: High throughput short read alignment via bi-directional BWT. In: Proc. IEEE International Conference on Bioinformatics and Biomedicine (BIBM), pp. 31–36 (2009)
6. Li, Z., Chen, Y., Mu, D., Yuan, J., Shi, Y., Zhang, H., Gan, J., Li, N., Hu, X., Liu, B., Yang, B., Fan, W.: Comparison of the two major classes of assembly algorithms: overlap-layout-consensus and de-Bruijn-graph. Brief Funct. Genomics 11(1), 25–37 (2012)
7. Ma, B., Tromp, J., Li, M.: PatternHunter: Faster and more sensitive homology search. Bioinformatics 18(3), 440–445 (2002)
8. Myers, E.W., Sutton, G.G., Delcher, A.L., Dew, I.M., Fasulo, D.P., Flanigan, M.J., Kravitz, S.A., Mobarry, C.M., Reinert, K.H., Remington, K.A., Anson, E.L., Bolanos, R.A., Chou, H.H., Jordan, C.M., Halpern, A.L., Lonardi, S., Beasley, E.M., Brandon, R.C., Chen, L., Dunn, P.J., Lai, Z., Liang, Y., Nusskern, D.R., Zhan, M., Zhang, Q., Zheng, X., Rubin, G.M., Adams, M.D., Venter, J.C.: A whole-genome assembly of Drosophila. Science 287(5461), 2196–2204 (2000)
9. Navarro, G., Raffinot, M.: Flexible Pattern Matching in Strings – Practical on-line search algorithms for texts and biological sequences. Cambridge University Press (2002)
10. Noé, L., Kucherov, G.: YASS: Enhancing the sensitivity of DNA similarity search. Nucleic Acid Research 33, W540–W543 (2005)
11. Ohlebusch, E., Gog, S.: Efficient algorithms for the all-pairs suffix-prefix problem and the all-pairs substring-prefix problem. Information Processing Letters 110(3), 123–128 (2010)
12. Simpson, J.T., Durbin, R.: Efficient de novo assembly of large genomes using compressed data structures. Genome Res. 22(3), 549–556 (2012)
13. Välimäki, N., Ladra, S., Mäkinen, V.: Approximate all-pairs suffix/prefix overlaps. Information and Computation 213, 49–58 (2012)
14. Vyverman, M., De Baets, B., Fack, V., Dawyndt, P.: Prospects and limitations of full-text index structures in genome analysis. Nucleic Acids Res. 40(15), 6993–7015 (2012)

Sequence Decision Diagrams*

Hind Alhakami[1], Gianfranco Ciardo[2], and Marek Chrobak[1]

[1] Dept. of Computer Science and Engineering, University of California, Riverside
[2] Dept. of Computer Science, Iowa State University

Abstract. Compact encoding of finite sets of strings is a classic problem. The manipulation of large sets requires compact data structures that allow for efficient set operations. We define *sequence decision diagrams* (SeqDDs), which can encode arbitrary finite sets of strings over an alphabet. SeqDDs can be seen as a variant of classic decision diagrams such as BDDs and MDDs where, instead of a fixed number of levels, we simply require that the number of paths and the lengths of these paths be finite. However, the main difference between the two is the target application: while MDDs are suited to store and manipulate large sets of constant-length tuples, SeqDDs can store arbitrary finite languages and, as such, should be studied in relation to finite automata. We do so, examining in particular the size of equivalent representations.

1 Introduction

Many data structures have been introduced to compactly encode finite sets of finite strings. *Substring indices* data structures, such as *tries*, suffix trees, suffix arrays, and DAWGs, exploit prefix sharing, suffix sharing, or both to achieve efficient storage of large sets. Beside compactness, the main purpose of *substring indices* data structures is to solve substring matching problem for multiple patterns in a given text with a time complexity proportional to the pattern size, not the whole text. These data structures allow for efficient matching, but updating them to add or delete strings is hard [1]. Additionally, the lack of efficient set manipulation algorithms for such data structures motivates work that leverages the benefits of *substring indices* while enabling efficient set manipulation.

In 2009, Loekito [7] introduced a new data structure, *sequence* BDD, SeqBDD, for short, that offers compact storage of finite languages. SeqBDDs are a half-relaxed variation of ZBDDs [8] where variables along *one-paths* may appear multiple times in any order. SeqBDDs inherit ZBDDs' efficient set manipulations, and also support algorithms to solve the substring matching problem.

Size complexity is crucial to decision diagrams, including SeqBDDs, due to two factors: first, decision diagrams are used to store efficiently an enormous amount of data; second, the time complexity of decision diagram algorithms is proportional to the size of the arguments, which is in turn sensitive to variable ordering. Since optimal variable ordering is an NP-complete problem [3], heuristics can only achieve a "good "variable ordering. Moreover, while sharing

---

* This work is supported in part by Ministry of Higher Education - Saudi Arabia, and National Science Foundation under grants CCF-1217314 and CCF-1442586.

E. Moura and M. Crochemore (Eds.): SPIRE 2014, LNCS 8799, pp. 149–160, 2014.
© Springer International Publishing Switzerland 2014

common suffixes as well as common prefixes contributes to the compactness of
SeqBDDs, embracing a binary representation degrades compactness [9].

We define *sequence decision diagrams* (SeqDDs) to encode arbitrary finite
languages. SeqDDs are somewhat analogous to a multi-valued variation of SeqBDDs, but are insensitive to variable ordering; in fact, they do not even associate variables or levels to nodes. Instead, they simply require that the number
of paths and the lengths of these paths be finite. We introduce two canonical
SeqDD definitions and discuss their compactness in relation to finite automata.
Canonical SeqDD promotes efficient algorithms for set manipulations and substring manipulations by exploiting node sharing and *memoization*. The rest of
the paper is organized as follows: Section 2 provides preliminaries. Section 3 introduces non-canonical and canonical SeqDDs. Section 4 discusses the relative
compactness of canonical SeqDDs. Section 5 introduces set and string manipulation algorithms. Section 6 provides preliminary applications of SeqDDs. Section
7 presents conclusions and future work.

2   Preliminaries

Finite automata are a well known data structure to describe regular languages.
While finite automata are memory efficient, their manipulation algorithms are
not guaranteed to provide minimized outputs even if their inputs are minimized.
On the other hand, decision diagrams have efficient manipulation algorithms but
most, for example BDDs [4] and MDDs [6], only target fixed-length languages.

2.1   Finite Automata

A finite automaton (FA) is a 5-tuple $(Q, \Sigma, \delta, q_0, F)$, with a finite set of states, a
finite alphabet, a transition function, a start state, and a set of accepting states.
Depending on the transition function, the FA is a *deterministic* FA (DFA, with
$\delta : Q \times \Sigma \to Q$) or a *non-deterministic* FA (NFA, with $\delta : Q \times \Sigma \cup \{\epsilon\} \to 2^Q$).
We also consider a *partial* DFA [2], a minimized DFA with partial transition
function $\delta : Q \times \Sigma \to Q \cup \{\emptyset\}$, obtained from the equivalent DFA by deleting all
states with no path to accepting states, as well as their incoming transitions.

2.2   Decision Diagrams

Binary decision diagrams (BDDs) are directed acyclic graph where each node is
associated with a boolean variable and encodes boolean functions over a structured boolean domain. Multi-valued decision diagrams (MDDs) generalize BDDs
by allowing nodes to have more than two outgoing edges, and provide a canonical representation of boolean functions over structured finite domains (we use
"MDDs" from now on, since BDDs are just a special case).

An *ordering* rule is enforced: assuming k domain variables $\{x_1, ..., x_k\}$, all
paths respect the order $x_k \prec x_{k-1} \prec \cdots \prec x_1 \prec x_0$, where x_0 is the range
variable associated with terminal nodes. Then, canonicity requires choosing a

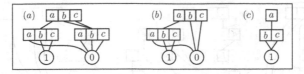

Fig. 1. Quasi (a), fully (b), and sparsely (c) reduced MDDs encoding $\mathcal{Y} = \{ab, ac\}$

reduction: *quasi*-reduced, only merge duplicate (i.e., isomorphic) nodes; *fully*-reduced, merge duplicate nodes and skip redundant (i.e., with identical children) nodes; or *sparsely*-reduced, merge duplicate nodes and omit nodes not reaching the **1**-terminal, and any edge pointing to them (Fig.1).

Decision diagrams excel at encoding sets that share many subsets, and their recursive structure enables effective use of dynamic programming through an *operation cache*, which virtually eliminates the need to recompute subproblems.

2.3 Notation

Given alphabet $\Sigma = \{s_1, \cdots, s_m\}$, with $m \in \mathbb{N}$, let Σ^* be the set of strings over Σ, i.e., $\Sigma^* = \{a_1 \cdots a_k : k \geq 0, \forall h, 1 \leq h \leq k, a_h \in \Sigma\}$. We introduce the following notation to discuss SeqDDs encoding a finite language $\mathcal{Y} \subset \Sigma^*$:

- If $\mathcal{Y} = \emptyset$, then $height(\mathcal{Y}) = \perp$, "undefined". Otherwise, the height of $\mathcal{Y}$ is the length of the longest string in it, $height(\mathcal{Y}) = \max\{|\sigma| : \sigma \in \mathcal{Y}\}$.
- $lengths(\mathcal{Y}) = \{k \in \mathbb{N} : \exists \sigma \in \mathcal{Y}, |\sigma| = k\}$, the set of all string lengths in $\mathcal{Y}$.
- For $k \in lengths(\mathcal{Y})$, $\mathcal{Y}_k = \{\sigma \in \mathcal{Y} : |\sigma| = k\}$, the strings of length k in $\mathcal{Y}$, and $\mathcal{Y}_{<k} = \{\sigma \in \mathcal{Y} : |\sigma| < k\}$, the strings of length less than k in $\mathcal{Y}$.
- For $a \in \Sigma$, $\mathcal{Y}/a = \{\sigma \in \Sigma^* : a \cdot \sigma \in \mathcal{Y}\}$, the strings that, preceded by a, form a string in $\mathcal{Y}$.
- For $k \in lengths(\mathcal{Y})$ and $a \in \Sigma$, $\mathcal{Y}_k/a = \{\sigma \in \Sigma^{k-1} : a \cdot \sigma \in \mathcal{Y}_k\}$, the strings that, preceded by a, form a string of length k in $\mathcal{Y}$.
- $||\mathcal{Y}|| = \sum_{\sigma \in \mathcal{Y}} |\sigma|$, the total number of symbols in $\mathcal{Y}$, not to be confused with $|\mathcal{Y}|$, the number of strings in $\mathcal{Y}$.

3 Definition of Sequence Decision Diagrams

We now define a class of decision diagrams to encode any finite subset of Σ^*.

Definition 1. A *sequence decision diagram* (SeqDD) is a directed acyclic finite graph with two *terminal* nodes, **0** and **1**, and such that each *nonterminal* node p has $m + 1$ outgoing edges, each labeled with a different element from $\Sigma \cup \{\epsilon\}$; we write $p[a] = q$ to indicate that the outgoing edge labeled with $a \in \Sigma \cup \{\epsilon\}$ points to node q, which can be a terminal or nonterminal node. □

Definition 2. The set of strings $\mathcal{X}(p)$ encoded by a SeqDD node p is:

$$\mathcal{X}(p) = \begin{cases} \emptyset, \text{ the empty set} & \text{if } p = \mathbf{0}, \\ \{\epsilon\}, \text{ the set containing only the empty string} & \text{if } p = \mathbf{1}, \\ \bigcup_{a \in \Sigma \cup \{\epsilon\}} \{a \cdot \sigma : \sigma \in \mathcal{X}(p[a])\} & \text{otherwise.} \end{cases}$$

□

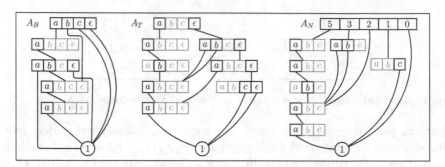

Fig. 2. A SeqDD$_B$, a SeqDD$_T$, and a SeqDD$_N$ encoding $\mathcal{Y} = \{aa, aaa, aabaa, baa, c, \epsilon\}$. Indices in gray point to terminal **0** (not represented for clarity).

Theorem 1. Given a finite set of strings $\mathcal{Y} \subset \Sigma^*$, there exists a SeqDD with a root (i.e., a node with no incoming edges) p satisfying $\mathcal{X}(p) = \mathcal{Y}$.
Proof. The proof is trivial and left to the reader.   □

As defined, SeqDDs are general non-canonical encoding of finite languages. Any set $\mathcal{Y} \subset \Sigma^*$ can be encoded by infinitely many SeqDDs because, if a node r encodes $\mathcal{Y}$, any node r' with $r'[a] = \mathbf{0}$ for each $a \in \Sigma$ and $r'[\epsilon] = r$ also encodes $\mathcal{Y}$, and the "insertion" of such "useless nodes" can be repeated at will (indeed, not just above the root, but anywhere along any path in the SeqDD). Thus, we now describe possible sets of restrictions to ensure canonicity. In any case:

- No *duplicate nodes* are allowed: the SeqDD cannot contain two nonterminal nodes p and q such that $p[a] = q[a]$ for every $a \in \Sigma \cup \{\epsilon\}$.
- No *empty nodes* are allowed: the SeqDD cannot contain a nonterminal node p such that $p[a] = \mathbf{0}$ for every $a \in \Sigma \cup \{\epsilon\}$.
- No ϵ-*nodes* are allowed: the SeqDD cannot contain a nonterminal node p such that $p[a] = \mathbf{0}$ iff $a \in \Sigma$.

Then, informally, canonicity is achieved by additionally "pushing" ϵ-edges (not pointing to **0**) toward the bottom, or toward the top, of the diagram (Fig. 2).

3.1   Definition of Canonical SeqDDs with ϵ at the Bottom

Definition 3. A SeqDD$_B$ is a SeqDD with no duplicate, empty, or ϵ-nodes where, for any nonterminal node p, either $p[\epsilon] = \mathbf{0}$ or $p[\epsilon] = \mathbf{1}$.   □

Theorem 2. Given a finite set of strings $\mathcal{Y} \subset \Sigma^*$, there exists a unique single-root SeqDD$_B$ whose root p satisfies $\mathcal{X}(p) = \mathcal{Y}$.
Proof. If $height(\mathcal{Y}) = \perp$, then $\mathcal{Y} = \emptyset$, and the canonicity restrictions imply that $p = \mathbf{0}$ is the only SeqDD$_B$ node encoding $\mathcal{Y}$. If $height(\mathcal{Y}) = 0$, then $\mathcal{Y} = \{\epsilon\}$, and the same restrictions imply that $p = \mathbf{1}$ is the only SeqDD$_B$ node encoding $\mathcal{Y}$. If $height(\mathcal{Y}) = k > 0$, assume the theorem holds for any $\mathcal{Y}'$ with $height(\mathcal{Y}') < k$. Clearly, $height(\mathcal{Y}/a) < k$ and, if $\epsilon \in \mathcal{Y}$, then $\mathcal{Y} = \{\epsilon\} \cup \bigcup_{a \in \Sigma} a \cdot \mathcal{Y}/a$, otherwise $\mathcal{Y} = \bigcup_{a \in \Sigma} a \cdot \mathcal{Y}/a$. Then, if $\epsilon \in \mathcal{Y}$, we can define node p, with $p[\epsilon] = \mathbf{1}$ and,

for each $a \in \Sigma$, $p[a] = q_a$, where q_a is the unique node encoding $\mathcal{Y}/a$ (by induction, q_a exist since $height(\mathcal{Y}/a) < k$). Note that we might have $\mathcal{Y}/a = \mathcal{Y}/b$ for $a \neq b$, this simply means that the two corresponding edges in p point to the same SeqDD$_B$ node (indeed nodes are shared across any of the descendants of p, to avoid duplicates). No other node q encoding $\mathcal{Y}$ can exist because it would have to differ from p in at least one index $a \in \Sigma$, while we must have $p[\epsilon] = q[\epsilon] = \mathbf{1}$. By inductive assumption, SeqDD$_B$'s $p[a]$ and $q[a]$ cannot encode the same set, that is, $\mathcal{X}(p[a]) = \mathcal{Y}/a \neq \mathcal{X}(q[a])$, thus there is a string $a \cdot \sigma'$ in $\mathcal{X}(p)$ and not in $\mathcal{X}(q)$, or vice versa. The case where $\epsilon \notin \mathcal{Y}$ is analogous, except that $p[\epsilon] = \mathbf{0}$. $\square$

3.2   Definition of Canonical SeqDDs with ϵ at the Top

For the alternative definition where we allow "ϵ at the top", it is easier to recast the definition of quasi-reduced MDDs [5] as a special case of SeqDDs.

Definition 4. A k-level MDD is the terminal node $\mathbf{1}$, if $k = 0$, or, if $k > 0$, it is a single-root SeqDD without duplicate, empty, or ϵ-nodes where the root p is such that $p[\epsilon] = \mathbf{0}$ and, for $a \in \Sigma$, $p[a]$ is a $(k-1)$-level MDD or $\mathbf{0}$. $\square$

Thus, the root of a k-level MDD encodes a nonempty set of strings of length k.

Definition 5. A k-level SeqDD$_T$ is a SeqDD without duplicate, empty, or ϵ-nodes whose root node p is such that, for $a \in \Sigma$, $p[a]$ is $\mathbf{0}$ or the root of a $(k-1)$-level MDD, while $p[\epsilon]$ is $\mathbf{0}$ or the root of an h-level SeqDD$_T$, $h < k$. $\square$

Thus, it is easy to prove by induction that the root p of a k-level SeqDD$_T$ encodes a nonempty set of strings of length k, $\bigcup_{a \in \Sigma} \mathcal{X}(q[a])$, plus a possibly empty set of strings of length less than k, $\mathcal{X}(q[\epsilon])$.

Theorem 3. Given a finite language $\mathcal{Y} \subset \Sigma^*$, there exists a unique single-root SeqDD$_T$ with root p such that $\mathcal{X}(p) = \mathcal{Y}$.
Proof. If $height(\mathcal{Y}) = \perp$, then $\mathcal{Y} = \emptyset$, and the canonicity restrictions imply that $p = \mathbf{0}$ is the only SeqDD$_T$ encoding $\mathcal{Y}$. If $height(\mathcal{Y}) = 0$, then $\mathcal{Y} = \{\epsilon\}$, and the same restrictions imply that $p = \mathbf{1}$ is the only SeqDD$_T$ encoding $\mathcal{Y}$. If instead $height(\mathcal{Y}) = k > 0$, assume that the theorem holds for any set $\mathcal{Y}'$ with $height(\mathcal{Y}') < k$. Since $\mathcal{Y} = \mathcal{Y}_{<k} \cup \bigcup_{a \in \Sigma} a \cdot \mathcal{Y}_k/a$, we can define node p such that, for $a \in \Sigma$, $p[a] = q_a$ with $\mathcal{X}(q_a) = \mathcal{Y}_k/a$, while $p[\epsilon] = q_\epsilon$ with $\mathcal{X}(q_\epsilon) = \mathcal{Y}_{<k}$. By inductive hypothesis, nodes q_a and q_ϵ are unique, as they all encode sets of height less than k and, since $\mathcal{Y}_k/a$ contains only strings of length $k - 1$, q_a is in particular the root of an MDD, i.e., $q_a[\epsilon] = \mathbf{0}$. Then, node p is also the only node encoding $\mathcal{Y}$ since any other node p' would have to differ from p in at least one child. If $p[\epsilon] \neq p'[\epsilon]$, there must exists a string σ of length less than k in $\mathcal{X}(p[\epsilon])$, thus $\mathcal{X}(p)$, and not in $\mathcal{X}(p'[\epsilon])$, thus $\mathcal{X}(p')$, or vice versa. If there is an $a \in \Sigma$ with $p[a] \neq p'[a]$, there must exists a string σ in $\mathcal{X}(p[a])$ and not in $\mathcal{X}(p'[a])$, so that $a \cdot \sigma$ is in $\mathcal{X}(p)$ and not in $\mathcal{X}(p')$, or vice versa ($a \cdot \sigma$ cannot possibly be in $\mathcal{X}(p'[\epsilon])$ as it is of length k). Either way, p' cannot encode the same set as p. $\square$

A SeqDD$_T$ relies on some concept of level for the nodes of the decision diagram. More specifically, a SeqDD$_T$ node encodes all the maximum-length strings in its children corresponding to elements of Σ and delegates the encoding of the shorter strings to its ϵ-child. A similar encoding for set $\mathcal{Y}$ partitions its strings according to their length, and uses a top node to make a decision based on the length of the string σ being searched, not on the first symbol of σ (Fig. 2). This leads us to a third, different in spirit but essentially equivalent, definition.

Definition 6. A SeqDD$_N$ is a set of "sparse" root nodes, each root r having a finite set $\mathcal{R}$ of outgoing edges labeled with different elements $k \in \mathbb{N}$, such that $r[k]$ points to a k-level MDD. The set encoded by r is $\bigcup_{k \in \mathcal{R}} \mathcal{X}(r[k])$.   □

4   Compactness of Canonical SeqDD Definitions

We now discuss the size of our SeqDDs, where the size of a SeqDD A is the number of edges it contains, $edges(A)$, rather than the number of nodes. Given the structural differences between a SeqDD$_B$ and a SeqDD$_T$, we compare them by thinking of them as finite automata. A closer look at a SeqDD$_B$ shows that it can easily be converted into a DFA (Theorem 4). On the other hand, a SeqDD$_T$ can be converted into a restricted type of NFA.

4.1   DFA Representation of SeqDD$_B$

Given a SeqDD$_B$ A_B encoding a finite language $\mathcal{Y} \subset \Sigma^*$, we can build an equivalent DFA $M = (Q, \Sigma, \delta, q_0, F)$. If $A_B = \mathbf{0}$ then $M = (\{q_0\}, \Sigma, \delta, q_0, \emptyset)$. Otherwise, we first define the states Q in terms of the nodes in A_B: every nonterminal node q in A_B corresponds to a state $q \in Q$, while node $\mathbf{1}$ in A_B corresponds to new state $f \in Q$ and node $\mathbf{0}$ corresponds to a new trap state $t \in Q$.

The initial state q_0 corresponds to A_B's root while the transition function $\delta : Q \times \Sigma \to Q$ is such that, for every $a \in \Sigma$ and edge $q[a] = p$ in A_B, there is a corresponding transition $\delta(q, a) = p$ and, if $q[\epsilon] = \mathbf{1}$, no transition is added, but q is added to the accepting states F. Lastly, state f is also added to F.

Theorem 4. Given a SeqDD$_B$ A_B encoding a finite language $\mathcal{Y} \subset \Sigma^*$, building an equivalent minimized DFA M requires linear time in the size of A_B.

Proof. The proof is direct from the translation algorithm above.   □

For memory efficiency, decision diagrams can be stored in a sparse form. In the case of a sparse SeqDD$_B$, this corresponds to a *partial* DFA, and the translation is analogous to the non-sparse version just discussed. From now on, we consider sparse representations for all canonical forms of SeqDD and for partial DFAs.

4.2   NFA Representation of SeqDD$_T$

To discuss the translation of a SeqDD$_T$ into an equivalent NFA, we first define RNFAs, a restricted version of NFAs, keeping in mind that our goal is to facilitate size comparisons between a SeqDD$_B$ and a SeqDD$_T$. To that end, our RNFA definition resembles the structure of SeqDD$_T$ while respecting the key characteristics of ordinary NFAs when encoding a finite language.

Definition 7. A restricted NFA (RNFA) is an acyclic NFA $N = (Q, \Sigma, \delta, Q_I, Q_F)$, where both Q_I and Q_F are singletons sets and, for each state $q \in Q$, the following condition holds: at most one outgoing ϵ-transition is allowed, and if $k = \max(lengths(L(q)))$ then all strings in $\bigcup_{a \in \Sigma} L(\delta(q, a))$ have length equal $k - 1$ and all strings in $L(\delta(q, \epsilon))$ have length at most $k - 1$. This value k is called the *level of* q. □

A *minimized* RNFA enforces the following restriction rules.

- No *duplicate states* are allowed: An RNFA cannot contains q and p such that $L(q) = L(p)$.
- No *empty states* are allowed: An RNFA cannot contain a state $q \in Q \setminus Q_I$ such that $L(q) = \emptyset$.
- No ϵ-*states* are allowed: An RNFA cannot contain a state $q \in Q \setminus Q_F$ such that $L(q) = \{\epsilon\}$.

Any RNFA can be converted to an equivalent minimized RNFA by adapting the bucket-sort based OBDD reduction algorithm proposed in [10]. The minimized RNFA for a given language is unique, the proof is omitted due to lack of space.

The following lemma affirms that RNFAs, like DFAs, can recognize any finite language (unlike DFAs, they obviously cannot accept any infinite language).

Lemma 1. *If* $\mathcal{Y} \subset \Sigma^*$ *is a finite language, there exists an RNFA* N *to accept* $\mathcal{Y}$.
Proof. The proof of existence is analogous to the one of Theorem 3. □

If SeqDD$_T$ A_T with a single root node r encodes a finite language $\mathcal{Y} \subset \Sigma^*$, the equivalent RNFA $T = (Q, \Sigma, \delta, Q_I, Q_F)$ is built as follows. Each nonterminal node q of A_T corresponds to a state $q \in Q$; terminal node **1** of A_T corresponds to a new state $\mathbf{1} \in Q$, and $F = \{\mathbf{1}\}$; finally, $Q_I = \{r\}$ (note that, if $r = \mathbf{0}$, we also must add r to Q). The transition function $\delta : Q \times \Sigma \cup \{\epsilon\} \to Q$ is such that, for every edge $q[a] = p$ in A_T with $a \in \Sigma \cup \{\epsilon\}$, there is a corresponding transition $\delta(q, a) = p$. Thus, in particular, if $r = \mathbf{0}$, then $T = (\{\mathbf{0}\}, \Sigma, \emptyset, \{\mathbf{0}\}, \{\mathbf{1}\})$, and the encoded language is $\mathcal{Y} = \emptyset$, while, if $A_T = \mathbf{1}$, then $T = (\{\mathbf{1}\}, \Sigma, \emptyset, \{\mathbf{1}\}, \{\mathbf{1}\})$ and the encoded language is $\mathcal{Y} = \{\epsilon\}$.

From the conversion process, it is easy to conclude that a canonical SeqDD size is bounded by the size of the corresponding FA in terms of number of transitions, plus the number of accepting states.

4.3   SeqDD Compactness Comparison by Means of Finite Automata

To study the relative compactness of canonical SeqDDs, we first discussed bounds on the number of states for equivalent DFAs and RNFAs; these are trivially reflected in similar bounds for SeqDD$_B$'s and SeqDD$_T$'s. To obtain bounds on the number of transitions, one could just multiply the state bounds by the alphabet size, but we are really interested in the actual number of edges for equivalent SeqDDs, thus partial FAs. This section shows that bounds similar to those for states hold also for edges.

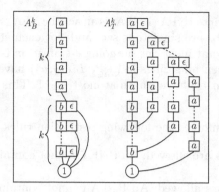

Fig. 3. Example of quadratic growth when translating SeqDD$_B$ into SeqDD$_T$

Theorem 5. Given a DFA $M = (Q, \Sigma, \delta_D, q_0, F)$ with n states encoding a finite language $\mathcal{Y} \subset \Sigma^*$, an equivalent minimized RNFA N has $O(n^2)$ states.
Proof. For each state $q \in Q$ and $k = 0, \ldots, height(\mathcal{Y})$, let $L(q, k) = L(q) \cap \Sigma^k$. Then, we build an equivalent RNFA N with states organized by level:

- Level 0 of the RNFA contains a single accepting state f.
- Level k contains a state $\langle q,k \rangle$ for each nonempty $L(q, k)$.
- The initial state of N is $\langle q_0, \max lengths(\mathcal{Y}) \rangle$.
- The transition function δ_N of N satisfies
  - For each state $\langle q,k \rangle$ with $k > 0$ in N and for each $a \in \Sigma$: $\langle p,k - 1 \rangle \in \delta_N(\langle q,k \rangle, a)$ iff $\delta_D(q, a) = p$.
  - For each state $\langle q,k \rangle$ in N, let h be the largest integer less than k such that state $\langle q,h \rangle$ exists in N; if such state exists, then $\langle q,h \rangle \in \delta_N(\langle q,k \rangle, \epsilon)$.

Note that the resulting RNFA might not be minimized, in the sense that it is possible that $\langle q,k \rangle$ and $\langle p,k \rangle$ encode the same language, in which case they should be merged. In any case, however, the number of states of the RNFA is at most equal to the number of states of the DFA times the maximum length of a string in $\mathcal{Y}$, which, again, is at most equal to the number of states. Thus the number of RNFA states is at most quadratic the number of DFA states. As the two automata obviously accept the same language $\mathcal{Y}$, the proof is complete. □

To show that the growth of of Theorem 5 is indeed possible, consider the family of languages $\mathcal{G} = \{\mathcal{G}_k : k \in \mathbb{N}\}$ over $\{a, b\}$. Let $\mathcal{G}_k = \{a^k b^k, a^k b^{k-1}, \cdots, a^k b, a^k\}$, so that $||\mathcal{G}_k|| = 3(k+1)k/2$. Then, the SeqDD$_T$ A_T^k encoding $\mathcal{G}_k$ contains $k^2 + 3k$ edges, while the SeqDD$_B$ A_B^k encoding $\mathcal{G}_k$ contains $3k$ edges (see Fig. 3).

Theorem 6. Given a minimized RNFA N with n states encoding a finite language $\mathcal{Y} \subset \Sigma^*$, an equivalent minimized DFA has at most $O(2^n)$ states.
Proof. The proof is immediate given the well known fact that an NFA-to-DFA conversion may result in an exponential increase in the number of states. □

Since RNFAs are a restricted form of NFAs, however, one may wonder whether an exponential growth can actually occur. To show that this is the case, consider the

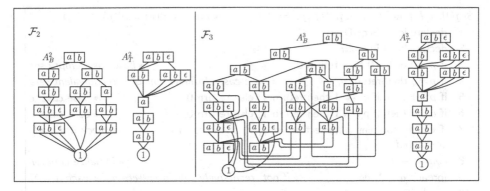

Fig. 4. Example of exponential growth when translating SeqDD$_T$ into SeqDD$_B$

family of languages $\{\mathcal{F}_k : k \in \mathbb{N}\}$ with $\mathcal{F}_k = \{xay : x, y \in \{a, b\}^*, |x| \le k, |y| = k\}$. Then, the SeqDD$_T$ A_T^k encoding $\mathcal{G}_k$ contains $7k - 1$ edges while the SeqDD$_B$ A_B^k encoding $\mathcal{G}_k$ contains $\Omega(2^k)$ edges (see Fig. 4). This is similar to the well-known construction that demonstrates the proof of Theorem 6.

5   Manipulation Algorithms for SeqDDs

We now consider two types of algorithms: *set manipulation algorithms* and *substring manipulation algorithms*. Those of the first type take two or more canonical SeqDDs with the same canonicity rule and perform set operations such as *union* or *intersection*. Those of the second type input a canonical SeqDD and a string, and select strings satisfying a criterion for matching a substring, changing a substring into another, or shorten or lengthen a string.

As with all decision diagram algorithms, we adopt a recursive style. SeqDD nodes are stored in a *unique table* to ensure canonicity. An *operation cache* ensures efficiency by virtually eliminating repeated computations. Each of the following *set manipulation algorithms* has been developed for SeqDD$_B$ and SeqDD$_N$ representations: union, intersection, set difference, symmetric set difference, and concatenation. For instance, the *Intersection* algorithm for two SeqDD$_B$'s traverses them top-down and builds the resulting SeqDD$_B$ bottom-up (see the pseudo-code in Fig. 5). SeqDD$_N$ set manipulation algorithms can be considered as shared MDD algorithms, since a SeqDD$_N$ is organized by the length of the strings encoded.

Various string manipulations can be performed. For example, the classical membership problem can be solved by a single trace, no longer than the *query size* + 1, starting from the root and ending in either terminal **1** or **0**. Set manipulation algorithms can also become handy in performing string manipulations; for instance, the membership problem is solved by a set intersection, and string replacement can be solved using a combination of set difference, intersection, and union. However, if we want to perform substring manipulations, the use of set manipulation algorithms becomes inefficient, hence we developed specific substring manipulation algorithms.

```
SeqDD_B  Intersection(SeqDD_B  p,  SeqDD_B  q)   • returns X(r) = X(p) ∩ X(q)
  1   declare local SeqDD_B r;
  2   declare local int count;
  3   if p=0 or q=0 then return 0;                           • base case: empty set
  4   if p=q then return p;          • base case: Intersection of two equivalent sets
  5   if p=1 then if q[ε]=1 then return 1; else return 0;        • base case: ε
  6   if q=1 then if p[ε]=1 then return 1; else return 0;        • base case: ε
  7   if Cache contains ⟨Intersection,{p,q}:r⟩ then return r;    • check if already
      computed
  8   count ← 0;                                          • initialize counter
  9   foreach a ∈ Σ do           • if not, recursively call Intersection for each a ∈ Σ
 10      r[a] ← Intersection(p[a],q[a]);
 11      if r[a]=0 then count ← count + 1;      • count edges pointing to terminal-0
 12   if count=|Σ| then r ← 0;             • potential empty-node or ε-node
 13   if p[ε]=1 and q[ε]=1 then                        • deal with ε case
 14      if r=0 or r=1 then r ← 1;
 15      else r[ε] ← 1;
 16   UniqueTableInsert(r);           • insert to unique table to ensure canonicity
 17   Cache ← ⟨Intersection,{p,q}:r⟩;  • record result in cache to avoid recomputation
 18   return r;
```

Fig. 5. SeqDD_B *Intersection* operation

The main advantage of using SeqDDs for substring manipulation lies in the ability to search or modify a set of strings at once, thanks to node sharing and *memoization*. For example, in a SeqDD_B, replacing the first occurrence of a substring t with t' is done once for all strings sharing a prefix that contains t. Moreover, a shared suffix is processed the first time we explore it; for other strings sharing that suffix the algorithm simply checks the *operation cache* for the result. A universal algorithm *replace* can replace, insert, or delete a specific substring: replacing ϵ by a string $t \neq \epsilon$ performs an insertion, while replacing t by ϵ performs a deletion. Of course, this can be refined by additionally providing to the algorithm specific substrings that must be found before and after the replacement location.

6   Applications of Sequence Decision Diagrams

Advancements in genome sequencing techniques along with their affordability have resulted in an increasing number of sequenced genomes. As a consequence, a concise representation that allows for efficient data manipulation is required to query, analyze, and retrieve this information. These processes are essential in various molecular biology problems.

SeqDD_B and SeqDD_N provide simple indexing data structures. Their compactness in regards to sequence indexing is summarized in Table 1. Given a string w of size x, it is well known that the size of a DAWG that encodes the

Table 1. Summary of the upper bound size of a SeqDD$_B$ or SeqDD$_N$ encodingthe set of all prefixes, suffixes, or subwords of a certain string of size x

Encoded set	DAWG size	SeqDD$_B$ size	SeqDD$_N$ size
Suffixes	$3x - 4$	$3x - 4$	$2x + 1$
Subwords	$3x - 4$	$3x - 4$	$(5x^2 + 3x + 6)/4$
Prefixes	x	x	$x^2 + 1$

set of suffixes / subwords of w is at most $3x - 4$ transitions, for $x > 2$ [2]. The size of a SeqDD$_B$ encoding w's suffixes (subwords) is bounded by $4x - 3$ ($5x - 6$) transitions. Technically, while a SeqDD$_B$ ϵ-transitions are shown in the figures as edges, in reality they can be encoded by a single bit, since an ϵ-transition can only point to the terminal state. Thus, the size of a SeqDD$_B$ is actually bounded by $3x - 4$ transition plus $x + 1$ bits when encoding the set of suffixes or $2x - 2$ bits when encoding the set of subwords given that all states are accepting. On the other hand, the size of a SeqDD$_N$ encoding subwords of w is bounded by $2x + \sum_{j=1}^{x} j + 3/2 \sum_{j=2}^{x-2} j$, which simplifies to $(5x^2 + 3x + 6)/4$ transitions.

Using SeqDD$_B$ or SeqDD$_N$ for indexing sequences allows for efficient manipulations. For instance, the membership problem requires time linear in the size of the query when handled one sequence at a time. Querying a large set of sequences at once could lead to substantial improvement in time complexity because decision diagrams exploit node sharing and *memoization*, if we build a SeqDD that encodes the query set and perform a simple intersection.

The longest common substring can be retrieved by intersecting the SeqDDs encoding the set of subwords of each sequence. Using SeqDD$_N$'s allows early pruning, but consumes space. To achieve better space efficiency, SeqDDs encoding the set of suffixes can be used along with a non-commutative variation of the intersection algorithm in Fig. 5, so that, when $p = 1$, the algorithm returns q. In this case, the longest common substring for more than two sequences is solved incrementally, thus SeqDD$_N$'s lose the advantages of early pruning. Note that both SeqDD intersection and its variation have time complexity proportional to the size of the smallest argument. A generalization of this problem is the DNA contamination problem.

The all-pairs suffix-prefix matching problem can be solved with multi-terminal SeqDD, a simple tweak to our original definition. Let $\mathcal{G} = \{s_1, s_2, \cdots, s_k\}$ be a set of strings, all pairs with matching prefix-suffix can be obtained by performing a prefix intersection between $\mathcal{Q}$ and p, where $\mathcal{Q}$ is a shared SeqDD with k handles, each pointing to a SeqDD q_i encoding the set of suffixes of s_i and p is a multi-terminal SeqDD encoding $\mathcal{G}$ with $k + 1$ terminal nodes corresponding to the 0-terminal and the k strings.

7   Conclusion

We introduced SeqDDs, multi-valued sequence decision diagrams, which can be seen as MDDs with no variable ordering but are nevertheless canonical. In fact, our SeqDDs do not have a notion of variables, hence any "size explosion" exclusively depends on the specific set to be encoded and on the canonization rule (we introduce two possibilities, $SeqDD_B$ and $SeqDD_T$). More importantly, SeqDDs are ideal for encoding finite sets of strings of arbitrary finite (but possibly different) lengths, that is, finite languages. $SeqDD_T$'s are analogous to shared MDDs, and may be best implemented by adding special nodes at the top level that makes a choice based on the string length; we call this version $SeqDD_N$. We study the compactness of our representations in terms of finite automata and show that there is no winner between the two versions: a $SeqDD_T/SeqDD_N$ can be quadratically larger than a $SeqDD_B$ for certain languages, but exponentially more compact for others; therefore, we are implementing algorithms for both versions. SeqDDs are useful for applications requiring compact storage and efficient manipulation of large sets of strings with high sharing rate. As future work, an edge-valued variation is a must for many applications, such as symbolic generation of probabilistic witnesses in CSL model checking.

References

1. Aoki, H., Yamashita, S., Minato, S.: An efficient algorithm for constructing a sequence binary decision diagram representing a set of reversed sequences. In: 2011 IEEE International Conference on Granular Computing (GrC), pp. 54–59 (2011)
2. Blumer, A., Blumer, J., Ehrenfeucht, A., Haussler, D., McConnell, R.: Building the minimal DFA for the set of all subwords of a word on-line in linear time. In: Paredaens, J. (ed.) ICALP 1984. LNCS, vol. 172, pp. 109–118. Springer, Heidelberg (1984)
3. Bollig, B., Wegener, I.: Improving the variable ordering of OBDDs is NP-complete. IEEE Trans. Comput., 993–1002 (1996)
4. Bryant, R.E.: Graph-based algorithms for boolean function manipulation. IEEE Trans. Comput., 677–691 (1986)
5. Ciardo, G., Lüttgen, G., Siminiceanu, R.I.: Saturation: An efficient iteration strategy for symbolic state space generation. In: Margaria, T., Yi, W. (eds.) TACAS 2001. LNCS, vol. 2031, pp. 328–342. Springer, Heidelberg (2001)
6. Kam, T., Villa, T., Brayton, R.K., Sangiovanni-Vincentelli, A.: Multi-valued decision diagrams: Theory and applications. Multiple-Valued Logic, 9–62 (1998)
7. Loekito, E., Bailey, J., Pei, J.: A binary decision diagram based approach for mining frequent subsequences. Knowledge and Information Systems, 235–268 (2010)
8. Minato, S.: Zero-suppressed BDDs for set manipulation in combinatorial problems. In: 30th Conference on Design Automation, pp. 272–277 (1993)
9. Requeno, J.I., Colom, J.M.: Compact representation of biological sequences using set decision diagrams. In: Rocha, M.P., Luscombe, N., Fdez-Riverola, F., Rodríguez, J.M.C. (eds.) 6th International Conference on PACBB. AISC, vol. 154, pp. 231–240. Springer, Heidelberg (2012)
10. Sieling, D., Wegener, I.: Reduction of OBDDs in linear time. Information Processing Letters, 139–144 (1993)

Shortest Unique Queries on Strings

Xiaocheng Hu[1], Jian Pei[2], and Yufei Tao[1]

[1] Chinese University of Hong Kong, New Territories, Hong Kong
[2] Simon Fraser University, Burnaby, Canada
{xchu,taoyf}@cse.cuhk.edu.hk, jpei@cs.sfu.ca

Abstract. Let D be a long input string of n characters (from an alphabet of size up to 2^w, where w is the number of bits in a machine word). Given a substring q of D, a *shortest unique query* returns a shortest unique substring of D that contains q. We present an optimal structure that consumes $O(n)$ space, can be built in $O(n)$ time, and answers a query in $O(1)$ time. We also extend our techniques to solve several variants of the problem optimally.

1 Introduction

Let D be a (long) string. Define $n = |D|$ where $|D|$ represents the length of D. Denote by $D[i]$ $(1 \le i \le n)$ the i-th character of D, and by $D[i:j]$ $(1 \le i \le j \le n)$ the substring of D starting at $D[i]$ and ending at $D[j]$. A string is *unique* if it has only one occurrence in D; otherwise, it is *repeating*. A substring $D[i_1:j_1]$ *contains* another $D[i_2:j_2]$ if $i_1 \le i_2$ and $j_1 \ge j_2$ hold at the same time.

In this paper, we study data structures on D that can efficiently answer the following query, which was recently proposed in [9], motivated by its fundamental nature in numerous applications in text retrieval and bioinformatics:

> **Shortest Unique Query:** Given a substring $q = D[x:y]$, such a query returns a substring of D with the minimum length among all the unique substrings of D containing q.

If $x = y$, we say that the query is a *point query*; otherwise, it is an *interval query*.

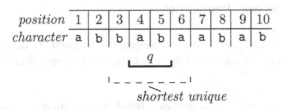

position	1	2	3	4	5	6	7	8	9	10
character	a	b	b	a	b	a	a	b	a	b

q

shortest unique

Fig. 1. An Example

Figure 1 shows a string D of length 10. Given $q = D[4:5] = \text{ab}$, a shortest unique query may return $D[3:6] = \text{baba}$ because its length 4 is the smallest among all the unique substrings containing q. To verify this, notice that (i) baba is unique because it

E. Moura and M. Crochemore (Eds.): SPIRE 2014, LNCS 8799, pp. 161–172, 2014.
© Springer International Publishing Switzerland 2014

has only one occurrence in D, whereas (ii) $D[3:5] = \text{bab}$ is repeating (it occurs also at $d[8:10]$), and so is $D[4:6] = \text{aba}$ (see $D[7:9]$). This implies no unique string of length at most 3 contains q. Note that, in general, a query result can be output with only 2 integers, which specify its starting and ending positions in D, respectively.

We make the standard assumption that each character of D fits in a machine word. If w is the number of bits in a word, this assumption implies that the alphabet where the characters of D are drawn can have a size up to 2^w. Unless otherwise stated, the default model of computation is RAM.

Existing Results. Previous research has focused exclusively on point queries. In their initial study [9], Pei et al. showed how to construct in $O(n^2)$ time an index of $O(n)$ size that answers a query in $O(1)$ time. Soon after that, Ileri et al. [6] and Tsuruta et al. [10] independently improved the construction time to $O(n)$. It is worth mentioning that $O(n)$ size is considered optimal in the sense that D itself requires $\Omega(n)$ words to store when the alphabet is large.

Our Results. We present the first study on interval queries. Our main result is a new structure of $O(n)$ space that can be built in $O(n)$ time, and answers a query in $O(1)$ time. In other words, we achieve the optimal efficiency as with the previous work, but on more general queries.

At this point, it seems fair to delve a bit into a crucial difference between designing a structure for point and interval queries. What makes point queries easy to handle is that *there are only n of them*! Therefore, the problem of indexing is more of a one-off computation problem: how to quickly compute the answers for all those n queries. Once this is done, one can simply store these answers in an array to allow constant query time. This idea, however, no longer works for interval queries because now we have $\Theta(n^2)$ of them. Therefore, there needs to be a major shift in the indexing strategy, calling for novel ideas.

The rest of the paper is organized as follows. In Section 2, we will clarify some basic facts relevant to this study. Then, Section 3 will present our structure for interval queries. Section 4 further demonstrates the usefulness of the proposed techniques by extending them (i) to answer queries with additional constraints, and (ii) to support interval queries in external memory optimally.

2   Basic Definitions and Properties

In this section, we pave the way for our subsequent discussion by defining several concepts related to minimal unique substrings and explaining some of their fundamental properties.

Definition 1. *Each integer $p \in [1, n]$ defines a* **left-fixed minimal unique substring** $MUS_{leftfix}(p)$ *as follows:*

- $MUS_{leftfix}(p) = nil$, *if $D[p:n]$ is repeating;*
- *otherwise, $MUS_{leftfix}(p) = D[p:z]$, where z is the smallest integer in $[p, n]$ such that $D[p:z]$ is unique.*

p	1	2	3	4	5	6	7	8	9	10
$MUS_{leftfix}(p)$	$D[1{:}3]$ =abb	$D[2{:}3]$ =bb	$D[3{:}6]$ =baba	$D[4{:}7]$ =abaa	$D[5{:}7]$ =baa	$D[6{:}7]$ =aa	$D[7{:}10]$ =abab	nil	nil	nil
$MUS_{rightfix}(p)$	nil	nil	$D[2{:}3]$ =bb	$D[2{:}4]$ =bba	$D[2{:}5]$ =bbab	$D[3{:}6]$ =baba	$D[6{:}7]$ =aa	$D[6{:}8]$ =aab	$D[6{:}9]$ =aaba	$D[7{:}10]$ =abab

Fig. 2. The left-fixed and right-fixed minimal unique substrings in Figure 1

In other words, $MUS_{leftfix}(p)$ is the shortest unique substring of D starting at $D[p]$. In the example of Figure 1, $MUS_{leftfix}(4)$, for instance, is $D[4:7]$ = abaa. Notice that $D[4:6]$ is repeating; and thus, $D[4:7]$ cannot be shortened on the right while still being unique. Viewed in another way, $D[4:7]$, $D[4:8]$ ..., $D[4:10]$ are all the unique substrings starting at $D[4]$; among them, $MUS_{leftfix}(4)$ is the shortest. See Figure 2 for the $MUS_{leftfix}(p)$ of all $p \in [1, 10]$.

The next definition is symmetric:

Definition 2. *Each integer $p \in [1, n]$ defines a* **right-fixed minimal unique substring** *$MUS_{rightfix}(p)$ as follows:*

- *$MUS_{rightfix}(p) = nil$, if $D[1:p]$ is repeating;*
- *otherwise, $MUS_{rightfix}(p) = D[z:p]$, where z is the largest integer in $[1,p]$ such that $D[z:p]$ is unique.*

The last row of Figure 2 shows the $MUS_{rightfix}(p)$ of all $p \in [1, 10]$ for our running example. Now we are ready to define the most important concept:

Definition 3. *A substring $D[i:j]$ is a* **minimal unique substring** *(MUS) if*

$$MUS_{leftfix}(i) = D[i:j] \text{ and } MUS_{rightfix}(j) = D[i:j].$$

In other words, $D[i:j]$ is an MUS if (i) it is unique, and (ii) it can be shortened on *neither* side while still being unique. We will use $\mathcal{M}$ to denote the set of MUS's in D. From Figure 2, one can verify easily that the $\mathcal{M}$ in our example is:

$$\mathcal{M} = \big\{ D[2:3] = \text{bb}, D[3:6] = \text{baba}, D[6:7] = \text{aa}, D[7:10] = \text{abab} \big\}. \quad (1)$$

$D[2:4]$ = bba, for example, is *not* an MUS because it can be shortened on the right into bb which is still unique.

Lemma 1. *The strings in $\mathcal{M}$ have distinct left endpoints, and distinct right endpoints.*

Proof. Suppose $D[i_1 : j_1]$ and $D[i_2 : j_2]$ are two different strings in $\mathcal{M}$ but $i_1 = i_2$. This means that they are both $MUS_{leftfix}(i_1)$. But only one string can be $MUS_{leftfix}(i_1)$, thus giving a contradiction. Similarly, it must hold that $j_1 \neq j_2$. □

It has been shown [10] that all the substrings defined earlier can be computed efficiently:

Lemma 2 ([10]). *All the left-fixed MUS's, right-fixed MUS's, and MUS's can be computed from D in $O(n)$ time.*

In general, a substring $D[i : j]$ requires only two integers to represent: integers i and j. Therefore, all the left-fixed MUS's, right-fixed MUS's, and MUS's can be stored in $O(n)$ words. This leads to the following useful fact:

Corollary 1. *In $O(n)$ time, we can compute a structure of $O(n)$ size that, given any substring $D[i : j]$, we can check whether it is unique in D in $O(1)$ time.*

Proof. Simply compute all the left-fixed MUS's using Lemma 2. Then, given a substring $D[i : j]$, declare that it is unique if and only if $j \geq z$, where z is such that $MUS_{leftfix}(i) = D[i : z]$. □

3   A Data Structure for Interval Queries

This section serves as a proof for our main result:

Theorem 1. *Given a data string of length n, we can pre-compute in $O(n)$ time an index structure that consumes $O(n)$ space, and answers any shortest unique query in $O(1)$ time.*

3.1   A 4-Candidate Lemma

Lemma 3. *The answer of the shortest unique query with substring $q = D[x : y]$ must be the shortest of the following 4 candidates:*

1. *$D[x : y]$ if it is unique*
2. *$MUS_{leftfix}(x)$*
3. *$MUS_{rightfix}(y)$*
4. *the shortest MUS containing q (breaking length ties arbitrarily). No such candidate exists if no MUS contains q.*

Proof. First of all, if $D[x : y]$ is unique, then clearly $D[x : y]$ is the answer because no string containing q can be any shorter. The following discussion focuses on the scenario where $D[x : y]$ is repeating.

Let $D[x' : y']$ be an answer to the query. If $x' = x$, then it must hold that $MUS_{leftfix}(x) = D[x' : y']$; otherwise, either $MUS_{leftfix}(x)$ or $D[x' : y']$ can be shortened on the right end while still being unique, which contradicts their definitions. Likewise, if $y' = y$, then $MUS_{rightfix}(y) = D[x' : y']$.

In the remaining scenario, $x' < x$ and $y' > y$. Suppose that $D[x' : y']$ was not an MUS, namely, it can be still be shortened either on the left or right while still being unique. However, as both $D[x' + 1 : y']$ and $D[x' : y' - 1]$ contain q, we have found a unique string containing q that is even shorter than $D[x' : y']$, which contradicts the definition of $D[x' : y']$. □

Whether Candidate 1—namely $D[x : y]$—is unique can be checked in constant time using an $O(n)$-space structure (see Corollary 1). Also, Candidates 2 and 3 can be obtained in constant time using an $O(n)$-space structure (see Lemma 2). It thus remains to give a structure for finding Candidate 4.

As before, let $\mathcal{M}$ be the set of MUS's of D. For each MUS $D[i:j]$ in $\mathcal{M}$, create an interval $[i, j]$. Denote by $\mathcal{I}$ the set of all the intervals created this way. For the example of Figure 1, we know from Equation 1 that

$$\mathcal{I} = \{[2,3], [3,6], [6,7], [7,10]\} \tag{2}$$

Lemma 4. *No two intervals in $\mathcal{I}$ can contain each other.*

Proof. Suppose, on the contrary, that $[i_1, j_1]$ and $[i_2, j_2]$ are two different intervals in I such that $[i_1, j_1]$ contains $[i_2, j_2]$. Recall that $D[i_1:j_1]$ and $D[i_2:j_2]$ are both MUS's of D. However, that $[i_1, j_1]$ contains $[i_2, j_2]$ indicates that we can shorten $D[i_1:j_1]$ to $D[i_2:j_2]$ which is still unique. This violates the definition of MUS. □

It is not hard to see that the problem ahead of us can be restated as:

Containment Min. Let $\mathcal{I}$ be a set of at most n intervals in the domain $[1, n]$ such that no two intervals contain each other (a requirement inherited from Lemma 4). Given an interval $[x, y]$ in the domain $[1, n]$, a *containment min query* returns the shortest one (breaking ties arbitrarily) among all the intervals in $\mathcal{I}$ containing $[x, y]$. We want to store $\mathcal{I}$ in a data structure to answer such queries efficiently.

3.2 The Proposed Structure

In this subsection, we will present a structure of $O(n)$ space that answers a containment min query in $O(1)$ time, which will complete our proof of Theorem 1.

Idea. Let $m = |\mathcal{I}|$. From now on, we will view $\mathcal{I}$ as an ordered set

$$\{I_1 = [i_1, j_1], I_2 = [i_2, j_2], ..., I_m = [i_m, j_m]\}$$

where $i_1 < i_2 < ... < i_m$, and therefore $j_1 < j_2 < ... < j_m$[1]. For any $a < b$, we say that I_a is on the *left* of I_b, and conversely, I_b is on the *right* of I_a. Given a subset $S \subseteq \mathcal{I}$, we say that it is a *consecutive subset* if $S = \{I_a, I_{a+1}, ...I_b\}$ for some a, b satisfying $1 \leq a \leq b \leq m$. We also regard the empty set $\emptyset$ as a consecutive subset.

For example, given the $\mathcal{I}$ in Equation 2, we have:

$$\{I_1 = [2,3], I_2 = [3,6], I_3 = [6,7], I_4 = [7,10]\}. \tag{3}$$

$\{I_3\}$ and $\{I_2, I_3, I_4\}$ are consecutive subsets, while $\{I_2, I_4\}$ is not. We observe:

Lemma 5. *For any $[x, y]$ in the domain $[1, n]$, the set of intervals of $\mathcal{I}$ containing $[x, y]$ must be a consecutive subset.*

Proof. Let a be the smallest integer such that $[i_a, j_a]$ contains $[x, y]$, and b be the largest integer such that $[i_b, j_b]$ contains $[x, y]$. For any integer $c \in [a, b]$, it holds that $i_c \leq i_b \leq x$ and $y \leq j_a \leq j_c$. In other words, $[i_c, j_c]$ contains $[x, y]$ as well. □

---

[1] Otherwise, there must be an interval containing another, which violates Lemma 4.

y	1	2	3	4	5	6	7	8	9	10
$\alpha(y)$	1	1	1	2	2	2	3	4	4	4

x	1	2	3	4	5	6	7	8	9	10
$\beta(x)$	nil	1	2	2	2	3	4	4	4	4

Fig. 3. Arrays α and β on the $\mathcal{I}$ in Equation 3

Algorithm 1. COMPUTING-α-ARRAY

Input: A set $\mathcal{I}$ of m intervals $I_1 = [i_1, j_1], ..., I_m = [i_m, j_m]$, sorted in ascending order of left point. The domain is $[1, n]$.

Output: Array α.

```
1  z ← 1
2  for y = 1 to n do
3      while j_z < y and z ≤ m do
4          z ← z + 1
5      if z ≤ m then
6          α(y) = z
7      else
8          α(y) = nil
9  return α
```

The above lemma motivates the following strategy for solving the containment min query. Given a query interval $[x, y]$, we will find the leftmost interval I_a in $\mathcal{I}$ containing $[x, y]$, and the rightmost interval I_b in $\mathcal{I}$ containing $[x, y]$. Then, the remaining task is to find the shortest interval among the consecutive subset $\{I_a, I_{a+1}, ..., I_b\}$, which is nothing but a standard *range min query* (RMQ)! We can index $\mathcal{I}$ using an RMQ structure [4,5] which uses $O(m) = O(n)$ space, can be constructed in $O(m)$ time, and answers an RMQ in $O(1)$ time.

Structure. It remains to explain how to design an index so that, given any $[x, y]$, we can derive the corresponding a and b in constant time. We resolve this issue with another key observation: a depends *only* on y! Formally, given a value $y \in [1, n]$, let us define $\alpha(y)$ as

- the smallest integer $z \in [1, m]$ such that $j_z \geq y$, if such a z exists;
- nil, otherwise.

In other words, I_z is the leftmost interval in $\mathcal{I}$ whose right endpoint is at least y. If such an interval exists, then $\alpha(y) = z$; otherwise, $\alpha(y) = nil$. The next lemma states the aforementioned observation formally:

Lemma 6. *Fix an integer $y \in [1, n]$. For any $x \in [1, y]$, all the following are true:*

1. *If $\alpha(y) = nil$, then $\mathcal{I}$ has no interval containing $[x, y]$.*
2. *If $I_{\alpha(y)}$ does not contain $[x, y]$, then $\mathcal{I}$ has no interval containing $[x, y]$.*
3. *If $I_{\alpha(y)}$ contains $[x, y]$, then it is the leftmost interval in $\mathcal{I}$ containing $[x, y]$.*

Proof. Statement 1 holds because when $\alpha(y) = nil$, all the intervals of $\mathcal{I}$ end strictly to the left of y.

Algorithm 2. CONTAINMENT-MIN

Input: A query interval $[x, y]$.
Output: The shortest interval in $\mathcal{I}$ containing $[x, y]$.
1   $a \leftarrow \alpha(y)$
2   $b \leftarrow \beta(x)$
3   **if** $a = nil$ or $b = nil$ **then**
4     $\lfloor$ **return** nil
5   **if** I_a *does not contain* $[x, y]$ **then**
6     $\lfloor$ **return** nil
7   perform an RMQ to retrieve the shortest interval among $I_a, I_{a+1}, ..., I_b$
8   **return** the above interval

To prove Statement 2, suppose on the contrary that there was an interval $[x_c, y_c]$ in $\mathcal{I}$ that contains $[x, y]$. It follows from the definition of $\alpha(y)$ that $c > \alpha(y)$. This means that $x_{\alpha(y)} < x_c \leq x$. On the other hand, from how $\alpha(y)$ is defined we know that $y_{\alpha(y)} \geq y$. Therefore, $[x_{\alpha(y)}, y_{\alpha(y)}]$ contains $[x, y]$, which contradicts the if-condition of the statement.

To prove Statement 3, suppose on the contrary that there was an interval $[x_c, y_c]$ in $\mathcal{I}$ containing $[x, y]$, and that this interval is on the left of $[x_{\alpha(y)}, y_{\alpha(y)}]$. Then, it follows that $y \leq y_c < y_{\alpha(y)}$, which contradicts the definition of $\alpha(y)$.   □

A similar observation holds on b—it depends only on x. Formally, given a value $x \in [1, n]$, define $\beta(x)$ as:

- the largest integer $z \in [1, m]$ such that $i_z \leq x$, if such a z exists;
- nil, otherwise.

In other words, $I_{\beta(x)}$ (if exists) is the rightmost interval in $\mathcal{I}$ whose left endpoint is at most x. Then, we have:

Lemma 7. *Fix an integer* $x \in [1, n]$. *For any* $y \in [x, n]$, *all the following are true:*

1. *If* $\beta(x) = nil$, *then* $\mathcal{I}$ *has no interval containing* $[x, y]$.
2. *If* $I_{\beta(x)}$ *does not contain* $[x, y]$, *then* $\mathcal{I}$ *has no interval containing* $[x, y]$.
3. *If* $I_{\beta(x)}$ *contains* $[x, y]$, *then it is the rightmost interval in* $\mathcal{I}$ *containing* $[x, y]$.

Proof. Symmetric to the proof of Lemma 6.   □

Figure 3 demonstrates all the $\alpha(y)$ and $\beta(x)$ values for the $\mathcal{I}$ of our running example shown in Equation 3. Using the two arrays, we can figure out in $O(1)$ time the values of a and b for any $[x, y]$ (recall that I_a and I_b are the leftmost and rightmost intervals of $\mathcal{I}$ containing $[x, y]$, respectively) using the previous two lemmas. Consider, for example, $x = 4$ and $y = 5$. Probing the α array gives us $\alpha(y) = 2$. Since $I_2 = [3, 6]$ contains $[x, y]$, we conclude from Lemma 6 that $a = 2$. Probing the β array gives us $\beta(x) = 2$. We thus conclude from Lemma 7 that $b = 2$.

Arrays α and β are all we need to complete our structure. Their space consumption is clearly $O(n)$. Furthermore, it is fundamental to compute them in $O(n)$ time. Algorithm 1 elaborates on the computation of α, whereas we omit the algorithm for β due to symmetry.

The above discussion results in our final query algorithm as shown in Algorithm 2. It is easy to see that the query time is $O(1)$.

4   Extensions

In this section, we discuss several extensional issues. First, in Sections 4.1 and 4.2, we will explain how to use the structure of Theorem 1 (without any modification) to answer two other useful queries, thus further demonstrating the power of our techniques. Then, Section 4.3 will present the I/O-efficient counterpart of Theorem 1.

4.1   Position Constrained Queries

In our current definition, the result of a shortest unique query can start and end anywhere in the data string D. Next, we formulate a variant where a query can specify the permissible ranges for the endpoints of its result:

Position Constrained Query. Such a query specifies (i) a substring $q = D[x : y]$, and (ii) two ranges $r_{start} = [s_1, s_2]$ and $r_{end} = [e_1, e_2]$ both in the domain $[1, n]$. It returns (if exists) a substring $D[i : j]$ with the minimum length such that
- $D[i : j]$ is unique
- $D[i : j]$ contains q
- $i \in [s_1, s_2]$ and $j \in [e_1, e_2]$.

Since $i \leq x$ and $j \geq y$ must always hold, it suffices to consider that $s_2 \leq x$ and $e_1 \geq y$.

For example, in Figure 1, consider a query with $q = D[4 : 5]$ (as shown) and $r_{start} = [3, 4]$ and $r_{end} = [8, 9]$. Then, $D[3 : 6]$ is no longer a legal answer because its right endpoint is not in r_{end}. Instead, the query should return $D[4 : 8] = \texttt{abaab}$.

Queries with $s_2 = x$ and $e_1 = y$. Let us first consider a special class of position constrained queries, where s_2 and e_1 always equal x and y, respectively. Interestingly, any query outside the class actually has the same result as a query inside the class, as explained later. Thus, solving this class of queries is the key.

Lemma 8. *Consider a position constrained query with $q = D[x : y]$, $r_{start} = [s, x]$, and $r_{end} = [y, e]$. Then:*

- *If $D[s : e]$ is repeating, the query has no result.*
- *Otherwise, the result is the shortest of the following 4 candidates:*
  1. *$D[x : y]$ if it is unique;*
  2. *$MUS_{leftfix}(x)$ if its right endpoint is in r_{end};*
  3. *$MUS_{rightfix}(y)$ if its left endpoint is in r_{start};*
  4. *The shortest MUS (breaking ties arbitrarily) that (i) contains q, (ii) has its left endpoint in r_{start}, and (iii) has its right endpoint in r_{end}. No such candidate exists if no MUS satisfies these conditions.*

Proof. The lemma's correctness follows from an argument almost identical to the one we used to prove Lemma 3.   □

With our experience with Lemma 3, it should be quite clear that we only need to clarify how to find Candidate 4, because all the other candidates and the necessary uniqueness checking can be done in $O(1)$ time under the $O(n)$ space budget. Furthermore, it is easy to see that the task of finding Candidate 4 boils down to the following problem:

Position Constrained Containment Min (PCCM). Let $\mathcal{I}$ be a set of $m \leq n$ intervals in the domain $[1, n]$ such that no two intervals contain each other (a requirement inherited from Lemma 4). Given intervals $[x, y]$, $[s, x]$, $[y, e]$ all in the domain $[1, n]$, a PCCM query returns the shortest interval in $\mathcal{I}$ (breaking ties arbitrarily) that (i) contains $[x, y]$, (ii) has its left endpoint in $[s, x]$, and (iii) has its right endpoint in $[y, e]$. We want to store $\mathcal{I}$ in a data structure to answer such queries efficiently.

The structure we need is exactly the one described in Section 3.2 for solving the containment min query, namely, the α and β arrays, and an RMQ index. A PCCM query is also answered by a single RMQ, which fetches the shortest interval in $\{I_a, I_{a+1}, ..., I_b\}$ for a pair of a and b carefully chosen as follows[2]:

$$a = \begin{cases} \alpha(y) & \text{if } s = 1 \text{ or } \beta(s-1) = nil \\ nil & \text{if } \beta(s-1) = m \\ \max\{\beta(s-1)+1, \alpha(y)\} & \text{otherwise} \end{cases}$$

$$b = \begin{cases} \beta(x) & \text{if } e = n \text{ or } \alpha(e+1) = nil \\ nil & \text{if } \alpha(e+1) = 1 \\ \min\{\alpha(e+1)-1, \beta(x)\} & \text{otherwise} \end{cases}$$

These values ensure that

- if $a = nil$ or I_a does not cover $[x, y]$, then the PCCM query has no answer;
- otherwise, $\{I_a, I_{a+1}, ..., I_b\}$ includes all and only the intervals of $\mathcal{I}$ containing $[x, y]$ whose left and right endpoints fall in $[s, x]$ and $[y, e]$, respectively.

The PCCM query algorithm is exactly the same as Algorithm 2 except that, at Lines 1 and 2, we should replace a and b with the ones given above.

General Queries. Now we consider position constrained queries with arbitrary $q = D[x : y]$, $r_{start} = [s_1, s_2]$, and $r_{end} = [e_1, e_2]$. As promised, each such query can be converted to one in the special class we have discussed:

Lemma 9. *To answer a positioned constrained query with $q = D[x : y]$, $r_{start} = [s_1, s_2]$, and $r_{end} = [e_1, e_2]$, we can simply return the result of the position constrained query with $q' = D[s_2 : e_1]$, $r'_{start} = [s_1, s_2]$, and $r'_{end} = [e_1, e_2]$.*

---

[2] We follow the convention that $\max\{v, nil\} = nil$ and $\min\{v, nil\} = nil$ for any integer v.

Proof. The lemma follows from the fact that the answer for the first query must contain $D[s_2 : e_1]$.   □

We thus conclude with:

Theorem 2. *Given a data string of length n, we can pre-compute in $O(n)$ time an index structure that consumes $O(n)$ space, and answers any position constrained query in $O(1)$ time.*   □

4.2 Find-All Queries

A shortest unique query may have more than one answer. For example, consider again $q = D[4 : 5] = \text{ab}$ in Figure 1. Besides $D[3 : 6]$, both $D[2 : 5] = \text{bbab}$ and $D[4 : 7] = \text{abaa}$ can be returned as a query result. Motivated by this, we define a new operation to retrieve all these possible results:

> **Find-All Query.** Given a substring $q = D[x : y]$, such a query returns *all* the substrings of D whose lengths are the minimum among the unique substrings of D containing q.

We will denote by k the number of substrings returned by a query (e.g., a find-all query with $q = D[4 : 5]$ returns $k = 3$ substrings). Next, we describe an algorithm that answers such a query in $O(k)$ time.

We achieve the purpose using position constrained queries. First, run a (normal) shortest unique query to get an answer string $D[i : j]$. Let $\ell = j - i + 1$ be the length of this string. The value i breaks the interval $[1, x]$ into two disjoint parts: $[1, i - 1]$ and $[i + 1, x]$. Now we can use two position constrained queries to find the next answers, if any. Due to symmetry, it suffices to explain how to do so for $[1, i - 1]$. We run a position constrained query with $q' = D[x : y]$, $r_{start} = [1, i - 1]$, and $r_{end} = [y, n]$. A crucial observation is that, if this query returns a string—say $D[i' : j']$—of length *greater* than ℓ, then we can assert that the original find-all query has no result substring that starts within $D[1 : i - 1]$. On the other hand, if $D[i' : j']$ indeed has length ℓ (note that its length cannot be shorter than ℓ), we have found another answer for the find-all query, after which we use i' to break $[1, i - 1]$ into even smaller intervals for recursion.

Algorithm 3 describes the above strategy in detail. To answer a find-all query, simply call FIND-ALL($D[x : y], [1, x], \ell$).

Lemma 10. *Our algorithm answers a find-all query in $O(k)$ time.*

Proof. Suppose that the j-th ($1 \leq j \leq k$) answer of the final-all query starts at position i_j, such that $1 \leq i_1 < i_2 < ... < i_k \leq x$. Clearly, these k positions break $[1, x]$ into at most $2k + 1$ disjoint parts: $[1, i_1 - 1], i_1, [i_1 + 1, i_2 - 1], ..., i_k, [i_k + 1, x]$. Our algorithm issues a position constrained query for each part. The query time then follows from Theorem 2.   □

Thus we have proved:

Theorem 3. *Given a data string of length n, we can pre-compute in $O(n)$ time an index structure that consumes $O(n)$ space, and answers any find-all query in $O(k)$ time, where k is the number of substrings reported.*

Algorithm 3. FIND-ALL $(D[x:y], [s_1, s_2], \ell)$

Input: $D[x:y]$ is a query substring, $[s_1, s_2]$ is an interval in the domain $[1, n]$, and ℓ is
the length of the shortest unique substrings containing $D[x:y]$.

Output: All the shortest unique substrings containing q whose left endpoints are in
$[s_1, s_2]$.

1 run a position constrained query with $q = D[x:y]$, $r_{start} = [s_1, s_2]$, and $r_{end} = [y, n]$
2 **if** *the query returns nil* **then**
3 | **return** $\emptyset$
4 $D[i:j] \leftarrow$ the string returned by the query
5 **if** *the length of* $D[i:j] > \ell$ **then**
6 | **return** $\emptyset$
7 $S_1 \leftarrow$ FIND-ALL$(D[x:y], [s_1, i-1], \ell)$
8 $S_2 \leftarrow$ FIND-ALL$(D[x:y], [i+1, s_2], \ell)$
9 **return** $\{D[i:j]\} \cup S_1 \cup S_2$

4.3 External Memory

The previous discussion has concentrated on the RAM model. In this section, we consider shortest unique queries in the standard *external memory* (EM) model [1]. Under this model, the machine is equipped with a disk that is formated into *blocks* of size B words, and with internal memory of $M \geq 2B$ words. An I/O exchanges a block of data between the disk and memory. The *space* of a structure is measured by the number of disk blocks it occupies, and the *time* of an algorithm is measured by the number of I/Os it performs.

The structure of Theorem 1 works *directly* in external memory. This means that one can simply store the structure by treating the disk as virtual memory. Given that the structure uses $O(n)$ words, the number of blocks it occupies is $O(n/B)$, where B is the number of words in a block. To answer a shortest unique query, one can simply apply the algorithm of Theorem 1 by again treating the disk as virtual memory. As the algorithm performs only $O(1)$ CPU calculation and probes $O(1)$ memory locations, its I/O cost is definitely bounded by $O(1)$.

Our structure can also be constructed efficiently. Remember that it has the following components:

- The $MUS_{leftfix}$ and $MUS_{rightfix}$ arrays (see Figure 2)
- The α and β arrays (Figure 3)
- An RMQ structure.

Both the $MUS_{leftfix}$ and $MUS_{rightfix}$ arrays can be built using the algorithm of [7] in $O(SORT(n))$ I/Os, provided that a *suffix array* [8] is given, where $O(SORT(n))$ is the number of I/Os needed to sort n elements. The suffix array itself can also be computed in $O(SORT(n))$ I/Os [3]. After the $MUS_{leftfix}$ and $MUS_{rightfix}$ arrays are ready, we can then obtain the set $\mathcal{M}$ of MUS's, sorted by left endpoint, in $O(SORT(n))$ I/Os. Then, the α and β arrays can be built using Algorithm 1 in $O(n/B)$ I/Os. An RMQ structure can also be created from $\mathcal{M}$ in $O(n/B)$ I/Os [2].

We now conclude with the last main result of this paper:

Theorem 4. *Given a data string of length n, we can pre-compute in $O(SORT(n))$ I/Os an index structure in external memory that occupies $O(n/B)$ blocks, and answers any shortest unique query in $O(1)$ I/Os.*

Acknowledgements. Xiaocheng and Yufei Tao were supported in part by projects GRF 4165/11, 4164/12, and 4168/13 from HKRGC. Jian Pei was supported by an NSERC Discovery grant and a BCIC NRAS Team project.

References

1. Aggarwal, A., Vitter, J.S.: The input/output complexity of sorting and related problems. CACM 31(9), 1116–1127 (1988)
2. Demaine, E.D., Landau, G.M., Weimann, O.: On cartesian trees and range minimum queries. In: Albers, S., Marchetti-Spaccamela, A., Matias, Y., Nikoletseas, S., Thomas, W. (eds.) ICALP 2009, Part I. LNCS, vol. 5555, pp. 341–353. Springer, Heidelberg (2009)
3. Dementiev, R., Kärkkäinen, J., Mehnert, J., Sanders, P.: Better external memory suffix array construction. ACM Journal of Experimental Algorithmics, 12 (2008)
4. Fischer, J., Heun, V.: Theoretical and practical improvements on the RMQ-problem, with applications to LCA and LCE. In: Lewenstein, M., Valiente, G. (eds.) CPM 2006. LNCS, vol. 4009, pp. 36–48. Springer, Heidelberg (2006)
5. Harel, D., Tarjan, R.E.: Fast algorithms for finding nearest common ancestors. SIAM J. of Comp. 13(2), 338–355 (1984)
6. İleri, A.M., Külekci, M.O., Xu, B.: Shortest unique substring query revisited. In: Kulikov, A.S., Kuznetsov, S.O., Pevzner, P. (eds.) CPM 2014. LNCS, vol. 8486, pp. 172–181. Springer, Heidelberg (2014)
7. Ilie, L., Smyth, W.F.: Minimum unique substrings and maximum repeats. Fundam. Inform. 110(1-4), 183–195 (2011)
8. Manber, U., Myers, E.W.: Suffix arrays: A new method for on-line string searches. SIAM J. of Comp. 22(5), 935–948 (1993)
9. Pei, J., Wu, W.C.-H., Yeh, M.-Y.: On shortest unique substring queries. In: ICDE, pp. 937–948 (2013)
10. Tsuruta, K., Inenaga, S., Bannai, H., Takeda, M.: Shortest unique substrings queries in optimal time. In: Geffert, V., Preneel, B., Rovan, B., Štuller, J., Tjoa, A.M. (eds.) SOFSEM 2014. LNCS, vol. 8327, pp. 503–513. Springer, Heidelberg (2014)

Online Multiple Palindrome Pattern Matching[*,**]

Hwee Kim and Yo-Sub Han

Department of Computer Science, Yonsei University
50, Yonsei-Ro, Seodaemun-Gu, Seoul 120-749, Republic of Korea
{kimhwee,emmous}@cs.yonsei.ac.kr

Abstract. A palindrome is a string that reads the same forward and backward. We say that two strings of the same length are pal-equivalent if for each possible center they have the same length of the maximal palindrome. Given a text T of length n and a set of patterns $P_1, \ldots, P_k$, we study the online multiple palindrome pattern matching problem that finds all pairs of an index i and a pattern P_j such that $T[i-|P_j|+1 : i]$ and P_j are pal-equivalent. We solve the problem in $O(m_k M)$ preprocessing time and $O(m_k n)$ query time using $O(m_k M)$ space, where M is the sum of all pattern lengths and m_k is the longest pattern length.

1 Introduction

A palindrome is a string that reads the same forward and backward. If a substring of a string is a palindrome, we say that the string has a palindromic substring or palindromic structure. It is crucial to find palindromes and identify similar palindromic structures in bio sequence analysis [8]. Many researchers examined the properties of palindromic structures in strings [2–6] and proposed efficient algorithms on palindromic structures [7, 10, 12]. We focus on the palindrome pattern matching problem introduced by I et al. [11]—they define two strings of the same length to be pal-equivalent if for each possible center they have the same length of the maximal palindrome. Given a text T of length n and a pattern P of length m, the palindrome pattern matching problem is to find all indices i such that $T[i-m+1 : i]$ and P are pal-equivalent. I et al. [11] presented two algorithms that solve the palindrome pattern matching for an arbitrary size alphabet: One solves the problem in $O(n + m)$ time and the other solves the problem in $O((n + m) \log \sigma + r)$ time, where σ is the alphabet size and r is the number of matching occurrences.

We notice that both algorithms by I et al. [11] require a preprocessing step of T, which makes algorithms unsuitable for an extremely large text or a stream text. This motivates us to consider the online pattern matching, where we should report the matching for each index i while reading T online. We tackle the

---

[*] This research was supported by the Basic Science Research Program through NRF funded by MEST (2012R1A1A2044562).

[**] Kim was supported by NRF (National Research Foundation of Korea) Grant funded by the Korean Government (NRF-2013-Global Ph.D. Fellowship Program).

E. Moura and M. Crochemore (Eds.): SPIRE 2014, LNCS 8799, pp. 173–178, 2014.
© Springer International Publishing Switzerland 2014

online multiple palindrome pattern matching based on a modification of the Aho-Corasick automaton [1]. For multiple patterns $P_1, \ldots, P_k$, our algorithm requires $O(m_k M)$ preprocessing time and runs in $O(m_k n)$ query time using $O(m_k M)$ space, where $M = \sum_{i=1}^{k} |P_i|$ and $m_k = \max(|P_i|)$.

2   Preliminaries

Given a finite set Σ of characters and a string w over Σ, let $|w|$ be the length of w and $w[i]$ be the symbol of w at position i, for $1 \le i \le |w|$. We define the empty string λ as a string of length 0. We use $w[i : j]$ to denote a substring $w[i]w[i+1] \cdots w[j]$, where $1 \le i \le j \le |w|$. A language over Σ is a set of strings over Σ. A finite-state automaton (FA) $\mathcal{A}$ is specified by $\mathcal{A} = (Q, \Sigma, \delta, s, F)$, where Q is a set of states, Σ is an alphabet, $\delta \subseteq Q \times \Sigma \times Q$ is a set of transitions, $s \in Q$ is the start state and $F \subseteq Q$ is a set of final states. A string w is accepted by $\mathcal{A}$ if there is a labeled path from s to a state in F such that the path spells out w. The language $L(A)$ of an FA $\mathcal{A}$ is the set of all strings accepted by $\mathcal{A}$. For more background knowledge in automata theory, the reader may refer to textbooks [9, 13].

For a string w, let w^R denote the reversed string of w. A string w is called a *palindrome* if $w = w^R$. The *radius* of a palindrome w is $\frac{|w|}{2}$. The *center* of a palindromic substring $w[i : j]$ of a string w is $\frac{i+j}{2}$. We call a palindromic substring $w[i : j]$ the *maximal palindrome* at the center $\frac{i+j}{2}$ if no other palindromes at the center $\frac{i+j}{2}$ have a larger radius than $w[i : j]$. Let $Pals(w)$ be the set of pairs of the center and the radius of all center-distinct maximal palindromes [10]. For two strings w and z of the same length, we say that w and z are *pal-equivalent* if $Pals(w) = Pals(z)$.

Definition 1 (Online Multiple Palindrome Pattern Matching). *Given a text T of length n and patterns $P_1, \ldots, P_k$ of length $m_1, \ldots, m_k$, find all pairs of an index i and a corresponding pattern P_j such that $Pals(P_j) = Pals(T[i-m_j+1 : i])$ after reading each character $T[i]$.*

For the online pattern matching, we call the time to preprocess the patterns *preprocessing time*, and the time to read the text to find matchings *query time*.

3   The Algorithm

The basic idea of the algorithm is to process multiple patterns at once with a single automaton based on the idea of the Aho-Corasick automaton [1]. Assume that given patterns $P_1, \ldots, P_k$ of length $m_1, \ldots, m_k$ are sorted by ascending order with respect to the length of the pattern and M is the sum of all pattern lengths. Before we design an algorithm, we have the following observation:

Observation 1. *For strings w, z and an index i, if there exists $(c, r) \in Pals(w)$ where $c \leq i$ and $c + r - 0.5 \geq i$, then $z[i] = z[2r-i]$. If there is no (c, r) satisfying the condition, then $z[i] \notin \{z[2r-i] \mid (c, r) \in Pals(w)$ and $c + r - 0.5 = i - 1\}$.*

Note that $z[i]$ is computed based on $z[l]$'s for $l < i$, instead of characters in w. Based on Observation 1, we define a *variable pattern* of a pattern P as follows:

Definition 2. *For a pattern P of length m over Σ of size t, a variable pattern P' is defined by an array $\mathbb{A}[m]$ of variables and an array $\mathbb{B}[m]$ of unequal conditions satisfying the following conditions:*

1. *$P'[i] = \mathbb{A}[l_i]$ for $1 \leq i, l_i \leq m$.*
2. *If there exists $(c, r) \in Pals(P)$ where $c \leq i$ and $c + r - 0.5 \geq i$, then $l_i = l_{2r-i}$, and thus, $P'[i] = P'[2r-i]$.*
3. *Otherwise, for all $j \in \{2r - i \mid (c, r) \in Pals(P)$ and $c + r - 0.5 = i - 1\}$, $\mathbb{B}[i] = j$ and $\mathbb{B}[j] = i$, and thus, $P'[i] \neq P'[j]$.*

Now we construct $P'_1, \ldots, P'_k$ simultaneously by Algorithm 1. All variable patterns share $\mathbb{A}$ while each variable pattern P'_j has a distinct array $\mathbb{B}[j][m]$ of unequal conditions in the algorithm. Fig. 1 shows an example of P' and $\mathbb{B}$.

$$P = \boxed{A \mid G \mid C \mid G \mid T \mid A}$$

$$P' = \boxed{\mathbb{A}[1]\mathbb{A}[2]\mathbb{A}[3]\mathbb{A}[2]\mathbb{A}[4]\mathbb{A}[5]}$$

$$\mathbb{A}[4] \neq \mathbb{A}[1], \mathbb{A}[2], \mathbb{A}[3], \mathbb{A}[5] \Rightarrow$$

$\mathbb{B}$	
1	2, 3, 4
2	1, 3, 4, 5
3	1, 2, 4
4	1, 2, 3, 5
5	2, 4

Fig. 1. A variable pattern P' and an array $\mathbb{B}$ of unequal conditions for $P = AGCGTA$

Based on Observation 1 and Definition 2, we establish the following result:

Lemma 1. *After running Algorithm 1, if there is a surjection of $\mathbb{A}$ to Σ where $\mathbb{A}[i] \neq \mathbb{A}[j]$ holds for all i, j such that $j \in \mathbb{B}[l][i]$, then $Pals(P'_l) = Pals(P_l)$. Moreover, given a string w such that $Pals(w) = Pals(P_l)$, there exists a surjection of $\mathbb{A}$ to Σ such that $P'_l = w$.*

Once we have $P'_1, \ldots, P'_k$, we can construct a special automaton $\mathcal{B} = (Q, \mathbb{A} \cup \{\#\}, \delta : Q \times \mathbb{A} \to Q, s, F, \Sigma, \mathbb{B}, \delta_f : Q \to Q, \mathcal{H} : Q \to 2^{\mathbb{A} \times (\mathbb{A} \cup \{\#\})}, \delta_p : Q \to Q)$. Note that five parameters—$\Sigma, \mathbb{B}, \delta_f, \mathcal{H}, \delta_p$—are added to the definition of a traditional FA. The automaton $\mathcal{B}$ simulates the Aho-Corasick algorithm [1], using $P'_1, \ldots, P'_k$ as patterns. In the Aho-Corasick algorithm, when there occurs a mismatch, the algorithm checks the longest suffix of the prefix of T read so far. The automaton $\mathcal{B}$ simulates the process by δ_f, and additionally, changes surjection of $\mathbb{A}$ to Σ according to $\mathcal{H}$. The suffix transition function δ_p contains transitions to find multiple matching occurrences on a single state. Algorithm 2 constructs

$\mathcal{B}$. We use a supplementary function StateForVP to return the state denoting the end of a given variable pattern. Fig. 2 shows an example of $\mathcal{B}$.

Algorithm 1. ConstructMultiVariablePattern

Input: Patterns $P_1, \ldots, P_k$ of length $m_1, \ldots, m_k$ over Σ of size t
Output: $P'_1, \ldots, P'_k, \mathbb{A}[m_k], \mathbb{B}[k][m_k]$

1   **for** $j \leftarrow 1$ **to** k **do**
2   compute $Pals(P_j)$ // we insert $(0.5, 0)$ to $Pals(P_j)$ for convenience
3      $c \leftarrow 0.5,\ d \leftarrow 0,\ s \leftarrow 0$
4   **for** $i \leftarrow 1$ **to** m_j **do**
5   find r such that $(c, r) \in Pals(P_j)$
6   **if** $d \geq i$ **then** $P'_j[i] \leftarrow P'_j[2r-i]$ **else**
7             $s \leftarrow s+1,\ P'_j[i] \leftarrow \mathbb{A}[s]$
8   **for each** $(c', r') \in Pals(P_j)$ **do**
9   **if** $c' + r' - 0.5 = i - 1$ **then**
10   add $2r' - i$ to $\mathbb{B}[j][i]$, add i to $\mathbb{B}[j][2r'-i]$
11   find r_1, r_2 such that $(c+0.5, r_1), (c+1, r_2) \in Pals(P_j)$
12         $d \leftarrow \max(d, c+r_1, c+r_2+0.5),\ r \leftarrow r+1$
13 **return** $P'_1, \ldots, P'_k, \mathbb{A}, \mathbb{B}$

Algorithm 2. ConstructMultiAutomaton

Input: Patterns $P_1, \ldots, P_k$ of length $m_1, \ldots, m_k$ over Σ of size t
Output: $\mathcal{B} = (Q, \mathbb{A} \cup \{\#\}, \delta, s, F, \Sigma, \mathbb{B}, \delta_f, \mathcal{H}, \delta_p)$

1   ConsturctMultiVariablePattern$(P_1, \ldots, P_k)$
2   add q_λ to Q and let $p_1, \ldots, p_k \leftarrow q_\lambda$
3   **for** $i \leftarrow 1$ **to** $m_k + 1$ **do**
4   **for each** P'_j where $i \leq m_j + 1$ **do**
5   let $P'_j[i] = \mathbb{A}[l]$ and $p_j = q_s$
6   **if** $i \neq m_j + 1$ **then** $\delta(q_s, \mathbb{A}[l]) \leftarrow q_{s \cdot l}$, add $q_{s \cdot l}$ to Q **if** $i = 2$ **then**
            $\delta_f(q_s) \leftarrow q_\lambda$, add $(\mathbb{A}[1] \leftarrow \#)$ to $\mathcal{H}(q_s)$ **else if** $i > 2$ **then**
7   find the smallest i' and corresponding j' such that
            $Pals(P'_{j'}[1 : i-i']) = Pals(P'_j[i' : i-1])$
8             $\delta_f(q_s) \leftarrow$ StateForVP$(P'_{j'}[1 : i-i'])$
9   **for** $g \leftarrow 1$ **to** $i - i'$ **do**
10   add $(\mathbb{A}[h] \leftarrow \mathbb{A}[h'])$ to $\mathcal{H}(q_s)$ for $P'_j[g] = \mathbb{A}[h]$ and
                $P'_j[g+i'-1] = \mathbb{A}[h']$
11   **for each** $\mathbb{A}[h]$ in $P'_j[1 : i-1]$ *without injective function in* $\mathcal{H}(q_s)$ **do**
            add $(\mathbb{A}[h] \leftarrow \#)$ to $\mathcal{H}(q_s)$
12   **if** $i = m_j$ **then** add $q_{s \cdot l}$ to F find the largest i' and corresponding j'
            such that $Pals(P'_{j'}[1 : i']) = Pals(P'_j[i-i'+1 : i])$
13   **if** $i' = m_{j'}$ **then** $\delta_p(p_j) \leftarrow$ StateForVP$(P'_{j'})$ $p_j \leftarrow q_{s \cdot l}$
14 **return** $(Q, \mathbb{A} \cup \{\#\}, \delta, q_\lambda, F, \Sigma, \mathbb{B}, \delta_f, \mathcal{H}, \delta_p)$

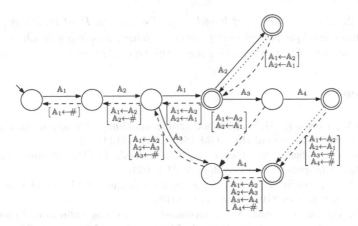

Fig. 2. An automaton $\mathcal{B}$ for $P_1 = AGA, P_2 = ACTG, P_3 = ATAT, P_4 = TCTGC$. Variables $\mathbb{A}[i]$ are written as $\mathbb{A}_i$ for better readability. Dashed transitions are failure transitions and dotted transitions are suffix transitions.

Algorithm 3. FindMultiPalindromeMatching

Input: Patterns $P_1, \ldots, P_k$ of length $m_1, \ldots, m_k$ over Σ of size t
Output: (i, P_j) such that $Pals(P_j) = Pals(T[i-m_j+1 : i])$

1 ConstructMultiAutomaton($P_1, \ldots, P_k$)
2 **for** $i \leftarrow 1$ **to** m_k **do** $\mathbb{A}[i] \leftarrow \#$ $q_l \leftarrow q_\lambda$ // current state
3 **for** $i \leftarrow 1$ **to** n **do**
4   **while** *one of the following conditions holds for all* $\mathbb{A}[j]$ *such that* $\delta(q_l, \mathbb{A}[j]) \neq \emptyset$
         1. $q_l \in F$
         2. $\mathbb{A}[j] \neq T[i], \#$
         3. $\mathbb{A}[j] = \#$ *and there exists* $j' \in \mathbb{B}[j][g]$ *such that* $\mathbb{A}[j'] = T[i]$ *and*
           $\delta(q_l, \mathbb{A}[j]) = StateForVP(P'_g[1 : |l|+1])$
5   **do**
6   **for each** $(\mathbb{A}[h] \leftarrow \mathbb{A}[h']) \in \mathcal{H}(q_l)$ **do** $\mathbb{A}[h] \leftarrow \mathbb{A}[h']$ $q_l \leftarrow \delta_f(q_l)$
7   **if** $\mathbb{A}[j] = \#$ **then** $\mathbb{A}[j] \leftarrow T[i]$ $q_l \leftarrow \delta(q_l, \mathbb{A}[j])$
8   **if** $q_l \in F$ **then return** $(i, P_{j'})$ where $StateForVP(P'_{j'}) = q_l$ $p_l \leftarrow q_l$
9   **while** $\delta_p(p_l) \neq \emptyset$ **do**
10         $p_l \leftarrow \delta_p(p_l)$
11   **return** $(i, P_{j'})$ where $StateForVP(P'_{j'}) = p_l$

Now we are ready to design an algorithm that solves the problem using $\mathcal{B}$. Algorithm 3 processes T in $\mathcal{B}$ and reports all matching end-indices and the corresponding matching patterns.

Lemma 2. *Algorithm 3 returns all pairs of an index i and a pattern P_j such that $Pals(P_j) = Pals(T[i-m_j+1 : i])$.*

Theorem 2. *Given a text T of length n and a pattern P of length m, we can solve the online multiple palindrome pattern matching problem with $O(m_k M)$ pre-processing time and $O(m_k n)$ query time using $O(m_k M)$ space.*

References

1. Aho, A.V., Corasick, M.J.: Efficient string matching: An aid to bibliographic search. Communications of the ACM 18(6), 333–340 (1975)
2. Allouche, J.-P., Baake, M., Cassaigne, J., Damanik, D.: Palindrome complexity. Theoretical Computer Science 292(1), 9–31 (2003)
3. Anisiu, M.-C., Anisiu, V., Kása, Z.: Total palindrome complexity of finite words. Discrete Mathematics 310(1), 109–114 (2010)
4. Brlek, S., Hamel, S., Nivat, M., Reutenauer, C.: On the palindromic complexity of infinite words. International Journal of Foundations of Computer Science 15(2), 293–306 (2004)
5. Droubay, X., Justin, J., Pirillo, G.: Episturmian words and some constructions of de luca and rauzy. Theoretical Computer Science 255(1-2), 539–553 (2001)
6. Glen, A., Justin, J., Widmer, S., Zamboni, L.Q.: Palindromic richness. European Journal of Combinatorics 30(2), 510–531 (2009)
7. Groult, R., Prieur, É., Richomme, G.: Counting distinct palindromes in a word in linear time. Information Processing Letters 110(20), 908–912 (2010)
8. Gusfield, D.: Algorithms on Strings, Trees, and Sequences: Computer Science and Computational Biology. Cambridge University Press (1997)
9. Hopcroft, J.E., Ullman, J.D.: Introduction to Automata Theory, Languages, and Computation. Addison–Wesley (1979)
10. Tomohiro, I., Inenaga, S., Bannai, H., Takeda, M.: Counting and verifying maximal palindromes. In: Chavez, E., Lonardi, S. (eds.) SPIRE 2010. LNCS, vol. 6393, pp. 135–146. Springer, Heidelberg (2010)
11. Tomohiro, I., Inenaga, S., Bannai, H., Takeda, M.: Palindrome pattern matching. Theoretical Computer Science 483, 162–170 (2013)
12. Manacher, G.: A new linear-time "on-line" algorithm for finding the smallest initial palindrome of a string. Journal of the ACM 22(3), 346–351 (1975)
13. Wood, D.: Theory of Computation. Harper & Row (1986)

Indexed Matching Statistics
and Shortest Unique Substrings

Djamal Belazzougui and Fabio Cunial

Helsinki Institute for Information Technology (HIIT)
Department of Computer Science, University of Helsinki, Finland*
name.surname@helsinki.fi

Abstract. The unidirectional and bidirectional matching statistics between two strings s and t on alphabet Σ, and the shortest unique substrings of a single string t, are the cornerstone of a number of large-scale genome analysis applications, and they encode nontrivial structural properties of s and t. In this paper we compute for the first time the matching statistics between s and t in $O((|s| + |t|) \log |\Sigma|)$ time and in $O(|s| \log |\Sigma|)$ bits of space, circumventing the need for computing the depths of suffix tree nodes that characterized previous approaches. Symmetrically, we compute for the first time the shortest unique substrings of a string t in $O(|t| \log |\Sigma|)$ time and in $O(|t| \log |\Sigma|)$ bits of space. A key component of our methods is an encoding of both the unidirectional and the bidirectional statistics that takes $2|t| + o(|t|)$ bits of space and that allows constant-time access to every position.

1 Introduction and Motivation

Let s and t be nonempty strings on alphabet $\Sigma = 1..\sigma$, let $\bar{s}$ and $\bar{t}$ be their reverse, and let $\$ = 0$ be a character not in Σ that is smaller than any character in Σ. In this paper we study the following concepts:

Definition 1. *Given two strings s and t and a threshold $\tau > 0$, the unidirectional matching statistics $\mathsf{MS}_{t,s,\tau}$ of t with respect to s is a vector of length $|t|$ that stores at index $i \in [0..|t| - 1]$ the length of the longest prefix of $t[i..|t| - 1]$ that occurs at least τ times in s.*

Definition 2. *Given a string t and a threshold $\tau > 0$, the unidirectional distinguishing statistics $\mathsf{DS}_{t,\tau}$ of t is a vector of length $|t|$ that stores at index $i \in [0..|t| - 1]$ the length of the shortest prefix of $t[i..|t| - 1]\$ that occurs at most τ times in t.*

We drop any subscript from $\mathsf{MS}_{t,s,\tau}$ and $\mathsf{DS}_{t,\tau}$ whenever s, t or τ are clear from the context. Note that $\mathsf{DS}_{t,\tau}[i] = \mathsf{MS}_{t,t,\tau+1}[i] + 1$ for every i and τ.[1] $\mathsf{MS}_{t,s,1}$ has

---

* This work was partially supported by Academy of Finland under grant 250345 (Center of Excellence in Cancer Genetics Research).

[1] MS and DS have often been regarded as different problems in the literature: we thank an anonymous reviewer for making this connection explicit.

E. Moura and M. Crochemore (Eds.): SPIRE 2014, LNCS 8799, pp. 179–190, 2014.
© Springer International Publishing Switzerland 2014

also been called *external matching* [23], and $DS_{t,1}$ has been called *distinguishing prefix* or *shortest unique substring* elsewhere [10]. By extension, $RS_t = DS_{t,1} - 1$ can be dubbed the *unidirectional repeating statistics* of t, since $RS_t[i]$ is the length of the longest substring that starts at position i and that occurs at least twice in t. RS_t has also been called *internal matching* elsewhere [23]. $MS_{t,s,1}$ and $DS_{t,1}$ are almost as old as the suffix tree itself [23], with first applications to file transmission [22]. We are also interested in the bidirectional versions of such concepts:

Definition 3. *Given two strings s and t and a threshold τ, the bidirectional matching statistics $BMS_{t,s,\tau}$ of t with respect to s is a vector of length $|t|$ that stores at index $i \in [0..|t| - 1]$ the length of the longest substring $t[x..y]$ with $x \le i \le y$ that occurs at least τ times in s.*

Bidirectional distinguishing and repeating statistics, denoted respectively by $BDS_{t,\tau}$ and $BRS_{t,\tau}$, can be defined in the same way. Computing $MS_{t,s,1}$ is a classical problem in string processing: the textbook solution scans t from left to right while navigating suffix links and child links in the suffix tree of s. Symmetrically, t can be scanned from right to left, while taking Weiner links and parent links in the (compressed) suffix tree of s [13]. Computing the depths of suffix tree nodes is the bottleneck of both approaches: such depths can be encoded either explicitly in $\Theta(|s| \log |s|)$ bits of space, and decoded in constant time [13], or implicitly on top of a compressed index, and decoded in $O(\log^\epsilon |s|)$ time [18]. In this paper we completely circumvent the need for computing the depths of suffix trees nodes, by indexing both s and $\bar{s}$ and by performing both a forward and a backward pass over t. This allows to compute $MS_{t,s,\tau}$ in $O((|s| + |t|) \log \sigma)$ time and $O(|s| \log \sigma)$ bits of space for the first time, for any τ.

$DS_{t,1}$ has been previously computed either in quadratic time using suffix trees [14], or in linear time using $O(|t| \log |t|)$ bits of space [10,20]. We adapt our MS algorithm to compute $DS_{t,\tau}$ for the first time in $O(|t| \log \sigma)$ bits of space and in $O(|t| \log \sigma)$ time, for any τ.

A key component of our methods is an efficient encoding of $MS_{t,s}$ and of DS_t, that takes $2|t| + o(|t|)$ bits of space and that allows to retrieve $MS[i]$ and $DS[i]$ in constant time for any i. This scheme uses ideas that have been previously applied to encode depths in compressed suffix trees [18]. Our index can represent $BMS_{t,s}$ and BDS_t using just $o(|t|)$ bits of additional space, while still supporting constant-time access to the statistics at any position. Note that $BMS_{t,s}$ has already been computed from $MS_{t,s}$ in the past [19], but it has been encoded in $O(|t| \log |t|)$ bits of space.

Before proceeding, we note that fast and succinct representations of MS and DS enable a number of large-scale applications. For example, the profiles of MS and DS can be used to discriminate between sequencing errors and single-nucleotide variations in large read collections [15], and DS has applications in primer design for PCR, in comparative genomics [8], and in summarizing the context of the occurrences of a pattern in a large text collection [14]. More interestingly, MS and DS encode a number of structural properties of s and t. For example, recall that a *repeat* of t is any string w that occurs at least twice in t.

A repeat w is *maximal* if both awb and cwd occur in s, with $\{a, b, c, d\} \subseteq \Sigma$, $a \neq c$ and $b \neq d$. An occurrence i of a repeat w in t is said to be *exposed* if $t[i..i + |w| - 1] = w$ and if no substring $t[i'..j']$ repeats in t, where $i' \leq i$, $j' \geq i + |w| - 1$, and $(i', j') \neq (i, i + |w| - 1)$. A *near-supermaximal repeat* of t is a repeat with at least one exposed occurrence. Clearly there is a near-supermaximal repeat exposed at position i in t if and only if $\mathsf{DS}_{t,1}[i] = \mathsf{DS}_{\bar{t},1}[|t| - i - \mathsf{DS}_{t,1}[i] + 1]$, and its length is $\mathsf{DS}_{t,1}[i] - 1$.

Symmetrically, recall that a *minimal absent word* of s is a string awb with $w \in \Sigma^*$ and $\{a, b\} \subseteq \Sigma$, such that both aw and wb occur in s, but awb does not occur in s. Clearly $t[j..j + \mathsf{MS}_{t,s,1}[j]]$ is a minimal absent word of s if and only if $\mathsf{MS}_{t,s,1}[j + 1] \geq \mathsf{MS}_{t,s,1}[j]$, in which case $t[j + 1..j + \mathsf{MS}_{t,s,1}[j] - 1]$ is a maximal repeat of s.

Recall also that a *maximal exact match* (MEM) between s and t is a triple (i, j, ℓ) where $0 \leq i < |s|$, $0 \leq j < |t|$, and $1 \leq \ell \leq \min\{|s|, |t|\}$, such that $s[i..i + \ell - 1] = t[j..j + \ell - 1]$, $s[i - 1] \neq t[j - 1]$ and $s[i + \ell] \neq t[j + \ell]$ (we assume that $s[-1] \neq t[-1]$ and that $s[|s|] \neq t[|t|]$). A *maximal unique match* (MUM) between s and t is a MEM (i, j, ℓ) such that string $s[i..i + \ell - 1]$ occurs exactly once in s and in t. MEMs, MUMs and their variants are used routinely in whole-genome alignment [11]. If $\mathsf{MS}_{t,s,1}[j] = \mathsf{MS}_{\bar{t},\bar{s},1}[|t| - j - \mathsf{MS}_{t,s,1}[j]]$, then there is a MEM $(i, j, \mathsf{MS}_{t,s,1}[j])$ at position j in t, and this is the longest MEM starting at j. This MEM is unique in t iff $\mathsf{MS}_{t,s,1}[j] \geq \mathsf{DS}_{t,1}[j]$, and it is unique in s iff $\mathsf{DS}_{t,s,1}[j]$ is defined, where $\mathsf{DS}_{t,s,\tau}$ is a binary version of distinguishing statistics. Note that $\mathsf{DS}_{t,s,1}$ can be implemented using $\mathsf{MS}_{t,s,2}$ and a bitvector $\mathtt{flag}$ such that $\mathtt{flag}[i] = 1$ iff $t[i..i + \mathsf{MS}_{t,s,1} - 1]$ occurs exactly once in s: $\mathsf{DS}_{t,s,1}[i]$ is defined if and only if $\mathtt{flag}[i] = 1$, in which case $\mathsf{DS}_{t,s,1}[i] = \mathsf{MS}_{t,s,2}[i] + 1$. We can thus decide in constant time whether a MUM starts at any position in t, and compute the length of such MUM.

Finally, given two positions $j > i$ in t, $\mathsf{MS}_{t,s,1}$ allows to compute the *average common substring* dissimilarity measure [21] between $t[i..j]$ and the whole s in $O(j - i)$ time, as well as the number k of factors in the relative LZ77 factorization of $t[i..j]$ with respect to s in $O(k)$ time. Besides having connections to Kolmogorov complexity and to the cross-entropy of finite-memory random sources [5], these measures are now the cornerstone of popular whole-genome comparison tools used in phylogenetics [21].

2   Computing and Indexing MS and DS

We denote by SA_s and BWT_s the suffix array and the Burrows-Wheeler transform of a string s, respectively. Recall that the *suffix array* $\mathsf{SA}_s[0, |s|]$ of s is the vector of indices such that $s[\mathsf{SA}_s[i], |s|]\$$ is the i-th smallest suffix of $s\$$ in lexicographical order, and the Burrows-Wheeler transform of s is the string $\mathsf{BWT}_s[0, |s|]$ satisfying $\mathsf{BWT}_s[i] = s[\mathsf{SA}_s[i] - 1]$ if $\mathsf{SA}_s[i] > 0$, and $\mathsf{BWT}_s[i] = \$$ otherwise. We define the *suffix array range*, or identically the *BWT range* $(i_w, j_w)_s$ of a substring w in string s, as the maximal interval $[i_w..j_w]$ in SA_s such that all the suffixes $s[\mathsf{SA}_s[i]..|s| - 1]$ for $i_w \leq i \leq j_w$ are prefixed by w.

We drop the subscript s from a range whenever the reference string is implied by the context. Incidentally, note that $\mathsf{RS}_t[i] = \max\{\mathsf{LCP}_t[j], \mathsf{LCP}_t[j+1]\}$, where $\mathsf{SA}[j] = i$ and LCP_t is the longest common prefix array of string t, i.e. $\mathsf{LCP}_t[j]$ is the longest prefix shared by suffixes $t[\mathsf{SA}_t[j]..|t|]\$$ and $t[\mathsf{SA}_t[j-1]..|t|]\$$ for all $j \in [1..|t|]$. Finally, we denote by $\neg s$ the complement of a bitstring s, i.e. $\neg s$ is the string t that satisfies $t[i] = 1 - s[i]$ for $0 \le i < |s|$.

It is clear that $\mathsf{MS}_{t,s,\tau}[i] \ge 0$, $\mathsf{DS}_{t,\tau}[i] \ge 1$ and $\mathsf{RS}_t[i] \ge 0$ for all i and τ. The key additional property on which most of this paper rests is that $\mathsf{MS}_{t,s,\tau}$, $\mathsf{DS}_{t,\tau}$ and RS_t are δ-*monotone sequences* [17]:

Definition 4. *Let $a = a_0 a_1 \ldots a_n$ and $\delta = \delta_1 \delta_2 \ldots \delta_n$ be two sequences of non-negative integers. Sequence a is said to be δ-monotone if $a_i - a_{i-1} \ge -\delta_i$ for all $i \in [1..n]$.*

In particular, $\mathsf{MS}_{t,s,\tau}[i] - \mathsf{MS}_{t,s,\tau}[i-1] \ge -1$, $\mathsf{DS}_{t,\tau}[i] - \mathsf{DS}_{t,\tau}[i-1] \ge -1$ and $\mathsf{RS}_t[i] - \mathsf{RS}_t[i-1] \ge -1$ for all $i \in [1..|t|-1]$. Other popular examples of δ-monotone sequences in string processing are the *permuted LCP array* [18], and the *longest previous factor array* used in the LZ77 factorization of a string t, with $\delta_i = 1$ for all $i \in [1..|t|-1]$, and the lengths of the partial matches when searching a text t for a string w with the KMP algorithm[2].

It is natural to represent $\mathsf{MS}_{t,s,\tau}$ in succinct space by encoding the consecutive offsets $\mathsf{MS}_{t,s,\tau}[i] - \mathsf{MS}_{t,s,\tau}[i-1]$. It turns out that the same data structure can answer queries on $\mathsf{MS}_{\bar{t},\bar{s},\tau}[i]$ as well.

Lemma 1. *There is a data structure that takes $2|t| + o(|t|)$ bits of space and that answers queries on $\mathsf{MS}_{t,s,\tau}[i]$ and on $\mathsf{MS}_{\bar{t},\bar{s},\tau}[i]$ for any i in constant time.*

Proof. We follow the approach described in [18]. Specifically, we build a sequence $\mathbf{ms}_{t,s,\tau}$ of $2|t|$ bits by appending, for each $i \in [0, |t|-1]$ in increasing order, the binary string:

$$\underbrace{00 \cdots\cdots\cdots 0}_{\substack{\mathsf{MS}_{t,s,\tau}[i] - \mathsf{MS}_{t,s,\tau}[i-1]+1 \\ \text{times}}} 1$$

where we set $\mathsf{MS}_{t,s,\tau}[-1] = 1$ for convenience. The resulting array contains either $2|t|$ or $2|t|-1$ bits: in the latter case, we append a final zero. Note that the number of zeros before the ith one in $\mathbf{ms}$ equals $i + \mathsf{MS}[i]$. Then, index $\mathbf{ms}$ to support select operations[3] in constant time using $2|t|+o(|t|)$ bits. We can thus compute $\mathsf{MS}[i]$ for any $i \in [0, |t|-1]$ by using the formula $\mathtt{select}(\mathbf{ms}, 1, i) - 2i$.

For a position $i \in [0, |t|-1]$, consider now the longest prefix of $\bar{t}[i..|t|-1]$ that occurs in $\bar{s}$, and assume that $i + \mathsf{MS}_{\bar{t},\bar{s},\tau}[i] = |t| - j$. Then substring $t[j..|t|-i-1]$ occurs in s, but substring $t[j-1..|t|-i-1]$ does not occur in s. Since the number of zeros before the $(j-1)$th one in $\mathbf{ms}$ is $j - 1 + \mathsf{MS}_{t,s,\tau}[j-1] < |t| - i$ and the

---

[2] Let i_x and i_{x+1} be the starting positions in t of two consecutive partial matches determined by KMP. In particular, let $t[i_x..i_x+a_x-1] = w[0..a_x-1]$ and $t[i_{x+1}..i_{x+1}+a_{x+1}-1] = w[0..a_{x+1}-1]$ for some positive maximal a_x and a_{x+1}. Then, $a_{x+1} - a_x \ge -\delta_{x+1}$, where δ_{x+1} is the shortest period of $w[0..a_x - 1]$.

[3] $\mathtt{select}(A, 1, i)$ is the position of the ith one in bitvector A, where i starts from zero.

number of zeros before the jth one in $\mathbf{ms}$ is $j + \mathrm{MS}_{t,s,\tau}[j] \geq |t| - i$, it follows that the ith zero from the right in $\mathbf{ms}$ is preceded by exactly $j = |t| - i - \mathrm{MS}_{\bar{t},\bar{s},\tau}[i]$ ones, therefore $\mathrm{MS}_{\bar{t},\bar{s},\tau}[i] = 2(|t| - i) - 1 - \mathtt{select}(\mathbf{ms}, 0, |t| - i - 1)$. $\square$

Corollary 1. $\mathbf{ms}_{\bar{t},\bar{s},\tau} = \neg\overline{\mathbf{ms}_{t,s,\tau}}$

Proof. Since $\mathrm{MS}_{\bar{t},\bar{s},\tau}[i] = 2(|t| - i) - 1 - \mathtt{select}(\mathbf{ms}_{t,s,\tau}, 0, |t| - i - 1)$ and at the same time $\mathrm{MS}_{\bar{t},\bar{s},\tau}[i] = \mathtt{select}(\mathbf{ms}_{\bar{t},\bar{s},\tau}, 1, i) - 2i$, the position of the ith one from the left in $\mathbf{ms}_{\bar{t},\bar{s},\tau}$ equals the position of the ith zero from the right in $\mathbf{ms}_{t,s,\tau}$ for all i. $\square$

We can encode $\mathrm{DS}_{t,\tau}$ in a similar bitvector $\mathbf{ds}_{t,\tau}$ with $2|t| + 1 + o(|t|)$ bits, by appending $\mathrm{DS}[i] - \mathrm{DS}[i-1] + 1$ zeros and a one for every $i \in [1..|t| - 1]$. We assume again $\mathrm{DS}[-1] = 1$, and we append a final zero if $\mathbf{ds}$ contains just $2|t|$ bits. As before $\mathrm{DS}[i] = \mathtt{select}(\mathbf{ds}, 1, i) - 2i$, but now $\mathrm{DS}_{\bar{t}}[i] = 2(|t| - i) + 1 - \mathtt{select}(\mathbf{ds}, 0, |t| - i)$. This implies that $2|t| + 1 - \mathtt{select}(\mathbf{ds}, 0, |t| - i) = \mathtt{select}(\mathbf{ds}_{\bar{t}}, 1, i)$, or in other words that $\mathbf{ds}_{\bar{t}} = 0 \cdot (\neg\mathbf{ds}_t[1..2|t|])$.

More generally, the encoding described in Lemma 1 can be used to index any δ-monotone sequence $a_0 \ldots a_{n-1}$ in $n + a_{n-1} + \sum_{i=1}^{n-1} \delta_i$ bits, by concatenating $a_i - a_{i-1} + \delta_i$ zeros followed by a one for every i. The number of zeros before the ith one in this bitvector is $a_i + \sum_{j=1}^{i} \delta_j$, thus it is necessary to keep all the prefix sums of δ to answer queries on a_i.

We are interested in applications where the reference string s is fixed, and we have to output either $\mathbf{ds}_{s,\tau}$, or $\mathbf{ms}_{t,s,\tau}$ in reply to a query containing t. It turns out that both bitvectors can be derived from the Burrows-Wheeler transform of s and of $\bar{s}$, augmented with the corresponding suffix tree topologies.

Theorem 1. *Let* BWT_s *and* $\mathrm{BWT}_{\bar{s}}$ *be the Burrows-Wheeler transform of a string* s *and of its reverse* $\bar{s}$, *indexed to support a backward step in* α *time. Assume that we have a representation of the suffix tree topology of* s *and of* $\bar{s}$ *that supports parent operations in* β *time. Given a string* t *and a threshold* τ, *we can compute* $\mathbf{ms}_{t,s,\tau}$ *in* $O(|t|(\alpha + \beta))$ *time, and in* $O(\log|s| + \log|t|)$ *bits of space in addition to the input and the output.*

Proof. We apply twice the algorithm described in [13]. First, we scan t from right to left, using BWT_s and the suffix tree topology of s to determine the runs of consecutive ones in $\mathbf{ms}$. Specifically, we build a bitvector $\mathbf{runs}[1..|t| - 1]$ where $\mathbf{runs}[i] = 1$ iff $\mathrm{MS}[i] = \mathrm{MS}[i-1] - 1$, i.e. iff there is no zero between the ith and the $(i-1)$th ones in $\mathbf{ms}$. Assume that we have the interval $(i_w, j_w)_s$ in BWT_s that corresponds to substring $w = t[k..k + \mathrm{MS}[k] - 1]$ (the single-character interval for $k = |t| - 1$ can be directly derived from the table $C[1..\sigma]$ used in backward search). We try to perform a backward step using symbol $a = t[k-1]$: if the step leads to an interval of size at least τ, we set $\mathbf{runs}[k] = 1$ and we update the BWT interval to $(i_{aw}, j_{aw})_s$. Otherwise, we set $\mathbf{runs}[i] = 0$, we update the BWT interval to the interval of the parent of the proper locus of w in the suffix tree of s, and we try another backward step with character a. We repeat these operations until a backward step leads to an interval of size at least τ. Note that, since we are not using the string depth operation, we don't know $\mathsf{MS}[k]$ for any $k < k^*$, where $k^* + \mathsf{MS}[k^*] = |t|$.

In the second phase we symmetrically scan t from left to right, using $\mathsf{BWT}_{\bar{s}}$, the suffix tree topology of $\bar{s}$, and vector **runs**, to build **ms**. Assume that we have the interval $(i_w, j_w)_{\bar{s}}$ in $\mathsf{BWT}_{\bar{s}}$ that corresponds to substring $w = t[k..h-1]$ such that $\mathsf{MS}[k] \geq h-k$ and $\mathsf{MS}[k-1] = h-k$ (again, we can derive the interval for $t[0]$ from the table $C[1..\sigma]$ used in backward search). We try to perform a backward step with symbol $t[h]$: if the step leads to an interval of size at least τ, we continue issuing backward steps with the following symbols of t, until we reach a position h^* in t such that a backward step with character $t[h^*]$ from the interval $(i_w, j_w)_{\bar{s}}$ of substring $w = t[k..h^*-1]$ leads to an interval of size less than τ. We thus know that $\mathsf{MS}[k] = h^* - k$, so we append $h^* - k - \mathsf{MS}[k-1] + 1 = h^* - h + 1$ zeros and a one to **ms**. Then, we iteratively replace $(i_w, j_w)_{\bar{s}}$ with the interval of the parent of the proper locus of w in the suffix tree of $\bar{s}$, and we try another backward step with symbol $t[h^*]$, until we reach an interval $(i_{w'}, j_{w'})_{\bar{s}}$ for which such backward step leads to an interval of size at least τ. Let this interval correspond to substring $w' = t[k'..h^*-1]$. Note that $\mathsf{MS}[k'] > \mathsf{MS}[k'-1] - 1$ and $\mathsf{MS}[x] = \mathsf{MS}[x-1] - 1$ for all $x \in [k+1..k'-1]$, therefore **runs**$[x]$ must be one for $x \in [k+1..k'-1]$ and **runs**$[k']$ must be zero, i.e. k' is the index of the first zero to the right of position k in **runs**. We can thus append $k' - k - 1$ ones to **ms** and repeat the process from substring $t[k'..h^*]$ and its interval in $\mathsf{BWT}_{\bar{s}}$.

If we store vector **runs** in the last $|t| - 1$ bits of **ms**, each iteration of the second phase of the algorithm overwrites only parts of **runs** that will not be used in following iterations. This is easy to see, and is left to the reader.   □

The two-pass approach of Theorem 1 completely avoids string-depth operations. Recall that computing the depth of a suffix tree node ultimately requires to decompress a position in the underlying compressed suffix array: if such array is encoded in $O(|s| \log \sigma)$ bits, the best known time complexity for this operation is $O(\log^\epsilon |s|)$. Complexity increases when the compressed suffix array is encoded in $|s| \log \sigma + o(|s|)$ bits or in $|s| \log \sigma (1 + o(1))$ bits. Note also that the algorithm in Theorem 1 uses either BWT_s or $\mathsf{BWT}_{\bar{s}}$ at every step, i.e. it does not need to keep their intervals synchronized. A similar result holds for $\mathsf{ds}_{s,\tau}$:

Theorem 2. *Let* BWT_s *and* $\mathsf{BWT}_{\bar{s}}$ *be the Burrows-Wheeler transform of a string* s *and of its reverse* $\bar{s}$, *indexed to support a backward step in* α *time. Assume that we have a representation of the suffix tree topology of* s *and of* $\bar{s}$ *that supports parent operations in* β *time. We can compute* $\mathsf{ds}_{s,\tau}$ *in* $O(|s|(\alpha + \beta))$ *time, and in* $O(\log |s|)$ *bits of space in addition to the input and the output.*

Proof. By applying almost verbatim the two-phase approach of Theorem 1. In the first phase we build vector **runs** by trying a backward step with character $a = s[k-1]$ from the interval in BWT_s of the longest string that starts at position k and that occurs more than τ times, i.e. from the interval of string $w = s[k..k + \mathsf{DS}_{s,\tau}[k] - 2]$. If this step leads to an interval of size greater than τ, we set **runs**$[k] = 1$ and we repeat from the BWT interval of aw. Otherwise, we set **runs**$[k] = 0$, we move to the interval of the parent of the proper locus of w in the suffix tree of s, and we try another backward step with character a. We repeat these operations until a backward step leads to an interval of size at most τ.

In the second phase, assume again that we know the interval in $\mathsf{BWT}_{\bar{s}}$ of the longest string that starts at position k and that occurs more than τ times, i.e. the interval of string $w = s[k..h-1]$ with $h = k + \mathsf{DS}_{t,\tau}[k] - 1$. We iteratively move to the interval of the parent of the proper locus of w in the suffix tree of $\bar{s}$ and we try a backward step with character $a = s[h]$, until such a step leads to an interval of size greater than τ. Thanks to $\mathtt{runs}$, we know that the interval from which the last backward step was taken corresponds to string $s[k'..h-1]$, where k' is the position of the first zero to the right of k in $\mathtt{runs}$. We can thus append $k' - k - 1$ ones to $\mathtt{ds}$ and move to string $s[k'..h]$. We then try backward steps with the characters at position $h+1, h+2, \dots$ until a backward step reaches an interval of size at most τ: this gives $\mathsf{DS}[k']$. $\qquad\square$

Corollary 2. *Given a string s, there is a data structure that allows to compute: (1) $\mathtt{ds}_{s,\tau}$ in $O(|s| \log \sigma)$ time and in $O(\log |s|)$ bits of space in addition to the input and the output; (2) $\mathtt{ms}_{t,s,\tau}$ for any string t, in $O(|t| \log \sigma)$ time and in $O(\log |s| + \log |t|)$ bits of space in addition to the input and the output. This data structure takes $2|s| \log \sigma + O(|s|)$ bits of space and can be built in $O(|s| \log \sigma)$ time using $O(|s| \log \sigma)$ bits of space.*

Proof. BWT_s and $\mathsf{BWT}_{\bar{s}}$ can be built in $O(|s| \log \log \sigma)$ time and $O(|s| \log \sigma)$ bits of space using the algorithm described in [9]. The wavelet trees on BWT_s and $\mathsf{BWT}_{\bar{s}}$ can then be built in $O(|s| \log \sigma)$ time. The suffix tree topologies of s and $\bar{s}$ can be built in $O(|s| \log \sigma)$ time using just the corresponding wavelet trees, using the approach described in [1,2]. $\qquad\square$

Corollary 3. *Given a string s, there is a data structure that allows to compute: (1) $\mathtt{ds}_{s,\tau}$ in $O(|s|)$ time and in $O(\log |s|)$ bits of space in addition to the input and the output; (2) $\mathtt{ms}_{t,s,\tau}$ for any string t, in $O(|t|)$ time and in $O(\log |s| + \log |t|)$ bits of space in addition to the input and the output. This data structure takes $2|s| \log \sigma + o(|s| \log \sigma)$ bits of space and can be built in randomized $O(|s|)$ time using $O(|s| \log \sigma)$ bits of space.*

Proof. Given a bitvector of length n, we can build in $O(n)$ time a data structure that supports constant-time select queries and that takes $2n + o(n)$ bits of space [4,12]. We achieve the claimed time complexity by plugging in Theorem 1 the index described in [3], which supports constant-time backward steps and takes $2|s| \log \sigma + o(|s| \log \sigma)$ bits of space. The latter is built in randomized $O(|s|)$ time and $O(|s| \log \sigma)$ bits of space, using the algorithm described in [1,2]. The suffix tree topologies are then built in $O(|s|)$ time from the indexes of [3]. $\qquad\square$

3   Computing and Indexing BMS and BDS

We start by generalizing the algorithm for computing BMS from MS described in [19], in order to make it work on user-defined *blocks*, rather than on positions, of the input string. We say that a pair (i, j) is a *matching statistics interval* of string t if $i \in [0..|t| - 1]$ and $j = i + \mathsf{MS}_{t,s,\tau}[i] - 1$. We say that an interval is *maximal* if it is not contained into any other interval, and we denote the list of all maximal intervals by $\mathcal{I}$. The following properties are immediate, and will be used extensively in the sequel.

Property 1. There is at most one interval in $\mathcal{I}$ that ends at any given position j in t. Given an interval (i, j), the maximal interval that contains (i, j) is $(j - \mathsf{MS}_{\bar{t},\bar{s},\tau}[|t| - j - 1] + 1, j)$.

Property 2. Let $(i_0, j_0), (i_1, j_1), \ldots, (i_m, j_m)$ be a subset of $\mathcal{I}$ sorted by increasing starting position. Then $\sum_{h=1}^{m} i_h - i_{h-1} \leq |t|$ and $\sum_{h=1}^{m} j_h - j_{h-1} \leq |t|$.

It is natural to compute $\mathsf{BMS}_{t,s,\tau}$ by scanning t and $\mathsf{MS}_{t,s,\tau}$ while keeping in memory the *active* subset of $\mathcal{I}$, i.e. the set of intervals in $\mathcal{I}$ that cover the current position in t. The following algorithm is an alternative to [19]:

Lemma 2 ([19]). $\mathsf{BMS}_{t,s,\tau}$ *can be computed from* $\mathsf{MS}_{t,s,\tau}$ *in* $O(|t|)$ *time and* $O(|t| \log |t|)$ *bits of space.*

Proof. Assume that we are at position k in t, and let I be the subset of $\mathcal{I}$ that contains all the intervals that start before position k and that cover position k, i.e. all intervals $(i, j) \in \mathcal{I}$ with $j = i + \mathsf{MS}[i] - 1$ and $i < k \leq j$. We implement I as a doubly-linked list, with a node for every distinct length of an interval in I. The node associated with length ℓ stores all the intervals of length ℓ in I as a doubly-linked list. We assume that I is sorted by decreasing length, and we keep it sorted during the algorithm using insertion sort. Moreover, we use array $\mathsf{ends}[0..|t| - 1]$ to store in $\mathsf{ends}[j]$ a pointer to the only interval in I that ends at j, if any. Let prev be a pointer to the *previous interval* $(i_{prev}, j_{prev}) = (k-1, k-2+\mathsf{MS}[k-1])$ if it belongs to I (prev is null if such interval does not belong to I), and let last be a pointer to the node of I that contains the interval $(i, j) \in I$ with maximum i. Let $\delta_k = \mathsf{MS}[k] - \mathsf{MS}[k - 1]$.

If $\delta_k = -1$, then $\mathsf{BMS}[k]$ is the length of the first node in I, list I does not change, and we set prev to null. Similarly, if $\mathsf{MS}[k] = 1$, then $\mathsf{BMS}[k]$ is the length of the first node in I, or one if I is empty, list I does not change, and we set prev to null. Otherwise, we proceed as follows. If $\mathsf{prev} \neq \emptyset$, we remove from I the interval pointed by prev and we set $\mathsf{ends}[j_{prev}]$ to null. Then, we add to I the current interval $(k, k + \mathsf{MS}[k] - 1)$, as follows.

Let ℓ be the length of node last. If $\mathsf{MS}[k] = \ell$ we add the new interval to last, otherwise we scan I linearly starting from the node associated with length ℓ, until we find the node associated with length $\mathsf{MS}[k]$ (we create a new node if no such node exists). This linear scan visits at most $|\mathsf{MS}[k] - \ell| = |j_h - j_{h-1} - i_h + i_{h-1}|$ nodes of I for every maximal interval $(i_h, j_h) \in \mathcal{I}$, thus it visits in total $\sum_{h=1}^{m} |j_h - j_{h-1} - i_h + i_{h-1}| \leq \sum_{h=1}^{m} j_h - j_{h-1} + i_h - i_{h-1} \leq 2|t|$ nodes of I. Note that this corresponds to charging the moves along I to the bits of ms.

After having added $(k, k+\mathsf{MS}[k]-1)$ to I, we update prev, last and $\mathsf{ends}[k+\mathsf{MS}[k] - 1]$ to point to the newly inserted interval. Once again, $\mathsf{BMS}[k]$ equals the length of the first node in I. Before moving to position $k + 1$, we remove from I the interval pointed by $\mathtt{ends}[k]$, if any. $\square$

Lemma 3. $\mathsf{BDS}_{t,\tau}$ *can be computed from* $\mathsf{DS}_{t,\tau}$ *in* $O(|t|)$ *time and* $O(|t| \log |t|)$ *bits of space.*

Proof. By applying the algorithm in Lemma 2 almost verbatim. Now, for every position k in t, I stores all the *minimal intervals* that start before k and cover k, i.e. all the intervals that start before k, cover k, and do not contain any other interval. I is sorted by increasing length. We insert interval $(k, k + \mathsf{DS}[k] - 1)$ in I for every position k in t, but if $\mathsf{DS}[k] - \mathsf{DS}[k-1] = -1$ we first remove the interval (i_{prev}, j_{prev}) pointed by $\mathtt{prev}$, since it is not minimal, and we set $\mathtt{ends}[j_{prev}]$ to null. We thus have the guarantee that there is at most one interval in I that ends at every position j of t. It is easy to see that the moves along I can still be charged to the bits of $\mathtt{ds}$. $\square$

Note that the notion of interval holds for any δ-monotone sequence $a = a_0 a_1 \dots a_n$, and the algorithms in Lemma 2 and 3 allow to compute, respectively, the length of a longest and of a shortest interval of a that covers every position $i \in [0..n]$ in the sequence, in time linear *in the size of the index of the sequence*. In particular, the algorithm in Lemma 2 can be applied to compute BRS_t, the bidirectional version of the repeating statistics, from the unidirectional RS_t.

Assume now that string t is the concatenation of m nonempty substrings, i.e. that $t = t_0 \cdot t_1 \cdots t_{m-1}$ with $t_i \in \Sigma^+$ for all $i \in [0..m-1]$. We call such substrings *blocks* in what follows, and we denote with $\beta(j)$ the block that contains position j in t. We assume that block boundaries are marked in a bitvector of size $|t|$, indexed to support rank operations[4]. We say that an interval *spans* block i if it starts before block i and if it ends after block i. The algorithm described in Lemma 2 can be adapted to compute the length of a longest matching statistics interval (or, symmetrically, of a shortest distinguishing statistics interval) that spans every block $i \in [1..m-2]$.

Lemma 4. *The length of a longest matching statistics interval that starts before block i and ends after block i, for every $i \in [1..m-2]$, can be computed from* $\mathsf{MS}_{t,s,\tau}$ *in* $O(|t|)$ *time.*

Proof. We apply again the algorithm in Lemma 2 almost verbatim. Now array $\mathtt{ends}$ has one position per block, and $\mathtt{ends}[k]$ points to the longest interval in I that starts before block k and ends inside block k.

Assume that we are processing block k. Let $r = \max\{\beta(i + \mathsf{MS}[i] - 1) : \beta(i) = k\}$ be the rightmost block that contains the ending position of an interval that starts inside block k. Similarly, let $\ell = \max\{\mathsf{MS}[i] : \beta(i) = k\}$ be the length of a longest interval that starts inside block k. We call any interval that starts inside block k and that ends inside block r a *farthest interval*, and any interval

---

[4] $\mathtt{rank}(A, 1, i)$ is the number of ones in bitvector A up to position i, included.

of maximum length ℓ a *longest interval*[5]. Let (i_r, j_r) be a farthest interval of maximum possible length, i.e. $j_r - i_r = \max\{\mathsf{MS}[i] : \beta(i) = k, \beta(i + \mathsf{MS}[i] - 1) = r\}$. This interval spans every block $h \in [k + 1..r - 1]$, and no interval that starts inside block k and spans the same blocks is longer, thus (i_r, j_r) will be active in the following iterations. Similarly, if $j_r - i_r < \ell$, let (i_ℓ, j_ℓ) be a longest interval. Clearly no interval that starts inside block k and that spans the same blocks as (i_ℓ, j_ℓ) is longer, thus (i_ℓ, j_ℓ) will be active in the following iterations.

When we process block k, we first remove from I the interval that ends inside block k, by following the pointer in $\mathsf{ends}[k]$, and we assign to block k the length of the first node in I. Then, we insert in I the interval (i_ℓ, j_ℓ) (if it exists), and then the interval (i_r, j_r). If pointer $\mathsf{ends}[\beta(\mathsf{j}_\ell)]$ is null, we store in $\mathsf{ends}[\beta(\mathsf{j}_\ell)]$ a pointer to (i_ℓ, j_ℓ). Otherwise, if the length of interval $\mathsf{ends}[\beta(\mathsf{j}_\ell)]$ is smaller than ℓ, we remove interval $\mathsf{ends}[\beta(\mathsf{j}_\ell)]$ from I and we store in $\mathsf{ends}[\beta(\mathsf{j}_\ell)]$ a pointer to (i_ℓ, j_ℓ). This is because the old interval spans the same blocks as the new interval, but it is shorter. We do the same for (i_r, j_r).

Insertions in I work in the same way as before, with last pointing to the interval $(i, j) \in I$ with maximum i. The total number of movements in I can be still bounded by $2|t|$, since we are inserting a subset of $\mathcal{I}$. $\qquad\square$

Lemma 5. *The length of a shortest distinguishing statistics interval that starts before block i and ends after block i, for every $i \in [1..m - 2]$, can be computed from $\mathsf{DS}_{t,\tau}$ in $O(|t|)$ time.*

Proof. By adapting the algorithm in Lemma 4. Assume that we are at block k, and let L be the list of all tuples $(i, j, j - i, \beta(j))$ where $j = i + \mathsf{DS}[i] - 1$, and $\beta(i) = k$. We sort L by increasing third component of each tuple (length), and then we stable-sort L by the last component of each tuple (ending block). Then, we scan the sorted L and we insert in I the first interval that we find associated with every block h, updating the corresponding pointers in ends if the previous interval is longer than the new interval. The total number of movements in I after insertions can be still bounded by $2|t|$, since we are inserting a subset of the minimal intervals we inserted in Lemma 3. $\qquad\square$

Once again, these blocked variants of BMS and BDS can be applied to any δ-monotone sequence (thus in particular to BRS) in time linear in the size of the index for such sequence.

As done with MS and DS, we would like to build succinct indexes that support *bidirectional* queries in constant time. To this end, we augment ms and ds by exploiting the following property, whose immediate proof is left to the reader:

Property 3. Given a position $j \in [0..|t| - 1]$, the position $i^* = \min\{i \in [0..j - 1] : j \leq i + \mathsf{MS}[i] - 1\}$ can be computed by $\mathsf{select}(\mathsf{ms}, 0, j) - j + 1$.

Note that the same property holds for DS, for RS, and for every δ-monotone sequence $a_0 a_1 \ldots a_n$ with $\delta_i = 1$ for all $i \in [1..n]$. However, the property does not generalize to δ-monotone sequences where δ_i is not constantly one for all $i \in [1..n]$.

---

[5] If all blocks have the same length, a longest interval spans every block in $[k+1..r-2]$.

Indeed, in such cases position j is covered by position $i < j$ iff the ith one in the index of the sequence is preceded by at least $j + 1 + \sum_{k=1}^{i}(\delta_k - 1)$ zeros.

Theorem 3. *There is a data structure that takes $2|t| + o(|t|)$ bits of space and that answers queries on $\mathrm{BMS}_{t,s,\tau}[i]$ for any i in constant time.*

Proof. We store `ms` in $2|t|$ bits, and we augment it with an index that takes $O(|t| \log \log |t| / \log |t|) \in o(|t|)$ bits, and that supports rank and select queries in constant time [7,16]. We assume that we can read any block of $\Theta(\log |t|)$ consecutive bits of `ms` in constant time. Moreover, we partition `ms` into $B = \lceil 2|t| / \log |t| \rceil$ blocks of size $\log |t|$ each. We use again the notation $\beta(i)$ to identify the block that contains position i in `ms`. For every block k, let (i_k, j_k) be a longest matching statistics interval of t such that $\beta(\mathtt{select}(\mathtt{ms}, 1, i_k)) = k$. We store the position of $\mathtt{select}(\mathtt{ms}, 1, i_k)$ inside each block in array $\mathtt{start}[0..B-1]$, using $2|t| \log \log |t| / \log |t|$ bits. Then, we build a range-maximum data structure RMQ on the pairs $(k, j_k - i_k)$ for $k \in [0..B-1]$, using $2(2|t| / \log |t|) + o(|t|) = 4|t| / \log |t| + o(|t| / \log |t|)$ bits of space [6].

Given a position j in t, we use Property 3 to compute i^*, the smallest $i < j$ with $j \leq i + \mathrm{MS}[i] - 1$. Let $p = \beta(\mathtt{select}(\mathtt{ms}, 1, i^*))$ and $q = \beta(\mathtt{select}(\mathtt{ms}, 1, j))$. Since all the matching statistics intervals that cover position j must start between i^* and j, we query RMQ with the pair $(p+1, q-1)$ to get the block $k \in [p+1..q-1]$ with longest interval in constant time, and we compute the length ℓ_1 of such interval by $\ell_1 = h - 2 \cdot \mathtt{rank}(\mathtt{ms}, 1, h)$ where $h = k \log |t| + \mathtt{start}[k]$. Finally, we load in constant time blocks p and q and we use the Four Russians technique to compute in constant time the lengths ℓ_2 and ℓ_3 of the longest matching statistics intervals that starts inside block p and q, respectively, using a precomputed table of size $o(|t|)$ bits. We finally return $\max\{\ell_1, \ell_2, \ell_3\}$. $\square$

Corollary 4. *There is a data structure that takes $2|t| + o(|t|)$ bits of space and that answers queries on $\mathrm{BDS}_t[i]$ for any i in constant time.*

Proof. By applying the approach in Theorem 3 verbatim, using `ds` instead of `ms` and a range-minimum rather than a range-maximum data structure. $\square$

References

1. Belazzougui, D.: Linear time construction of compressed text indices in compact space. In: Proceedings of the 46th ACM Symposium on Theory of Computing. ACM (2014)
2. Belazzougui, D.: Linear time construction of compressed text indices in compact space. ArXiv preprint ArXiv:1401.0936 (2014)
3. Belazzougui, D., Navarro, G.: Alphabet-independent compressed text indexing. ACM Transactions on Algorithms 10(4) (2014)
4. Clark, D.: Compact Pat Trees. PhD thesis, University of Waterloo, Canada (1996)
5. Farach, M., Noordewier, M., Savari, S., Shepp, L., Wyner, A., Ziv, J.: On the entropy of DNA: Algorithms and measurements based on memory and rapid convergence. In: Proceedings of the Sixth Annual ACM-SIAM Symposium on Discrete Algorithms, pp. 48–57 (1995)

6. Fischer, J., Heun, V.: Space-efficient preprocessing schemes for range minimum queries on static arrays. SIAM Journal on Computing 40(2), 465–492 (2011)
7. Golynski, A.: Optimal lower bounds for rank and select indexes. Theoretical Computer Science 387(3), 348–359 (2007)
8. Haubold, B., Pierstorff, N., Möller, F., Wiehe, T.: Genome comparison without alignment using shortest unique substrings. BMC Bioinformatics 6(1), 123 (2005)
9. Hon, W.-K., Sadakane, K., Sung, W.-K.: Breaking a time-and-space barrier in constructing full-text indices. SIAM J. Comput. 38(6), 2162–2178 (2009)
10. İleri, A.M., Külekci, M.O., Xu, B.: Shortest unique substring query revisited. In: Kulikov, A.S., Kuznetsov, S.O., Pevzner, P. (eds.) CPM 2014. LNCS, vol. 8486, pp. 172–181. Springer, Heidelberg (2014)
11. Kurtz, S., Phillippy, A., Delcher, A.L., Smoot, M., Shumway, M., Antonescu, C., Salzberg, S.L.: Versatile and open software for comparing large genomes. Genome Biology 5(2), R12 (2004)
12. Munro, J.I.: Tables. In: Chandru, V., Vinay, V. (eds.) FSTTCS 1996. LNCS, vol. 1180, pp. 37–42. Springer, Heidelberg (1996)
13. Ohlebusch, E., Gog, S., Kügel, A.: Computing matching statistics and maximal exact matches on compressed full-text indexes. In: Chavez, E., Lonardi, S. (eds.) SPIRE 2010. LNCS, vol. 6393, pp. 347–358. Springer, Heidelberg (2010)
14. Pei, J., Wu, W.-H., Yeh, M.-Y.: On shortest unique substring queries. In: 2013 IEEE 29th International Conference on Data Engineering (ICDE), pp. 937–948. IEEE (2013)
15. Philippe, N., Salson, M., Commes, T., Rivals, E.: CRAC: An integrated approach to the analysis of RNA-seq reads. Genome Biology 14(3), R30 (2013)
16. Raman, R., Raman, V., Satti, S.R.: Succinct indexable dictionaries with applications to encoding k-ary trees, prefix sums and multisets. ACM Transactions on Algorithms (TALG) 3(4), 43 (2007)
17. Robertson, M.M.: A generalization of quasi-monotone sequences. Proceedings of the Edinburgh Mathematical Society (Series 2) 16(01), 37–41 (1968)
18. Sadakane, K.: Compressed suffix trees with full functionality. Theory of Computing Systems 41(4), 589–607 (2007)
19. Schnattinger, T., Ohlebusch, E., Gog, S.: Bidirectional search in a string with wavelet trees and bidirectional matching statistics. Inf. Comput. 213, 13–22 (2012)
20. Tsuruta, K., Inenaga, S., Bannai, H., Takeda, M.: Shortest unique substrings queries in optimal time. In: Geffert, V., Preneel, B., Rovan, B., Štuller, J., Tjoa, A.M. (eds.) SOFSEM 2014. LNCS, vol. 8327, pp. 503–513. Springer, Heidelberg (2014)
21. Ulitsky, I., Burstein, D., Tuller, T., Chor, B.: The average common substring approach to phylogenomic reconstruction. Journal of Computational Biology 13(2), 336–350 (2006)
22. Weiner, P.: The file transmission problem. In: Proceedings of the National Computer Conference and Exposition, June 4-8, pp. 453–453. ACM (1973)
23. Weiner, P.: Linear pattern matching algorithms. In: Switching and Automata Theory, pp. 1–11. IEEE (1973)

I/O-Efficient Dictionary Search with One Edit Error

Chin-Wan Chung[1], Yufei Tao[2], and Wei Wang[3]

[1] Korean Advanced Institute of Science and Technology, Daejeon, Korea
[2] Chinese University of Hong Kong, New Territories, Hong Kong
[3] University of New South Wales, Sydney, Australia
chungcw@kaist.edu, taoyf@cse.cuhk.edu.hk,
weiw@cse.unsw.edu.au

Abstract. This paper studies the *1-error dictionary search* problem in external memory. The input is a set D of strings whose characters are drawn from a constant-size alphabet. Given a string q, a query reports the ids of all strings in D that are within 1 edit distance from q. We give a structure occupying $O(n/B)$ blocks that answers a query in $O(1 + \frac{m}{wB} + \frac{k}{B})$ I/Os, where n is the total length of all strings in D, m is the length of q, k is the number of ids reported, w is the size of a machine word, and B is the number of words in a block.

1 Introduction

In this paper, we consider the *1-error dictionary search* problem defined as follows. The input D—the *dictionary*—is a set of strings whose characters are drawn from a constant-size alphabet Σ. Each string is associated with a distinct integer as its id. Given a string q with characters in Σ, a query reports the ids of those strings $s \in D$ with $dist(s, q) \leq 1$, where $dist(s, q)$ is the *edit distance* between s and q (a.k.a. the *Levenshtein distance*).[1]

The problem is fundamental to search engines that aim to tolerate typos in the keywords entered by users. It is well known [12,13] that between 80%-95% of the typos in practice are 1-edit-distance errors, thus providing a strong motivation to support 1-error dictionary search efficiently.

Computation Model. We study the problem in the standard *external memory* (EM) model [1]. Under this model, a machine has a disk that is an infinite sequence of *blocks*, each of which contains B words. Computation takes place only in (internal) memory, whose size M (in number of words) is at least $2B$. An I/O exchanges a block of data between the disk and memory. The *time* of an algorithm is the number of I/Os performed. The *space* of a structure is the number of blocks in the shortest prefix of the disk containing all the bits of the structure. Denote by w the number of bits in a machine word. We make the standard assumption that $w = \Omega(\lg B)$ (otherwise, the machine cannot even encode the address space of all the words in memory).

We will also make frequent use of the following notations:

---

[1] Specifically, $dist(s, q)$ is the smallest number of the following edit operations needed to convert s to q: (i-ii) inserting or deleting a character, and (iii) replacing a character with another one.

E. Moura and M. Crochemore (Eds.): SPIRE 2014, LNCS 8799, pp. 191–202, 2014.
© Springer International Publishing Switzerland 2014

Table 1. Comparison of our and previous results in external memory

space	query	update	source	remarks
$O(n/B)$	$O(m+k)$	$O(l)$ amortized (ins. only)	[5]	assume that data and query strings are equally long
$O(\frac{t}{B} + \frac{n}{wB})$	$O(m+k)$	-	[3]	
$O(\frac{n}{B} \lg n)$	$O(\frac{m}{B} + \frac{k}{B} + \lg n \lg \lg_B n)$	-	[11]	designed for the more general "full-text indexing" problem
$O(n/B)$	$O(1 + \frac{m}{wB} + \frac{k}{B})$	$O(l)$ expected	**New**	

- n: the total length of all the strings in D;
- t: the number of strings in D;
- m: the length of a query string;
- k: the number of qualifying strings for a query.

We define $\lg_b x = \max\{1, \log_b x\}$ with $b = 2$ if omitted.

Previous Results. The 1-error dictionary search problem was first studied in internal memory. Belazzougui [3] presented a static structure of $O(t + n/w)$ space that answers a query in $O(m + k)$ time. Belazzougui and Venturini [4] showed that the space can be reduced if the dictionary has small entropy, but their result does not improve that of [3] in general. Brodal and Gasieniec [5] considered a special instance of the problem, where all the data and query strings have the same length m (i.e., a query with string q essentially retrieves the strings whose *hamming distances* are within 1 from q). For this instance, they gave a structure of space $O(n)$ that answers a query in $O(m + k)$ time; in addition, their structure is semi-dynamic: it supports the insertion of a length-l data string in $O(l)$ amortized time.

The above structures, when applied in external memory, incur $O(m + k)$ I/Os answering a query. This, unfortunately, is at least a factor of B away from what we would like to achieve. Currently, the most I/O-efficient structure is due to Hon et al. [11]. Their structure, which is intended for a more general problem called *full-text indexing*, uses $O(\frac{n}{B} \lg n)$ blocks, and answers a query in $O(\frac{m}{B} + \frac{k}{B} + \lg n \cdot \lg \lg_B n)$ I/Os. It does not support updates efficiently.

It is worth mentioning that, while in this paper we focus on queries with 1-edit-distance errors, progress has also been made in the past decade towards tolerating a larger number of errors. Interested readers may refer to [6,7,15] for entry points into the literature.

Our Results. We give a new structure for 1-error dictionary search that uses $O(n/B)$ space, answers a query in $O(1 + \frac{m}{wB} + \frac{k}{B})$ I/Os, and supports the insertion and deletion of a length-l string in $O(l)$ expected I/Os (see Table 1 for a comparison). With B set to 1, our structure also works in the RAM model directly, and provides a new tradeoff between space and query time there.

Remarks. Henceforth, we will focus on a binary alphabet $\Sigma = \{0, 1\}$. Our techniques can be extended to any constant-size alphabet, without affecting the claimed complexities. Clarification will be duly made when this is not straightforward.

2 Exact Matching

In this section, we revisit the (precise) *dictionary search* problem, whose solution will be useful later. Specifically, the input is the same dictionary D as in 1-error dictionary search. Given a string q, we want to report its id in D if $q \in D$, or declare its absence otherwise. This problem is also commonly known as *exact matching*.

Given a string s, we denote by $|s|$ the length of s. For an $i \in [1, |s|]$, let s_i be the i-th character of s. Given i, j with $1 \le i \le j \le |s|$, let $s_{i..j}$ be the substring of s starting and ending at s_i and s_j, respectively. Specially, if $i > j$, $s_{i..j}$ is an empty string.

2.1 Preliminaries

Let us first review a result on perfect hashing. Given an integer $x > 0$, $[x]$ represents the set $\{0, 1, ..., x - 1\}$. Consider a set $S \subseteq [U]$ for some positive integer U. Set $n = |S|$. Let f be a function from $[U]$ to $[2n]$. We say that f is *perfect* on S if it is injective with respect to S, namely, for any two different integers x_1, x_2 of S, $f(x_1) \neq f(x_2)$. Furthermore, f is *stable* if, for each $x \in S$, $f(x)$ remains the same until x is deleted from S.

Lemma 1 ([8]). *Let U be a positive integer at most 2^{wB-1}. For any $S \subseteq [U]$, we can store a stable perfect function f of S using $O(1 + \frac{n}{wB} \lg \lg \frac{U}{n})$ blocks, where $n = |S|$. For any $x \in [U]$, $f(x)$ can be computed in constant I/Os. If an integer is inserted or deleted in S, the representation of f can be updated in constant I/Os expected.*

We will also need the next fact about the string B-tree:

Lemma 2 ([10]). *Let D be a set of t strings (each with an integer id), and n be their total length. We can store D in a string B-tree of $O(1 + \frac{t}{B} + \frac{n}{wB})$ space such that, given a string q of length m, using $O(\lg_B t + \frac{m}{wB})$ I/Os, we can report the id of q in D or declare the absence of q in D. A string of length l can be inserted and deleted in D with $O(\lg_B t + \frac{l}{wB})$ I/Os.*

2.2 A New Structure

We consider a slightly more general version of exact matching. Suppose that each string $s \in D$ carries an arbitrary *information field* that occupies $O(1)$ words. We want to support a *probe operation*: given a string q, decide whether $q \in D$ and if so, return the information field of q. The rest of the subsection serves as a proof for:

Theorem 1. *A set D of strings (each with an integer id) can be stored in a structure of $O(1 + \frac{t}{B} + \frac{n}{wB})$ blocks, where t is the number of strings in D and n is their total length, such that a probe operation can be performed in $O(1 + \frac{m}{wB})$ I/Os, where m is the length of the query string. To insert/delete a string with length l in D, the structure can be updated in $O(1 + \frac{l}{wB})$ expected I/Os.*

We say that a string is *short* if its length is at most $wB - 2$; otherwise, it is *long*. The two types of strings are processed separately.

Short Strings. Suppose that D has only short strings. We maintain a stable perfect function f on D by interpreting each string in D as an integer in $[2^{wB-2}]$. By Lemma 1, this demands $O(1 + \frac{t}{wB} \lg(wB))$ space, which is $O(1 + \frac{t}{B})$ because $w = \Omega(\lg B)$. Let I be an array of size $2t$ for storing information fields. For each string $s \in D$, its information field is stored in $I[f(s)]$.

To store the strings of D, we divide $[2t]$ into $\lceil 2t/B \rceil$ disjoint intervals of length B except possibly the last one. Refer to each interval as a *chunk*. All the at most B strings of D mapped to the same chunk by f are managed by a string B-tree, each of which occupies $O(1 + \frac{n'}{wB})$ space where n' is the total length of the strings it manages (Lemma 2). All the string B-trees use $O(1 + \frac{t}{B} + \frac{n}{wB})$ space in total.

To perform a probe with a short string q, we search the string B-tree of chunk $f(q)$ in $O(\lg_B B + \frac{|q|}{wB}) = O(1)$ I/Os (Lemma 2). If found, we return $I[f(q)]$ with another $O(1)$ I/Os; otherwise, q is not in D.

To insert a short string s in D, we first calculate $f(s)$ and update $I(f(s))$ in $O(1)$ expected I/Os (Lemma 1). Then, insert s in the string B-tree responsible for the chunk covering $I(f(s))$. By Lemma 2, the insertion cost is $O(\lg_B B + \frac{wB}{wB}) = O(1)$ I/Os. Using standard rebuilding techniques [14], we can resize I in $O(1)$ worst-case time per insertion. A deletion can be handled similarly.

Long Strings. Now, consider that D has only long strings. We sometimes regard a long string s of length l as a *blocked string* $\tilde{s}$ of length $\lceil l/(wB - 2) \rceil$, where each character of $\tilde{s}$ comes from an alphabet $\tilde{\Sigma}$ of size 2^{wB-2}. Specifically, if we chop s into blocks of size $wB - 2$ (possibly except the last block), character $\tilde{s}_j$ corresponds to the j-th block, where $1 \le j \le \lceil l/(wB - 2) \rceil$. Denote by $\tilde{D}$ the set of blocked strings obtained from D.

Let T be a trie built on $\tilde{D}$. For an internal node u in T, let $child(u)$ be the set of its child nodes (note that $|child(u)|$ can be up to 2^{wB-2}). As each node in $child(u)$ is a character in $\tilde{\Sigma}$, it can be regarded as a short string. To allow efficient navigation, we create an aforementioned structure (for short strings) on $child(u)$ such that given any character $\tilde{\sigma} \in \tilde{\Sigma}$, we can tell in constant time whether $\tilde{\sigma} \in child(u)$ and if so, also the address of the child $\tilde{\sigma}$. For each string $s \in D$, its information field is stored at the node of T whose root-to-leaf path corresponds to $\tilde{s}$.

The short-string structure on an internal node u of T consumes $O(1 + \frac{|child(u)|}{B} + \frac{|child(u)|wB}{wB}) = O(|child(u)|)$ space, i.e., $O(1)$ blocks per child. Hence, the entire space usage of T is asymptotically the number of nodes in T. Since each $s \in D$ necessitates $O(|s|/(wB))$ nodes, T has $O(n/(wB))$ nodes. It thus follows that our structure occupies $O(n/(wB))$ space overall.

To perform a probe, we simply search T with $\tilde{q}$. The cost is clearly $O(|\tilde{q}|) = O(|q|/(wB))$. To insert a long string s of length l, we first obtain its blocked string $\tilde{s}$ in $O(l/(wB)) = O(|\tilde{s}|)$ I/Os. Then, insert $\tilde{s}$ in T using $O(|\tilde{s}|)$ I/Os. Specifically, at the node of $\tilde{s}_i$ ($i \ge 1$) in T, we can identify the node of $\tilde{s}_{i+1}$ in constant time (by exploiting the short-string structure on $\tilde{s}_i$), or create it in constant time if it does not exist. The deletion algorithm is analogous.

3 One-Error Dictionary Search

This section serves as a proof for our main result:

Theorem 2. *A set D of strings (each with an integer id) can be stored in a structure of $O(n/B)$ blocks, where n is the total length of all the strings in D, such that a 1-error dictionary search query can be answered in $O(1 + \frac{m}{wB} + \frac{k}{B})$ I/Os, where m is the length of the query string, and k is the number of qualifying strings. To insert/delete a string with length l in D, the structure can be updated in $O(l)$ expected I/Os.*

We will focus on finding those strings $s \in d$ with 1 edit distance from a query string q (the string with 0 edit distance, if exists, can be found by exact matching). There are only 3 possibilities: an insertion, deletion, or a replacement of a character turns s to q—in these cases s is said to be an *insertion, deletion* and *replacement match*, respectively. We will concentrate on insertion matches in Section 3.1-3.3. The other types of matches can be reported using similar techniques, which are omitted from this extended abstract due to the space limit.

3.1 Signature Edits

We define an *insertion match* s to be an *appending match* if we can add a character at the end of s to turn it into q. Otherwise, s is a *non-appending match*. For instance, given $q = 11110$, $s = 1111$ is an appending match, while $s = 1110$ is a non-appending match. We will focus on reporting non-appending matches, because there is at most one appending match, as can be found by exact matching after trimming the last character of q.

Consider an insertion that turns a non-appending match s to q. Denote the insertion as (s, i, c) if it adds character c before s_i for some $i \in [1, |s|]$. Recall that there can be multiple such insertions. We define (s, i, c) to be a *signature insertion* if $c \neq s_i$. It turns out that only one insertion can be signature:

Lemma 3. *Every non-appending match s has a unique signature insertion that turns s into q.*

Proof. We first prove that s can be turned into q by a signature insertion. Since s is a non-appending match, there exists an $i \in [1, |s|]$ such that (s, i, c) turns s into q. If $c \neq s_i$, then this insertion is signature. Otherwise, let $j > i$ be the smallest integer such that $s_{j-1} \neq s_j$. Such j definitely exists; otherwise, we can append c to s to turn it into q, contradicting the fact that s is non-appending. Thus, (s, j, c) is a signature insertion.

We now prove that there is only one signature insertion. Assume that s has two: (s, i_1, c_1) and (s, i_2, c_2) with $i_1 < i_2$, both of which convert s to q. From (s, i_1, c_1), we know $q_{i_1} = c_1 \neq s_{i_1}$. However, from (s, i_2, c_2), we know that $q_{1..(i_2-1)} = s_{1..(i_2-1)}$, implying that $q_{i_1} = s_{i_1}$, giving a contradiction. $\square$

3.2 Short Strings

In this subsection, we describe a structure for a dictionary D that has only short strings (i.e., length at most $wB - 2$). Consider a string $s \in D$. Obviously, s has $|s|$ signature insertions: for each $i \in [1, |s|]$, the i-th signature insertion adds the opposite of s_i before s_i. Let $\mathcal{N}^+(s)$ be the set of strings obtained by applying those signature insertions on s, respectively. We have:

Lemma 4. $|\mathcal{N}^+(s)| = |s|$ *and* $\mathcal{N}^+(s)$ *is exactly the set of strings for which s is a non-appending match.*

Proof. $|\mathcal{N}^+(s)| = |s|$ is because any two signature insertions turn s into different strings. It is obvious that s is a non-appending match of every string in $\mathcal{N}^+(s)$. Finally, by Lemma 3, every string of which s is non-appending match belongs to $\mathcal{N}^+(s)$.

□

Structure. Let us first make a *disjoint-neighbor assumption*: for any two $s \neq s'$ in D, $\mathcal{N}^+(s)$ and $\mathcal{N}^+(s')$ are disjoint. Let D^+ be the union of $\mathcal{N}^+(s)$ of all $s \in D$. By Lemma 4, D^+ can have at most $\sum_{s \in D} |s| = n$ strings. Clearly, a query string q has a non-appending match in D if and only if $q \in D^+$. Next we utilize this fact to find the non-appending match of q efficiently (there is only one match under the disjoint-neighbor assumption).

We maintain a stable perfect function h^+ on D^+ using Lemma 1 (notice that all strings of D^+ have length at most $wB - 1$). The representation of h^+ occupies $O(1 + \frac{n}{wB} \lg(wB)) = O(n/B)$ space. Recall that h^+ maps each string $s^+ \in D^+$ to a distinct integer in $[2n]$. We create an array Δ^+ of size $2n$ to record signature insertions. Specifically, for each $s^+ \in D^+$, $\Delta^+[h^+(s^+)]$ stores a pair $(id(s), i)$ if (s, i, c) is the signature insertion generating s^+, where $id(s)$ gives the id of s. Each cell of Δ^+ can be stored in a word. Overall, Δ^+ uses $O(n/B)$ blocks. Finally, we build a structure of Theorem 1 on D so that exact matching in D can be done efficiently. The total space of our structure is therefore $O(n/B)$.

Given a string q, we search for its non-appending match as follows. First, locate cell $\Delta^+[h^+(q)]$ in constant I/Os. If nothing exists in the cell, $q \notin D^+$ and hence, has no non-appending match. Instead, suppose that $\Delta^+[h^+(q)]$ contains a pair $(id(s), i)$. We obtain a string q' from q by deleting q_i, and perform exact matching with q' in D using constant I/Os. If s is found, we return its id; otherwise, q has no non-appending match.[2] The total query time is $O(1)$.

Update. We discuss only insertions because a reverse procedure supports deletions. To insert a (short) string s, we first insert it in the exact matching structure (on D) using $O(1)$ expected I/Os (Theorem 1). Then, keeping s memory-resident, we can generate each string $s^+ \in \mathcal{N}^+(s)$ in memory. For each s^+ with signature insertion (s, i, c), calculate $h^+(s^+)$ and store $(id(s), i)$ at $\Delta^+[h^+(s^+)]$ in constant expected I/Os (Lemma 1). Therefore, the total insertion cost is $O(|\mathcal{N}^+(s)|) = O(l)$ I/Os expected.

Eliminating the Disjoint-Neighbor Assumption. We first need to solve a related problem we call *find-all-any*. Let r be an integer satisfying $0 \leq r \leq w$, and define $\phi = \lg(wB) + r$. Let $g \leq CwB/\phi$ for any constant $C > 0$, and $S_1, ..., S_g$ be g sets of integers in $[wB]$. Each integer in any S_i ($1 \leq i \leq g$) is associated with an arbitrary *information field* of r bits. We want to maintain a structure to support the following operations efficiently:

---

[2] The exact matching with q' is for detecting the scenario where $q \neq s^+$ but $h^+(q) = h^+(s^+)$.

- given an $i \in [1, g]$, insert/delete an integer in S_i, as well as its information field.
- *find-all*(*i*): given an $i \in [1, g]$, return the information fields of all the integers in S_i.
- *find-any*(*i*): given an $i \in [1, g]$, return an arbitrary integer in S_i and its information field.

Lemma 5. *Let* $L = \sum_{i=1}^{g} |S_i|$. *There is a structure of* $O(1 + \frac{\phi L}{wB})$ *space supporting an insertion/deletion in* $O(1)$ *expected I/Os, find-any in* $O(1)$ *I/Os, and find-all in* $O(1 + \frac{\phi |S_i|}{wB})$ *I/Os.*

Proof. We say that S_i is *small* if $|S_i| \leq wB/(2\phi)$, and *big* if $|S_i| \geq wB/\phi$. S_i is neither small nor big when $wB/(2\phi) < |S_i| < wB/\phi$. We follow the invariant that all small sets are together managed by a B-tree U. In U, the integers (of the small sets) are first sorted by the sets they come from, and then by their values. In other words, each integer corresponds to a composite key (set-id, value), which fits in $O(\lg(wB))$ bits. The information field of an integer is stored in the same leaf node with the integer. Each leaf node contains $\Theta(wB/\phi)$ integers. It will be guaranteed that U manages $O((wB/\phi)^2)$ integers, and thus occupies $O(\frac{(wB/\phi)^2}{wB/\phi}) = O(wB)$ blocks. Hence, each block pointer within U can be stored in $O(\lg(wB))$ bits. We therefore can set the fanout of U to $\Theta(wB/\lg(wB))$ so that U has only constant levels.

Each big set S_i is stored in a *big structure*, which consists of a hash structure (e.g., [9]) and a linked list. The hash structure is created on the integers of S_i, whereas the linked list contains these integers (i.e., each integer has two copies: in the hash structure and linked list, respectively) as well as their information fields. The ordering in the linked list is not important, as long as each block accommodates $\Theta(wB/\phi)$ integers and their information fields. At all times, for each integer, we let its copies in the hash structure and the linked list keep pointers to each other. This allows us to reach the copy in the linked list in constant I/Os.

When S_i is neither big nor small, it can be stored either in U or a big structure. Hence, U can index $O(g(wB/\phi)) = O((wB/\phi)^2)$ integers.

Next, we explain how to insert/delete an integer x in S_i. First, if S_i is currently stored in U, we insert/delete x in U using $O(1)$ I/Os. Otherwise, update the hash structure and linked list on S_i in $O(1)$ expected I/Os. In the linked list, we sometimes need to split a block or merge two blocks to ensure $\Theta(wB/\phi)$ integers per block. Each split/merge incurs $O(wB/\phi)$ I/Os to correct the pointers between the hash structure and linked list. By standard techniques (e.g., as in updating a B-tree), this happens only after $\Omega(wB/\phi)$ updates, so that each update is charged only constant I/Os for the split/merge. S_i may become big when it is in U, or conversely, become small when it is in a big structure. In either case, we perform an *overhaul* to move the entire S_i into a big structure or U respectively using $O(wB/\phi)$ I/Os. As at least $wB/(2\phi)$ updates must have occurred between two overhauls, each update is charged only constant I/Os for an overhaul. Therefore, overall an update requires constant expected I/Os amortized. The amortization can be easily removed using lazy rebuilding techniques of [14] (see also [2]).

The other two operations can be supported efficiently. To perform *find-all*(*i*), we simply search U or scan the linked list on S_i, depending on where S_i is stored. In either case, the cost is $O(1 + \frac{\phi |S_i|}{wB})$, recalling that U has $O(1)$ levels. Finally, to perform *find-any*(i), we simply return the first integer of S_i in U or its linked list. The cost is clearly $O(1)$. $\qquad\square$

Next, we remove the disjoint-neighbor assumption. Define D^+ again as the union of $\mathcal{N}^+(s)$, i.e., having discarded all duplicates. Divide $[2n]$ into $\lceil 2n/B \rceil$ intervals (a.k.a. *chunks*) of size B. Consider a string $s^+ \in D^+$, which may now belong to the $\mathcal{N}^+(s)$ of several $s \in D$. In other words, multiple signature insertions may have generated s^+. Let $E^+(s^+)$ be the set of those signature insertions.

Recall that every signature insertion has the form (s, i, c). A crucial observation is that all signature insertions in $E^+(s^+)$ differ in their values of i. To prove this, first notice that no two signature insertions in $E^+(s^+)$ can have come from the same s, according to Lemma 3. Next, assume that $E^+(s^+)$ had signature insertions (s_1, i, c_1) and (s_2, i, c_2) with $s_1 \neq s_2$. Notice that c_1 must be identical to c_2, because both of them were equal to $(s^+)_i$. It thus follows that s_1 had to be the same as s_2, giving a contradiction.

Because of the previous observation, we will sometimes regard $E^+(s^+)$ as a set of integers: $\{i \mid (s, i, c) \in E^+(s^+)\}$. Further, we associate i with an information field $id(s)$ where s is the unique string such that $(s, i, c) \in E^+(s^+)$. Clearly, i and $id(s)$ can be stored in $\lg(wB)$ and w bits, respectively.

For each chunk, we store at most B sets of integers, namely, a set $E^+(s^+)$ for every s^+ whose $h^+(s^+)$ is covered by the chunk. We index these sets with a find-all-any structure of Lemma 5. To see that this is possible, set $r = w$ and hence $\phi = \lg(wB) + w$, in which case a find-all-any structure can manage up to $g = CwB/\phi = C\frac{wB}{\lg(wB)+w}$ sets where $C > 0$ can be any constant. In other words, it can indeed be used to manage B sets because $B = O(\frac{wB}{\lg(wB)+w})$ (applying $w = \Omega(\lg B)$).

Therefore, by Lemma 5, the find-all-any structure of a chunk uses $O(1 + \frac{L}{B})$ space where L is the total size of those sets. As there are $O(n/B)$ chunks, the structures of all chunks require $O(n/B)$ space altogether. We also build all the structures as were necessary when the disjoint-neighbor assumption was made. The only difference is that, at each cell $\Delta^+[h^+(s^+)]$, we store an arbitrary pair $(id(s), i)$ where $(s, i, c) \in E^+(s^+)$.

A query is answered as before, except that after finding the match of q' in D, we report all the string ids in $E^+(h^+(q))$ using a *find-all* operation which performs $O(1 + \frac{k}{B})$ I/Os by Lemma 5. This is correct because, by definition, all the signature insertions in $E^+(h^+(q))$ generate exactly q. To insert a length-l string s, we proceed as before, but also perform an insertion in the find-all-any structure of a relevant chunk for each signature insertion (s, i, c). The algorithm of deleting a string s is also the same as before, except that if s happens to be the string whose id is in cell $\Delta^+[h^+(s^+)]$, we perform a *find-any* to extract another element from $E^+(s^+)$ to fill in the cell. By Lemma 5, the above changes incur only constant expected I/Os per signature insertion, and hence, $O(l)$ time in total.

Now we have arrived at:

Lemma 6. *We can store D in a structure of $O(n/B)$ blocks such that all insertion matches of a short query string with length at most wB can be performed in $O(1 + \frac{k}{B})$ I/Os, where k is the number of qualifying strings. To insert/delete a short string with length l in D, the structure can be updated in $O(l)$ expected I/Os.*

Remark. When Σ is not binary, the space consumption of our structure increases by a factor linear to $|\Sigma|$, and hence, remains $O(n/B)$ when $|\Sigma| = O(1)$. The major change is that, for each string s and each $i \in [1, |s|]$, there are $|\Sigma| - 1$ signature insertions, namely, (s, i, c) for every possible $c \neq s_i$ in Σ. Accordingly, $\mathcal{N}^+(s)$ has size $|s|(|\Sigma| - 1) = O(|s|)$. Furthermore, the length of a short string should be defined instead as $C \cdot wB$ for some appropriate $0 < C < 1$, making sure that a short string fits in one block (each character of Σ is now encoded by $\log_2 |\Sigma| = \Theta(1)$ bits). The other changes are obvious and therefore omitted.

3.3 Long Strings

This section explains the structure for *long* strings, each of which has length at least wB. We divide these strings by their lengths, and manage each group of strings with the same length separately. Our discussion below concentrates on a particular length l. For convenience, let $\rho = wB - 2$, and we will assume that $l = \lambda\rho - 1$ for some integer $\lambda \geq 2$. If not, we pad enough (at most $\rho - 1$) 0's at the end of each string to make the property hold. The padding can increase the space, query, and update costs of our structure by no more than a constant factor, as will be clear shortly.

Searching the Blocked Strings. Following the notation in Section 2.2, given a long string s, we denote its corresponding blocked string as $\tilde{s}$. Recall that $\tilde{s}$ is obtained by chopping s into blocks of size ρ. We may regard each character of $\tilde{s}$ as being in an alphabet $\tilde{\Sigma}$ of size 2^ρ, or equivalently, as a binary string of length at most ρ, whichever is more convenient. When the second interpretation is adopted, s_i is a bit in $\tilde{s}_j$ where $i \in [1, |s|]$ and $j = \lceil i/\rho \rceil$.

We denote by $\overline{s}$ as the reverse string of s (e.g., if $s = 10010$ then $\overline{s} = 01001$). Just like s, $\overline{s}$ also has its blocked string $\tilde{\overline{s}}$.

Consider a string q of length $l + 1$. Let s be a non-appending match of q. Note that both $\tilde{s}$ and $\tilde{q}$ have length λ (as long as $\rho \geq 2$). According to Lemma 3, there exists a unique signature insertion (s, i, c) that converts s to q. Let $j = \lceil i/\rho \rceil$. Depending on the value of j, we define a binary string $q^\star(j)$:

- If $j = \lambda$, then $q^\star(j) = \tilde{q}_\lambda$. In this case, $|q^\star(j)| = \rho$.
- Otherwise, $q^\star(j)$ is the string obtained by appending the first character of $\tilde{q}_{j+1}$ to $\tilde{q}_j$ (viewing both $\tilde{q}_j$ and $\tilde{q}_{j+1}$ in binary form). In this case, $|q^\star(j)| = \rho + 1$.

For example, suppose $\rho = 3$ and $q = 110100$; thus, if $j = 1$, then $q^\star(j) = 1101$, whereas if $j = 2$ then $q^\star(j) = 100$. The next lemma gives a crucial observation.

Lemma 7. *Let s be a non-appending match of q with signature insertion (s, i, c). Define $j = \lceil i/\rho \rceil$. All the following are true:*

  (i) $\tilde{s}_{1..(j-1)} = \tilde{q}_{1..(j-1)}$
  (ii) $\tilde{\overline{s}}_{1..(\lambda-j)} = \tilde{\overline{q}}_{1..(\lambda-j)}$
  (iii) $\tilde{s}_j$ *is a non-appending match of* $q^\star(j)$.

Proof. By the definition of i, it holds that $s_{1..(i-1)} = q_{1..(i-1)}$, and $s_{i..l} = q_{(i+1)..(l+1)}$. The latter suggests $\overline{s}_{1..(l-i+1)} = \overline{q}_{1..(l-i+1)}$.

(i) Notice that $\rho(j-1) \leq i-1$. Thus, $\tilde{s}_{1..(j-1)} = \tilde{q}_{1..(j-1)}$ holds because they (in binary form) are prefixes of the same length of $s_{1..(i-1)}$ and $q_{1..(i-1)}$, respectively.

(ii) First observe that $\rho(\lambda - j) \leq l - i + 1$: since $\rho\lambda = l + 1$, this is equivalent to showing $i \leq j\rho$, which is true by the definition of j. Hence, $\bar{\tilde{s}}_{1..(\lambda-j)} = \bar{\tilde{q}}_{1..(\lambda-j)}$ holds because they are prefixes of the same length of $\bar{s}_{1..(l-i+1)}$ and $\bar{q}_{1..(l-i+1)}$, respectively.

(iii) We focus on $j < \lambda$ because the case $j = \lambda$ is obvious. We will abbreviate $q^\star(j)$ simply as $q^\star$. Since $\tilde{s}_j$ is clearly an insertion match of $q^\star$, what remains to prove is that it is *not* an appending match. The remainder of the proof will regard $\tilde{s}_j$ as a binary string. Let $i' = i - \rho(j-1)$, namely, $(\tilde{s}_j)_{i'}$ is the same character as s_i.

Now assume that $\tilde{s}_j$ was an appending match of $q^\star$. Let c' be the last character of $q^\star$. It thus follows that $\tilde{s}_j : c' = q^\star$ (where ":" means concatenation), implying $(\tilde{s}_j)_{i'} = q^\star_{i'}$. On the other hand, the fact that (s, i, c) turns s to q implies $c = q^\star_{i'}$. It thus follows that $c = (\tilde{s}_j)_{i'} = s_i$, violating the definition of (s, i, c). $\square$

Structure. Let $\tilde{D}(l)$ be the set of blocked strings $\tilde{s}$ of all $s \in D$ with length l. Likewise, let $\bar{\tilde{D}}(l)$ be the set of blocked strings $\bar{\tilde{s}}$ of all such s. Regarding each blocked string as consisting of characters from $\tilde{\Sigma}$ (of size 2^ρ), we build a trie T on $\tilde{D}(l)$ and another trie $\bar{T}$ on $\bar{\tilde{D}}(l)$.

Consider a blocked string $\tilde{s} \in \tilde{D}(l)$, and the corresponding $\bar{\tilde{s}} \in \bar{\tilde{D}}(l)$. Recall that both $\tilde{s}$ and $\bar{\tilde{s}}$ have length λ. For each $j \in [1, \lambda]$, we do the following. First, identify the node u in T corresponding to $\tilde{s}_{1..(j-1)}$, and the node $\bar{u}$ in $\bar{T}$ corresponding to $\bar{\tilde{s}}_{1..(\lambda-j)}$. Insert $\tilde{s}_j$ to a set $S(u, \bar{u})$, which is therefore a set of short strings (when viewed in binary form). We build a structure of Lemma 6 on $S(u, \bar{u})$; let us represent this structure as $P^+(u, \bar{u})$, referred to as a P^+-*structure*.

Finally, we need a structure that, given any $(u, \bar{u})$, returns the beginning address of $P^+(u, \bar{u})$ in constant time. For this purpose, it suffices to maintain a standard dynamic hash structure (e.g., [9]) on the pair of addresses of u and $\bar{u}$; we will refer to it as the P^+-*lookup structure*.

Update. The algorithm for inserting a string of length l follows directly from the above description. The cost is $O(\frac{l}{wB} \cdot wB) = O(l)$ expected by Lemma 6 and our earlier discussion on maintaining a trie on blocked strings. Deletion is analogous.

Query. Given a string q of length $l + 1$, we find its non-appending matches as follows. For each $j \in [1, \lambda]$, we identify the node u in T corresponding to $\tilde{q}_{1..(j-1)}$, and the node $\bar{u}$ in $\bar{T}$ corresponding to $\bar{\tilde{q}}_{1..(\lambda-j)}$. Report all the non-appending matches of $q^\star(j)$ in $S(u, \bar{u})$. The algorithm's correctness is established by the next lemma.

Lemma 8. *Every non-appending match of q is reported exactly once.*

Proof. Lemma 7 shows that every non-appending match of q must be reported at least once. Next we prove that no non-appending match s can be reported twice. For this purpose, let (s, i, c) be the signature insertion that converts s to q, and let $j^\star = \lceil i/\rho \rceil$. We will prove that, for any $j' \neq j^\star$, our algorithm does *not* report s when $j = j'$. In other words, s is reported only when $j = j^\star$.

Suppose on the contrary that s was reported at j', i.e., $\tilde{s}_{j'}$ is a non-appending match of $q^\star(j')$. We can easily rule out the possibility of $j' < j^\star$. This is because, by the definition of i and $j^\star$, $\tilde{s}_{j'}$ must be equivalent to $\tilde{q}_{j'}$ for every $j' < j^\star$, implying that $\tilde{s}_{j'}$ is an appending match of $q^\star(j')$.

Consider $j' > j^\star$. By the definition of (s, i, c), we know $q_i = c$. Since s is reported by our algorithm when $j = j'$, $\tilde{s}_{j'-1}$ and $\tilde{q}_{j'-1}$ must correspond to the same node in trie T, implying that $s_{1..(\rho(j'-1))} = q_{1..(\rho(j'-1))}$. As $i \leq \rho j^\star \leq \rho(j'-1)$, we have $s_i = q_i$, namely, $s_i = c$, thus violating the definition of (s, i, c). $\square$

Analysis. To bound the space of our structure, let t' be the number of length-l strings in D, and $n' = t'l$ be the total length of these strings. The analysis in Section 2.2 shows that each of T and $\tilde{T}$ uses $O(\frac{t'}{B} + \frac{n'}{wB}) = O(\frac{n'}{wB})$ space.

Next we discuss the space of the P^+-structures. Consider a pair $(u, \overline{u})$ whose $S(u, \overline{u})$ is non-empty. Note that there are at most $O(n'/(wB))$ such $(u, \overline{u})$ because a string $s \in D$ of length l can necessitate $O(l/(wB))$ such pairs. Let $t(u, \overline{u})$ be the number of strings in $S(u, \overline{u})$, and $n(u, \overline{u})$ be their total length. By Lemma 6, $P^+(u, \overline{u})$ occupies $O(n(u, \overline{u})/B)$ blocks. Therefore, the space of all the P^+-structures is at most

$$\sum_{(u, \overline{u}) \text{ with non empty } S(u, \overline{u})} O\left(1 + \frac{n(u, \overline{u})}{B}\right)$$

$$= O\left(\frac{n'}{wB} + \frac{n'}{B}\right) = O(n'/B).$$

Finally, the space of the P^+-lookup structure is linear to the number of non-empty sets $S(u, \overline{u})$; as there are $O(n'/(wB))$ non-empty sets, the P^+-lookup structure occupies $O(n'/(wB^2))$ space. A query searches $O(l/(wB))$ P^+-structures. Combining Lemmas 6 and 8 proves that the query time is $O(\frac{l}{wB} + \frac{k}{B})$ overall.

We thus have established:

Lemma 9. *We can store D in a structure of $O(n/B)$ blocks such that all insertion matches of a length-m query string can be found in $O(1 + \frac{m}{wB} + \frac{k}{B})$ I/Os, where k is the number of qualifying strings. To insert/delete a long string with length l in D, the structure can be updated in $O(l)$ expected I/Os.*

Remark. The double-trie idea is due to [5], but the challenge in our contexts is to make the idea work on blocked strings. For this purpose, it is crucial to derive Lemmas 7 and 8, both of which rely on the notion of signature edits and the separation of non-appending matches. The two lemmas, notion, and separation are where our contributions lie.

Acknowledgements. Chin-Wan Chung was supported in part by Defense Acquisition Program Administration and Agency for Defense Development under the contract UD140022PD, Korea. Yufei Tao was supported in part by projects GRF 4165/11, 4164/12, and 4168/13 from HKRGC. Wei Wang was partly funded by ARC DP130103401 and DP130103405.

References

1. Aggarwal, A., Vitter, J.S.: The input/output complexity of sorting and related problems. CACM 31(9), 1116–1127 (1988)
2. Arge, L., Vitter, J.S.: Optimal external memory interval management. SIAM J. of Comp. 32(6), 1488–1508 (2003)
3. Belazzougui, D.: Faster and space-optimal edit distance "1" dictionary. In: Annual Symp. on Combinatorial Pattern Matching, pp. 154–167 (2009)
4. Belazzougui, D., Venturini, R.: Compressed string dictionary look-up with edit distance one. In: Kärkkäinen, J., Stoye, J. (eds.) CPM 2012. LNCS, vol. 7354, pp. 280–292. Springer, Heidelberg (2012)
5. Brodal, G.S., Gasieniec, L.: Approximate dictionary queries. In: Hirschberg, D., Meyers, G. (eds.) CPM 1996. LNCS, vol. 1075, pp. 65–74. Springer, Heidelberg (1996)
6. Chan, H.-L., Lam, T.-W., Sung, W.-K., Tam, S.-L., Wong, S.-S.: A linear size index for approximate pattern matching. In: Lewenstein, M., Valiente, G. (eds.) CPM 2006. LNCS, vol. 4009, pp. 49–59. Springer, Heidelberg (2006)
7. Cole, R., Gottlieb, L.-A., Lewenstein, M.: Dictionary matching and indexing with errors and don't cares. In: STOC, pp. 91–100 (2004)
8. Demaine, E.D., auf der Heide, F.M., Pagh, R., Pătraşcu, M.: De dictionariis dynamicis pauco spatio utentibus (*lat.* on dynamic dictionaries using little space). In: Correa, J.R., Hevia, A., Kiwi, M. (eds.) LATIN 2006. LNCS, vol. 3887, pp. 349–361. Springer, Heidelberg (2006)
9. Dietzfelbinger, M., Karlin, A.R., Mehlhorn, K., auf der Heide, F.M., Rohnert, H., Tarjan, R.E.: Dynamic perfect hashing: Upper and lower bounds. SIAM J. of Comp. 23(4), 738–761 (1994)
10. Ferragina, P., Grossi, R.: The string B-tree: A new data structure for string search in external memory and its applications. JACM 46(2), 236–280 (1999)
11. Hon, W.-K., Lam, T.W., Shah, R., Tam, S.-L., Vitter, J.S.: Cache-oblivious index for approximate string matching. Theoretical Computer Science 412(29), 3579–3588 (2011)
12. Kukich, K.: Techniques for automatically correcting words in text. ACM Comp. Surv. 24(4), 377–439 (1992)
13. Navarro, G.: A guided tour to approximate string matching. ACM Comp. Surv. 33(1), 31–88 (2001)
14. Overmars, M.H.: The Design of Dynamic Data Structures. Springer (1987)
15. Tsur, D.: Fast index for approximate string matching. Journal of Discrete Algorithms 8(4), 339–345 (2010)

Online Pattern Matching for String Edit Distance with Moves*

Yoshimasa Takabatake[1], Yasuo Tabei[2], and Hiroshi Sakamoto[1]

[1] Kyushu Institute of Technology
{takabatake,hiroshi}@donald.ai.kyutech.ac.jp
[2] PRESTO, Japan Science and Technology Agency
tabei.y.aa@m.titech.ac.jp

Abstract. Edit distance with moves (EDM) is a string-to-string distance measure that includes substring moves in addition to ordinal editing operations to turn one string to the other. Although optimizing EDM is intractable, it has many applications especially in error detections. Edit sensitive parsing (ESP) is an efficient parsing algorithm that guarantees an upper bound of parsing discrepancies between different appearances of the same substrings in a string. ESP can be used for computing an approximate EDM as the L_1 distance between characteristic vectors built by node labels in parsing trees. However, ESP is not applicable to a streaming text data where a whole text is unknown in advance. We present an online ESP (OESP) that enables an online pattern matching for EDM. OESP builds a parse tree for a streaming text and computes the L_1 distance between characteristic vectors in an online manner. For the space-efficient computation of EDM, OESP directly encodes the parse tree into a succinct representation by leveraging the idea behind recent results of a dynamic succinct tree. We experimentally test OESP on the ability to compute EDM in an online manner on benchmark datasets, and we show OESP's efficiency.

1 Introduction

Streaming text data appears in many application domains of information retrieval. Social data analysis faces a problem for analyzing continuously generated texts. In computational biology, recent sequencing technologies enable us to sequence individual genomes in a short time, which resulted in generating a large collection of genome data. There is therefore a strong incentive to develop a powerful method for analyzing streaming texts on a large-scale.

Edit distance with moves (EDM) is a string-to-string distance measure that includes substring moves in addition to insertions and deletions to turn one string to the other in a series of editing operations. The distance measure is motivated in error detections, e.g., insertions and deletions on lossy communication channels [9], typing errors in documents [4] and evolutionary changes in

---

* This work was supported by JSPS KAKENHI(24700140,26280088) and the JST PRESTO program.

E. Moura and M. Crochemore (Eds.): SPIRE 2014, LNCS 8799, pp. 203–214, 2014.
© Springer International Publishing Switzerland 2014

Table 1. Summary of recent pattern matching methods for EDM. The table summaries upper bound for the approximation ratio of EDM, computation time and space for each method. The space for ESP and OESP is presented in bits. N is the length of an input string; σ is the alphabet size; n is the number of variables in CFG; $\alpha \in (0,1]$ is a parameter for a hash table; $\lg^*$ is the iterated logarithm; $\lg$ stands for $\log_2$.

	Appro. ratio	Time	Space	Algorithm
SNN [12]	$O(\lg N \lg^* N)$	$O(N^{O(1)} + N\mathrm{polylog}(N))$	$O(N^{O(1)})$	Offline
Shapira and Storer [16]	$O(\lg N)$	$O(N^2)$	$O(N \lg N)$	Offline
ESP [3]	$O(\lg N \lg^* N)$	$O(N \lg^* N/\alpha)$	$N \lg \sigma$ $+n(\alpha + 3)\lg(n + \sigma)$	Offline
OESP	$O(\lg^2 N)$	$O(\frac{N \lg N \lg n}{\alpha \lg \lg n})$	$n(\alpha + 1)\lg(n + \sigma)$ $+n \lg(\alpha n) + 5n + o(n)$	Online

biological sequences [5]. Computing an optimum solution of EDM is intractable, since the problem is known to be NP-complete [12]. Therefore, researchers have paid considerable efforts to develop efficient approximation algorithms that are only applicable to an offline case where a whole text is given in advance (Table 1). Early results include the reversal model [8,1] which takes a substring of unrestricted size and replaces it by its reverse in one operation. Muthukrishnan and Sahinalp [12] proposed an approximate nearest neighbor considered as a sequence comparison with block operations. Recently, Shapira and Storer proposed a polylog time algorithm with $O(\lg N \lg^* N)$ approximation ratio for the length N of an input text.

Edit sensitive parsing (ESP) [3] is an efficient parsing algorithm developed for approximately computing EDM between strings in an offline setting. ESP builds from a given string a parse tree that guarantees upper bounds of parsing discrepancies between different appearances of the same substring, and then it represents the parse tree as a vector each dimension of which represents the frequency of the corresponding node label in a parse tree. L_1 distance between such characteristic vectors for two strings can approximate the EDM. Although ESP has an efficient approximation ratio $O(\lg N \lg^* N)$ and runs fast in $O(N \lg^* N/\alpha)$ time for a parameter $\alpha \in (0,1]$ for hash tables, its applicability is limited to an offline case. For applications in web mining and Bioinformatics, computing an EDM of massive streaming text data has ever been an important task. An open challenge, which is receiving increased attention, is to develop a scalable online pattern matching for EDM.

We present an online pattern matching for EDM. Our method is an online version of ESP named *online ESP (OESP)* that (i) builds a parse tree for a streaming text in an online manner, (ii) computes characteristic vectors for a substring at each position of the streaming text and a query, and (iii) computes the L_1 distance between each pair of characteristic vectors. The working space of our method does not depend on the length of text but the size of a parse tree. To make the working space smaller, OESP builds a parse tree from a streaming text and directly encodes it into a succinct representation by leveraging the idea behind recent results of an online grammar compression [11,10] and a dynamic succinct tree [14]. Our representation includes a novel succinct representation of a tree named *post-order unary degree sequence (POUDS)* that is built by the post-order traversal of a tree and a unary degree encoding. To guarantee the approximate EDM computed by OESP, we also prove an upper bound of the approximation ratio between our approximate EDM and the exact EDM.

Experiments using standard benchmark texts revealed OESP's efficiencies.

2 Preliminaries

2.1 Basic Notation

Let Σ be a finite alphabet forming texts, and $\sigma = |\Sigma|$. Σ^* denotes the set of all texts over Σ, and Σ^ℓ denotes the set of all texts of length ℓ over Σ, i.e. $\Sigma^\ell = \{S \in \Sigma^* || S| = \ell\}$. We assume a recursively enumerable set $\mathcal{X}$ of variables such that $\Sigma \cap \mathcal{X} = \phi$ and all elements in $\Sigma \cup \mathcal{X}$ are totally ordered. A sequence of symbols from $\Sigma \cup \mathcal{X}$ is called a string. The length of string S is denoted by $|S|$, and the cardinality of a set C is similarly denoted by $|C|$. A pair and triple of symbols from $\Sigma \cup \mathcal{X}$ are called digram and trigram, respectively. Strings x and z are said to be the prefix and suffix of the string $S = xyz$, respectively, and x, y, z are called substrings of S. The i-th symbol of S is denoted by $S[i]$ ($1 \le i \le |S|$). For integers i and j with $1 \le i \le j \le |S|$, the substring of S from $S[i]$ to $S[j]$ is denoted by $S[i, j]$. N denotes the length of a text S and it can be variable in an online setting.

2.2 Context-Free Grammar

A *context-free grammar (CFG)* is a quadruple $G = (\Sigma, V, D, Z_s)$ where V is a finite subset of $\mathcal{X}$, D is a finite subset of $V \times (V \cup \Sigma)^*$ of production rules, and $Z_s \in V$ represents the start variable. D is also called a *phrase dictionary*. Variables in V are called nonterminals. The set of strings in Σ^* derived from Z_s by G is denoted by $L(G)$. A CFG G is called *admissible* if for any $Z \in \mathcal{X}$ there is exactly one production rule $Z \to \gamma \in D$. We assume $|\gamma| = 2$ or 3 for any production rule $Z \to \gamma$.

The parse tree of G is represented as a rooted ordered tree with internal nodes labeled by variables in V and leaves labeled by elements in Σ, and the label sequence of its leaves are equal to an input string. Any internal node $Z \in V$ in a parse tree corresponds to a production rule in the form of $Z \to \gamma$ in D. The height of Z is the height of the subtree whose root is Z.

2.3 Phrase and Reverse Dictionaries

For a set V of production rules, a *phrase dictionary* D is a data structure for directly accessing the phrase $S \in (\Sigma \cup V)^*$ for any given $Z \in V$ if $Z \to S \in D$. A *reverse dictionary* $D^{-1} : (\Sigma \cup V)^* \to V$ is a mapping from a given sequence of symbols to a variable. D^{-1} returns a variable Z associated with a string S if $Z \to S \in D$; otherwise, it creates a new variable $Z' \notin V$ and returns Z'. For example, if $D = \{Z_1 \to abc, Z_2 \to cd\}$, $D^{-1}(a, b, c)$ returns Z_1, while $D^{-1}(b, c)$ creates Z_3 and returns it.

2.4 Problem Definition

In order to describe our method we first review the notion of EDM. The EDM $d(S, Q)$ between two strings S and Q is the minimum number of edit operations defined below to transform S into Q:

1. Insertion: A character a at position i in S is inserted, which generates $S[1, i - 1]aS[i]S[i + 1, N]$,
2. Deletion: A character a at position i in S is deleted, which generates $S[1, i - 1]S[i + 1, N]$,
3. Replacement: A character at position i is replaced by a, which generates $S[1, i - 1]aS[i + 1, N]$,
4. Substring move: A substring $S[i, j]$ is moved and inserted at the position k, which generates $S[1, i - 1]S[j + 1, k - 1]S[i, j]S[k, N]$.

Problem 1 (Online pattern matching for EDM). *For a streaming text $S \in \Sigma^*$, a query $Q \in \Sigma^*$, and a distance threshold $k \geq 0$, find all $i \in [1, |S|]$ such that the EDM between a substring $S[i, i + |Q|]$ and Q is at most k, i.e. $d(S[i, i + |Q|], Q) \leq k$.*

Cormode and Muthukrishnan [3] presented an offline algorithm for computing EDM. In their algorithm, a special type of derivation tree called ESP is constructed for approximately computing EDM. We present an online variant of ESP. Our algorithm approximately solves Problem 1 and is composed of two parts: (i) an online construction of a parse tree space-efficiently and (ii) an approximate computation of EDM from the parse tree. Although our method is an approximation algorithm, it guarantees an upper bound for the exact EDM. We now discuss the two parts in the next section.

3 Online Algorithm

OESP builds a special form of CFG and directly encodes it into a succinct representation in an online manner. Such a representation can be used as space-efficient phrase/reverse dictionaries, which resulted in reducing the working space. In this section, we first present a simple variant of ESP in order to introduce the notion of *alphabet reduction* and *landmark*. We then detail OESP and approximate computations of the EDM in an online manner. In the next section, we present an upper bound of the approximate EDM for the exact EDM.

3.1 ESP

Given an input string $S \in \Sigma^*$, we decompose the current S into digrams WX or trigrams WXY associated with variables as production rules, and iterate this process while $|S| > 1$ for the resulting S.

In each iteration, ESP uniquely partitions S into maximal non-overlapping substrings such that $S = S_1 S_2 \cdots S_\ell$ and each S_i is categorized into one of three types, i.e., type1: a repetition of a symbol, type2: a substring not including a type1 substring and of length at least $\lceil \lg |S| \rceil$, and type3: a substring being neither type1 nor type2 substrings.

At one iteration of parsing S_i, ESP builds two kinds of subtrees from digram WX and trigram WXY, respectively. The first type is a 2-tree corresponding to a production rule in the form of $Z \to WX$. The second type is a 3-tree corresponding to $Z \to WXY$.

ESP parses S_i according to its type. In case S_i is a type1 or type3 substring, ESP performs the left aligned parsing where 2-trees are built from left to right in S_i and a 3-tree is built for the last three symbols if $|S_i|$ is odd, as follows:

- If $|S_i|$ is even, ESP builds $Z \to S_i[2j-1, 2j]$, $j = 1, ..., |S_i|/2$,
- Otherwise, it builds $Z \to S_i[2j-1, 2j]$ for $j = 1, ..., (\lfloor |S_i|/2 \rfloor - 1)$, and builds $Z \to S_i[2j-1, 2j+1]$ for $j = \lfloor |S_i|/2 \rfloor$.

In case S_i is type2, ESP further partitions $S_i = s_1 s_2...s_\ell$ $(2 \leq |s_j| \leq 3)$ by the *alphabet reduction* described below, and builds $Z \to s_j$ for $j = 1, ..., \ell$.

After parsing all S_i to S_i', ESP continues this process for the resulted string by concatenating all S_i' $(i = 1, ..., \ell)$ at the next level.

Alphabet Reduction: Alphabet reduction is a procedure for partitioning a string of type2 into digrams and trigrams. Given S of type2, consider each $S[i]$ represented as binary integers. Let p be the position of the least significant bit in which $S[i]$ differs from $S[i-1]$, and let $bit(p, S[i]) \in \{0, 1\}$ be the value of $S[i]$ at the p-th position, where p starts at 0. Then, $L[i] = 2p + bit(p, S[i])$ is defined for any $i \geq 2$. Since S contains no repetition (i.e., S is type2), the string L defined by $L = L[2]L[3] \ldots L[|S|]$ is also type2. We note that if the number of different symbols in S is m, denoted by $[S] = m$, clearly $[L] \leq 2 \lg m$. Then, $S[i]$ is called *landmark* if (i) $L[i]$ is maximal such that $L[i] > \max\{L[i-1], L[i+1]\}$ or (ii) $L[i]$ is minimal such that $L[i] < \min\{L[i-1], L[i+1]\}$ and not adjacent to any other maximal landmark.

Because L is type2 and $[L] \leq \lg |S|$, any substring of S longer than $\lg |S|$ must contain at least one landmark. After deciding all landmarks, if $S[i]$ is a landmark, we replace $S[i-1, i]$ by a variable X and update the current dictionary with $X \to S[i-1, i]$. After replacing all landmarks, the remaining substrings are replaced by the left aligned parsing.

3.2 Post-order CFG

OESP builds a post-order partial parse tree (POPPT) and directly encodes it into a succinct representation. A partial parse tree defined by Rytter [15] is the ordered tree formed by traversing a parse tree in a depth-first manner and pruning out all descendants under every node of nonterminal symbols appearing no less than twice.

Definition 1 (POPPT and POCFG [11]). *A post-order partial parse tree (POPPT) is a partial parse tree whose internal nodes have post-order variables. A post-order CFG (POCFG) is a CFG whose partial parse tree is a POPPT.*

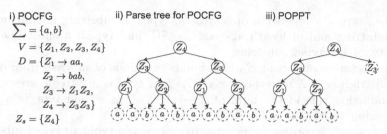

Fig. 1. Example of a POCFG, the parse tree of a POCFG, a post-order partial parse tree (POPPT)

Note that the number of nodes in the POPPT is at most $3n$ for a POCFG of n variables, because the right-hand sides consist of digrams or trigrams in the production rules and the numbers of internal nodes and leaves are n and at most $2n$, respectively.

Examples of a POCFG and POPPT are shown in Figure 1-i) and iii), respectively. The POPPT is built by traversing the parse tree in Figure 1-ii) in depth-first manner and pruning out all the descendants under the node having the second X_3. The resulted POPPT in Figure 1-iii) consists of internal nodes having post-order variables.

A major advantage of POPPT is that we can directly encode it into a succinct representation which can be used as a phrase dictionary. Such a representation enables us to reduce the working space of OESP by using it in a combination with a reverse dictionary.

3.3   Online construction of a POCFG

OESP builds from a given input string a POCFG that guarantees upper bounds of parsing discrepancies between the same substrings in the string. The basic idea of OESP is to (i) start from symbols in an input text, (ii) replace as many as possible of the same digrams or trigrams in common substrings by the same nonterminal symbols, and (iii) iterate this process in a bottom-up manner until it generates a complete POCFG. The POCFG is built in an online manner and the POPPT corresponding to it consists of nodes having two or three children.

OESP builds two types of subtrees in a POPPT from strings XY and WXY. The first type is a 2-tree corresponding to a production rule in the form of $Z \rightarrow XY$. The second type is a 3-tree corresponding to a production rule in the form of $Z \rightarrow WXY$.

OESP builds a 2-tree or 3-tree from a substring of a limited length. Let u be a string of length m. A function $\mathcal{L} : (\Sigma \cup V)^m \times [m] \rightarrow \{0, 1\}$ classifies whether or not the i-th position of u has a landmark, i.e., the i-th position of u has a landmark if $\mathcal{L}(u, i) = 1$. $\mathcal{L}(u, i)$ is computed from a substring $u[i - 1, i + 2]$ of length four. OESP builds a 3-tree from a substring $u[i + 1, i + 3]$ of length three if the i-th position of u does not have a landmark; otherwise, it builds a 2-tree from a substring $u[i + 2, i + 3]$ of length two. The landmarks on a string are

---

Algorithm 1. Online construction of ESP. D is phrase dictionary, D^{-1} is reverse dictionary, and q_k is queue at level k.

```
 1: function OESP
 2:     D := ∅; initialize queues q_k
 3:     while reading a new character c from an input text do
 4:         PROCESSSYMBOL(q_1, c)
 5:     end while
 6: end function
 7: function PROCESSSYMBOL(q_k, X)
 8:     q_k.enqueue(X)
 9:     if q_k.size() = 4 then
10:         if L(q_k, 2) = 0 then                                    ▷ Build a 2-tree
11:             Z := D^{-1}(q_k[3], q_k[4]); D := D ∪ {Z → q_k[3]q_k[4]}
12:             PROCESSSYMBOL(q_{k+1}, Z)
13:             q_k.dequeue(); q_k.dequeue()
14:         end if
15:     else if q_k.size() = 5 then                                  ▷ Build a 3-tree
16:         Z := D^{-1}(q_k[3], q_k[4], q_k[5]); D := D ∪ {Z → q_k[3]q_k[4]q_k[5]}
17:         PROCESSSYMBOL(q_{k+1}, Z)
18:         q_k.dequeue(); q_k.dequeue(); q_k.dequeue()
19:     end if
20: end function
```

---

decided such that they are synchronized in long common subsequences to make the parsing discrepancies as small as possible.

The algorithm uses a set of queues, $q_k, k = 1, ..., m$, where q_k processes the string at k-th level of a parse tree of a POCFG and builds 2-trees and 3-trees at each k. Since OESP builds a balanced parse tree, the number m of these queues is bounded by $\lg N$. In addition, landmarks are decided on strings of length at most four, and the length of each queue is also fixed to five. Algorithm 1 consists of the functions OESP and PROCESSSYMBOL.

The main function is OESP which reads new characters from an input text and gives them to the function PROCESSSYMBOL one by one. The function PROCESSSYMBOL builds a POCFG in a bottom-up manner. There are two cases according to whether or not a queue q_k has a landmark. For the first case of $\mathcal{L}(q_k, 2) = 0$, i.e. q_k does not have a landmark, the 2-tree corresponding to a production rule $Z \to q_k[3]q_k[4]$ in a POCFG is built for the third and fourth elements $q_k[3]$ and $q_k[4]$ of the k-th queue q_k. For the other case, the 3-tree corresponding to a production rule $Z \to q_k[3]q_k[4]q_k[5]$ is built for the third, fourth and fifth elements $q_k[3]$, $q_k[4]$ and $q_k[5]$ of the k-th queue q_k. In both cases, the reverse dictionary D^{-1} returns a nonterminal symbol replacing a sequence of symbols. The generated symbol Z is given to the higher q_{k+1}, which enables the bottom-up construction of a POCFG in an online manner.

The computation time and working space depend on implementations of phrase and reverse dictionaries. The phrase dictionary for a POCFG of n variables can be implemented using a standard array of at most $3n \lg (n + \sigma)$ bits of space and $O(1)$ access time. In addition, the reverse dictionary can be implemented using a chaining hash table and a phrase dictionary implemented as an array. Thus, the working space of OESP using these data structures is at most $n(4 + \alpha) \lg (n + \sigma)$ bits. In the following subsections, we present space-efficient representations of phrase/reverse dictionaries.

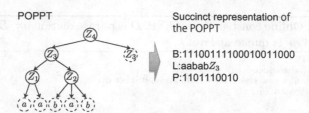

Fig. 2. Succinct representation of a POCFG for a phrase dictionary

3.4   Compressed Phrase Dictionary

OESP directly encodes a POCFG into a succinct representation that consists of bit strings B, P and a label sequence L. A bit string B is built by traversing a POPPT and putting c 0s and 1 for a node having c children in the post-order. The final 0 in B represents the super node. We shall call the bit string representation of a POPPT *posterior order unary degree sequence (POUDS)*. To dynamically build a tree and access any node in the POPPT, we index B by using the *dynamic range min/max tree* [14]. Our POUDS supports two tree operations: $child(B, i, j)$ returns the j-th child of a node i; $num\_child(B, i)$ returns the number of children for a node i. They are computed in $O(\lg m / \lg \lg m)$ time while using $2m + o(1)$ bits of space for a tree having m nodes.

A bit string P is built by traversing a POPPT and putting 1 for a leaf and 0 for an internal node in the post-order. P is indexed by the rank/select dictionary [6,13]. The label sequence L stores symbols of leaves in a POPPT.

We can access any element in L as a child of a node i in the following. First, we compute $c = num\_child(B, i)$ and children nodes $p = child(B, i, j)$ for $j \in [1, c]$. Then, we can compute the positions in L corresponding to the positions of these children as $q = rank_1(P, p)$ that returns the number of occurrences of 1 in $P[0, p]$ in $O(1)$ time. We obtain leaf labels as $L[q]$. For a POCFG of n nonterminal symbols, we can access the right-hand side of symbols from the left-hand side of a symbol of a production rule in $O(\lg n / \lg \lg n)$ time while using at most $n \lg (n + \sigma) + 5n + o(n)$ bits of space.

3.5   Compressed Reverse Dictionary

We implement a reverse dictionary using a chaining hash table that has a load factor $\alpha \in (0, 1]$ in a combination with a phrase dictionary. The hash table has αn entries and each entry stores a list of integers i representing the left-hand side X_i of a rule. For the rule $X_i \to S$, the hash value is computed from the right-hand side S. Then, the list corresponding to the hash value is scanned to search for X_i while checking elements referred to as S in a phrase dictionary. Thus, the expected access time is $O(1/\alpha)$. The space for a POCFG with n nonterminal symbols is $\alpha n \lg(n + \sigma)$ bits for the hash table and $n \lg(n + \sigma)$ bits for the lists, which resulted in $n(\alpha + 1) \lg(n + \sigma)$ bits in total.

A crucial observation in OESP is that indexes i for nonterminal symbols X_i are created in a strictly increasing order. Thus, we can organize each list in a hash table as a strictly increasing sequence of the indexes of nonterminal symbols.

We insert a new index i into a list in the hash table, and we append it at the end of the list. Each list in the hash table consists of a strictly increasing sequence of indexes. To make each index smaller, we compute the difference between an index i and the previous one j, and we encode it by the delta code, which resulted in the difference $i - j$ being encoded in $1 + \lfloor \lg(i - j) \rfloor + 2 \lfloor \lg \lfloor 1 + \lg(i - j) \rfloor \rfloor$ bits. For all n nonterminal symbols, the space for the lists is upper bounded by $n(1 + \lg(\alpha n) + 2 \lg \lg(\alpha n))$. bits The space for the hash table is $\alpha n \lg(n + \sigma + n(1 + \lg(\alpha n) + 2 \lg \lg(\alpha n))$ bits in total, resulting in $\alpha n \lg(n + \sigma) + n(1 + \lg(\alpha n))$ bits by multiplying the original α by a constant.

Since the reverse dictionary is implemented using the chaining hash and the phrase dictionary, its total space is at most $n(\alpha + 1) \lg(n + \sigma) + n(5 + \lg(\alpha n)) + o(n)$ bits. We can obtain the following result.

Lemma 1. *For a string length N, OESP builds a POCFG of n nonterminal symbols and its phrase/reverse dictionaries in $O(\frac{N \lg n}{\alpha \lg \lg n})$ expected time using at most $n(\alpha + 1) \lg(n + \sigma) + n \lg(\alpha n) + 5n + o(n)$ bits of space.*

3.6 Online Pattern Matching with EDM

We approximately solve problem 1 by using OESP. First, the parse tree is computed from a query Q by OESP. Let $T(Q)$ be a set of node labels in the parse tree for Q. We then compute a vector $V(Q)$ each dimension $V(Q)(e)$ of which represents the frequency of the corresponding node label e in $T(Q)$.

OESP builds another parse tree for a streaming text S in an online manner. $T(S)[i, i + |Q|]$ is a set of node labels included in the subtree corresponding to a substring $S[i, i + |Q|]$ from i to $i + |Q|$ in $T(S)$. $V(S)[i, i + |Q|]$ can be constructed for each $i \in [1, |S| - |Q|]$ by adding the node labels corresponding to $S[i, i + |Q|]$ and subtracting the node labels not included in $T(S)[i, i + |Q|]$ from $V(S)[i, i + |Q|]$, which can be performed in $\lg |S|$ time.

L_1-distance approximates the EDM between $V(S)[i, i + |Q|]$ and $V(Q)$, and it is computed as $\|V(S)[i, i + |Q|] - V(Q)\| = \sum_{e \in (T(S)[i,i+|Q|] \cup T(Q))} |V(S)[i, i + |Q|](e) - V(Q)(e)|$. We obtain the results with respect to computational time and space for computing the L_1 distance from lemma 1 as follows.

Theorem 1. *For a streaming text S of length N, OESP approximately solves the problem 1 in $O(\frac{N \lg N \lg n}{\alpha \lg \lg n})$ expected time using at most $n(\alpha + 1) \lg(n + \sigma) + n \lg(\alpha n) + 5n + o(n)$ bits of space.*

4 Upper Bound of Approximation

We present an upper bound of the approximate EDM in this section.

Theorem 2. $\|V(S) - V(Q)\| = O(\lg^2 m) d(S, Q)$ *for any $S, Q \in \Sigma^*$ and $m = \max\{|S|, |Q|\}$.*

Proof. Let $e_1, e_2, \ldots, e_d$ be a shortest series of editing operations such that $S_{k+1} = S_k(e_k)$ where $S_1 = S$, $S_d(e_d) = Q$, and $d = d(S, Q)$. It is sufficient

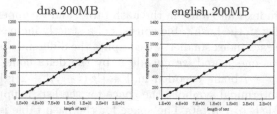

Fig. 3. Computation time in seconds for the length of text

to prove the assumption: there exists a constant c such that $\|V(S) - V(Q)\| \leq c \lg^2 m$ for $R(e) = S$. $S(i)$ denotes the string resulted by the i-th iteration of ESP where $S(0) = S$. Let p_i, q_i be the smallest integers satisfying $S(i)[p_i] \neq Q(i)[p_i]$ and $S(i)[|S|(i) - q_i] \neq Q(i)[|Q(i)| - q_i]$, respectively. We show that $q_i - p_i \leq \lg m + 1$ for each height i. This derives $\|V(S) - V(Q)\| \leq 2 \lg m (\lg m + 1)$ because $i \leq \lg m$.

We begin with the case that e is an insertion of a symbol. Clearly, it is true for $i = 0$ since $q_0 - p_0 \leq 1$. We assume the hypothesis on some height i. Let $S(i)[p']$ be the closest landmark from $S(i)[p_i]$ with $p' < p_i$ and $S(i)[q']$ be the closest landmark from $S(i)[q_i]$ with $q_i < q'$. For the next height, let $S(i+1) = S_1 S_2 S_3$ such that the tail of S_1 derives $S(i)[p_i]$ and the tail of S_2 derives $S(i)[q_i]$, and let $Q(i+1) = Q_1 Q_2 Q_3$ such that $|Q_1| = |S_1|$ and $|Q_3| = |S_3|$. On any iteration of ESP, the left aligned parsing is performed from a landmark to its closest landmark. It follows that, for S_1, $S_1[j] = Q_1[j]$ except their tails, for S_2, $|S_2| \leq \lfloor \frac{1}{2}(q_i - p_i) \rfloor \leq \lfloor \frac{1}{2}(\lg m + 1) \rfloor$, and for S_3, we can estimate $S_3[j] = Q_3[j]$ for any $j > \lfloor \frac{1}{2} \lg m \rfloor$. Thus, $q_{i+1} - p_{i+1} \leq 1 + \lfloor \frac{1}{2}(\lg m + 1) \rfloor + \lfloor \frac{1}{2} \lg m \rfloor \leq \lg m + 1$. Since $d(S, Q) = d(Q, S)$, this bound is true for the deletion of any symbol. The case that e is a replacement is similar.

Moreover, the bound holds for the case of insertion or deletion of any string of length at most $\lg m$. Using this, we can reduce the case of move operation of a substring u as follows. Without loss of generality, we assume u is a type2 substring and let $u = xyz$ such that x/z are the shortest prefix/suffix of u that contain a landmark, respectively. Then, we note that the y inside of u is transformed to a same string for any occurrence of u. Therefore, the case of moving u from S to obtain Q is reduced to the case of deleting x, z at some positions and inserting them into other positions. Since $|x|, |z| \leq \lg m$, the case of moving u is identical to the case of inserting two symbols and deleting two symbols, i.e., $\|V(S) - V(Q)\| \leq 8 \lg m (\lg m + 1)$.

From theorem 1 and 2, we obtain the following main theorem.

Theorem 3. *EDM is $O(\lg^2 N)$-approximable by the proposed online algorithm with $O(\frac{N \lg N \lg n}{\alpha \lg \lg n})$ expected time and $n(\alpha + 1) \lg (n + \sigma) + n(5 + \lg (\alpha n)) + o(n)$ bits of space.*

Proof. By the theorem 2, we obtain the bound $\|V(S[i, i + |Q|]) - V(Q)\| = O(\lg^2 |Q|) d(S[i, i + |Q|], Q)$ for any $i \in [1, |S| - |Q|]$. The time complexity is proved by the theorem 1. Thus, for the strings S and Q with $N = |S| \geq |Q|$, the result is concluded.

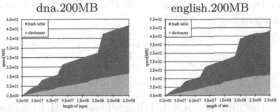

Fig. 4. Working space of dictionary and hash table for the length of text

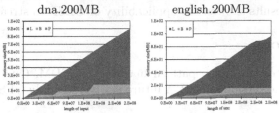

Fig. 5. Working space of a POUDS (B), a label sequence (L) and a bit string (P) which organizes a dictionary

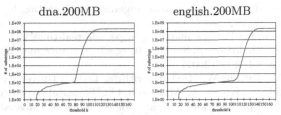

Fig. 6. The number of substrings whose EDM to a query is no more than each threshold

5   Experiments

We evaluated OESP on one core of an eight-core Intel Xeon CPU E7-8837 (2.67GHz) machine with 1024GB memory. We used two standard benchmark texts dna.200MB and english.200MB downloadable from `http://pizzachili.` `dcc.uchile.cl/texts.html`. We sampled texts of length 100 from these texts as queries. We also used computation time and working space as evaluation measures.

Figure 3 shows computation time for increasing the length of text. The computation time increased linearly for the length of text.

Figure 4 shows working space for increasing the length of text. The space of dictionary was much smaller than that of hash table. The dicionary used 115MB for dna.200MB and 121MB for english.200MB, while the hash table used 368MB for dna.200MB and 382MB for english.200MB.

Figure 5 shows space of a POUDS, a label sequence and a bit string organizing a dictionary for increasing the length of text. The space of dictionary and bit string was much smaller than that of the label sequence for dna.200MB and english.200MB. Table 2 details those space.

Table 2. Space for POUDS B, label sequence P and bit string P organizing a dictionary on dna.200MB and english.200MB.

	L[MB]	B[MB]	P[MB]
dna.200MB	89.95	17.62	7.73
english.200MB	95.72	14.99	8.22

Figure 6 shows the number of substring whose EDM to a query is at most a threshold. There were thresholds where the number of substrings dramatically increases. The results showed the applicability of OESP to streaming texts.

6   Conclusion

We have presented an online pattern maching for EDM. Our method named OESP is an online version of ESP. A future work is to apply OESP to real world streaming texts.

References

1. Bafna, V., Pevzner, P.A.: Genome rearrangements and sorting by reversals. SIAM Jour. on Comp. 25, 272–289 (1996)
2. Clifford, R., Sach, B.: Pattern matching in pseudo real-time. JDA 9, 67–81 (2011)
3. Cormode, G., Muthukrishnan, S.: The string edit distance matching problem with moves. TALG 3, 2:1–2:19 (2007)
4. Crochemore, M., Rytter, W.: Text Algorithms. Oxford University Press (1994)
5. Durbin, R., Eddy, S., Krogh, A., Mitchison, G.: Biological sequence analysis: Probabilistic models of proteins and nucleic acids. Cambridge University Press (1998)
6. Jacobson, G.: Space-efficient static trees and graphs. In: Proc. of FOCS, pp. 549–554 (1989)
7. Jalsenius, M., Porat, B., Sach, B.: Parameterized matching in the streaming model. In: STACS, pp. 400–411 (2013)
8. Kececioglu, J., Sankoff, D.: Exact and approximation algorithms for the inversion distance between two chromosomes. In: Apostolico, A., Crochemore, M., Galil, Z., Manber, U. (eds.) CPM 1993. LNCS, vol. 684, pp. 87–105. Springer, Heidelberg (1993)
9. Levenshtein, V.I.: Binary codes capable of correcting deletions, insertions and reversals. Soviet Physics Doklady 10, 707–710 (1996)
10. Maruyama, S., Tabei, Y.: Fully-online grammar compression in constant space. In: Proc. of DCC, pp. 218–229 (2014)
11. Maruyama, S., Tabei, Y., Sakamoto, H., Sadakane, K.: Fully-online grammar compression. In: Kurland, O., Lewenstein, M., Porat, E. (eds.) SPIRE 2013. LNCS, vol. 8214, pp. 218–229. Springer, Heidelberg (2013)
12. Muthukrishnan, S., Sahinalp, S.C.: Approximate nearest neighbors and sequence comparison with block operations. In: Proc. of STOC, pp. 416–424 (2000)
13. Navarro, G., Providel, E.: Fast, small, simple rank/select on bitmaps. In: Klasing, R. (ed.) SEA 2012. LNCS, vol. 7276, pp. 295–306. Springer, Heidelberg (2012)
14. Navarro, G., Sadakane, K.: Fully-functional static and dynamic succinct trees. TALG (2012) (accepted); A preliminary version appeared in SODA 2010 (2010)
15. Rytter, W.: Application of Lempel-Ziv factorization to the approximation of grammar-based compression. Theor. Comp. Sci. 302(1-3), 211–222 (2003)
16. Shapira, D., Storer, J.A.: Edit distance with move operations. JDA 5, 380–392 (2007)

K^2-Treaps: Range Top-k Queries in Compact Space*

Nieves R. Brisaboa[1], Guillermo de Bernardo[1], Roberto Konow[2,3], and Gonzalo Navarro[2]

[1] Databases Lab., Univ. of A. Coruña, Spain
{brisaboa,gdebernardo}@udc.es
[2] Dept. of Computer Science, Univ. of Chile
{rkonow,gnavarro}@dcc.uchile.cl
[3] Escuela de Informática y Telecomunicaciones, Univ. Diego Portales, Chile

Abstract. Efficient processing of top-k queries on multidimensional grids is a common requirement in information retrieval and data mining, for example in OLAP cubes. We introduce a data structure, the K^2-treap, that represents grids in compact form and supports efficient prioritized range queries. We compare the K^2-treap with state-of-the-art solutions on synthetic and real-world datasets, showing that it uses 30% of the space of competing solutions while solving queries up to 10 times faster.

1 Introduction

Top-k queries on multidimensional weighted point sets ask for the k heaviest points in a range. This type of query arises most prominently in data mining and OLAP processing (e.g., find the sellers with most sales in a time period) and in GIS applications (e.g., find the cheapest hotels in a city area), but also in less obvious document retrieval applications [16]. In the example of sales, one coordinate is the seller id, which are arranged hierarchically to allow queries for sellers, stores, areas, cities, states, etc., and the other is time (in periods of hours, days, weeks, etc.). Weights are the amounts of sales made by a seller during a time slice. Thus the query asks for the k heaviest points in some range $Q = [x_1, x_2] \times [y_1, y_2]$ of the grid.

Data mining and information systems such as those mentioned above usually handle huge amounts of data and may have to serve millions of queries per second. Representing this steadily increasing amount of data space-efficiently can make the difference between maintaining the data in main memory or having to resort to external memory, which is orders of magnitude slower.

We introduce a new compact data structure that performs fast range top-k queries on multidimensional grids and is smaller than state-of-the-art compact

---

* Funded by Millennium Nucleus Information and Coordination in Networks ICM/FIC P10-024F, by a Conicyt scholarship, by MICINN (PGE and FEDER) TIN2009-14560-C03-02 and TIN2010-21246-C02-01, by CDTI, MEC and AGI EXP 00064563/ITC-20133062, and by Xunta de Galicia (with FEDER) GRC2013/053.

E. Moura and M. Crochemore (Eds.): SPIRE 2014, LNCS 8799, pp. 215–226, 2014.
© Springer International Publishing Switzerland 2014

data structures. Our new representation, called K^2-treap, is inspired by two previous data structures: the K^2-tree [5] and the treap [18]. The K^2-tree is a compressed and self-indexed structure initially designed to represent Web graphs and later used in other domains as an efficient and compact representation of binary relations. The treap is a binary tree that satisfies the invariants of a binary search tree and a heap at the same time, which is useful for prioritized searches. Our results show that the K^2-treap answers queries up to 10 times faster, while using just 30% of the space, of state-of-the-art alternatives.

2   Basic Concepts

Rank and select on bitmaps. Let $B[1, n]$ be a sequence of bits, or bitmap. We define operations $rank_b(B, i)$ as the number of occurrences of $b \in \{0, 1\}$ in $B[1, i]$, and $select_b(B, j)$ as the position in S of the jth occurrence of b. B can be represented using $n + o(n)$ bits [15], so that both operations are solved in constant time.

Wavelet trees and discrete grids. An $n \times m$ grid with n points, exactly one per column (i.e., x values are unique), can be represented using a *wavelet tree* [10, 12]. This is a perfect balanced binary tree of height $\lceil \lg m \rceil$ where each node corresponds to a contiguous range of values $y \in [1, m]$ and represents the points falling in that y-range, sorted by increasing x-coordinate. The root represents $[1, m]$ and the two children of each node split its y-range in half, until the leaves represent a single y-coordinate. Each internal node stores a bitmap, which tells whether each point corresponds to its left or right child. Using $rank$ and $select$ queries on the bitmaps, the wavelet tree uses $n \lg m + o(n \log m)$ bits, and can count the number of points in a range in $O(\log m)$ time, because the query is decomposed into bitmap ranges on at most 2 nodes per wavelet tree level. Any point can be tracked up (to find its x-coordinate) or down (to find its y-coordinate) in $O(\log m)$ time as well.

K^2-trees. The K^2-tree [5] is a data structure to compactly represent sparse binary matrices (which can also be regarded as point grids). The K^2 tree subdivides the matrix into K^2 submatrices of equal size. The submatrices are considered left-to-right and top-to-bottom, and each is represented with a bit, set to 1 if the submatrix contains at least one non-zero cell. Each node whose bit is 1 is recursively decomposed, subdividing its submatrix into K^2 children, and so on. The subdivision ends when a fully-zero submatrix is found or when we reach the individual cells. The K^2-tree can answer range queries with multi-branch top-down traversal of the tree, following only the branches that overlap the query range. While it has no good worst-case time guarantees, in practice times are competitive. The worst-case space, if t points are in an $n \times n$ matrix, is $K^2 t \log_{K^2} \frac{n^2}{t}(1 + o(1))$ bits. This can be reduced to $t \lg \frac{n^2}{t}(1 + o(1))$ if the bitmaps are compressed. This is similar to the wavelet tree space, but in practice K^2-trees use much less space when the points are clustered.

The K^2-tree is stored in two bitmaps: T stores the bits of all the levels except the last one, in a level-order traversal, and L stores the bits of the last level (corresponding to individual cells). Given a node at position p in T, its children are be located from position $rank_1(T, p) \cdot K^2$ in $T : L$. This property enables K^2-tree traversals using just T and L.

Treaps and Priority Search Trees. A *treap* [18] is a binary search tree with nodes having two attributes: *key* and *priority*. The treap maintains the binary search tree invariants for the keys and the heap invariants for the priorities, that is the key of a node is larger than those in its left subtree and smaller than those in its right subtree, whereas its priority is not smaller than those in its subtree. The treap does not guarantee logarithmic height, except on expectation if priorities are independent of keys [13]. The *priority search tree* [14] is somewhat similar, but it is balanced. In this case, a node is not the one with highest priority in its subtree, but that element is stored in addition to the element at the node. The element stored separately is also removed from the subtree. Priority search trees can be used to solve 3-sided range queries on n-point grids, returning t points in time $O(t + \log n)$.

3 Related Work

Navarro et al. [17] introduced compact data structures for various queries on two-dimensional weighted points, including range top-k queries. They enhance the bitmaps of each node as follows: Let $x_1, \ldots, x_r$ be the points represented at a node, and $w(x)$ be the weight of point x. Then a range maximum query (RMQ) data structure built on $w(x_1), \ldots, w(x_r)$ is stored together with the bitmap. Such a structure uses $2r + o(r)$ bits and finds the position of the maximum weight in any range $[w(x_i), \ldots, w(x_j)]$ in constant time [8] and without accessing the weights themselves. Therefore, the total space becomes $3n \lg m + o(n \log m)$ bits.

To solve top-k queries on a grid range $Q = [x_1, x_2] \times [y_1, y_2]$, we first traverse the wavelet tree to identify the $O(\log m)$ bitmap intervals where the points in Q lie (a counting query would, at this point, just add up all the bitmap interval lengths). The heaviest point in Q in each bitmap interval is obtained with an RMQ, but we need to obtain the actual priorities in order to find the heaviest among the $O(\log m)$ candidates. The priorities are stored sorted by x- or y-coordinate, so we obtain each one in $O(\log m)$ time by tracking the point with maximum weight in each interval. Thus a top-1 query is solved in $O(\log^2 m)$ time. For a top-k query we must maintain a priority queue of the candidate intervals, and each time the next heaviest element is found, we remove it from its interval and reinsert in the queue the two resulting subintervals. The total query time is $O((k + \log m) \log(km))$.

It is possible to reduce the time to $O((k + \log m) \log^\epsilon m)$ time and $O(\frac{1}{\epsilon} n \log m)$ bits, for any constant $\epsilon > 0$ [16], but the space usage is much higher, even if linear.

4 The K^2-treap

In one dimension, an RMQ structure using $2n + o(n)$ bits [8] is sufficient to answer range top-k queries in $O(k \log k)$ or $O(k \log \log n)$ time, using the algorithm just described on a single interval. However, a similar RMQ structure for two or more dimensions needs $\Omega(mn \log m)$ bits [9] (on dense grids), and therefore it is better to directly look for representations of the data points that can also answer range top-k queries. The idea is to combine a K^2-tree with a treap data structure. If keys are $[1, n]$, treaps can be stored in $2n + o(n)$ bits plus the priorities [11], whereas priority search trees cannot. In two and more dimensions, however, this advantage vanishes. Therefore, our data structure combines the K^2-tree with a priority search tree, which is more convenient for its balancing guarantees.

4.1   Data Structure

Consider a matrix $M[n \times n]$ where each cell can either be empty or contain a weight in the range $[0, d-1]$. We consider a quadtree-like recursive partition of M into K^2 submatrices, the same performed in the K^2-tree with binary matrices. We build a conceptual K^2-ary tree similar to the K^2-tree, as follows: the root of the tree will store the coordinates of the cell with the maximum weight of the matrix, and the corresponding weight. Then the cell just added to the tree is marked as *empty*, deleting it from the matrix. If many cells share the maximum weight, we pick anyone of them. Then, the matrix is conceptually decomposed into K^2 equal-sized submatrices, and we add K^2 child nodes to the root of the tree, each representing one of the submatrices. We repeat the assignment process recursively for each child, assigning to each of them the coordinates and value of the heaviest cell in the corresponding submatrix and removing the chosen points. The procedure continues recursively for each branch until we find a completely empty submatrix (either because the matrix did not contain any weights in the region or because the cells with weights have been "emptied" during the construction process) or we reach the cells of the original matrix.

Fig. 1 shows an example of K^2-treap construction, for $K = 2$. At the top of the image we show the state of the matrix at each level of decomposition. $M0$ represents the original matrix, where the maximum value is highlighted. The coordinates and value of this cell are stored in the root of the tree. In the next level of decomposition (matrix $M1$) we find the maximum values in each quadrant (notice that the cell assigned to the root has already been removed from the matrix) and assign them to the children of the root node. The process continues recursively, subdividing each matrix into K^2 submatrices. The cells chosen as local maxima are highlighted in the matrices corresponding to each level of decomposition, except in the last level where all the cells are local maxima. Empty submatrices are marked in the tree with the symbol "-".

The data structure is represented in three parts: The location of local maxima, the weights of the local maxima, and the tree topology.

Local maximum coordinates: The conceptual K^2-treap is traversed levelwise, reading the sequence of cell coordinates from left to right in each level. The sequence of coordinates at each level ℓ is stored in a different sequence $coord[\ell]$. The coordinates at each level ℓ of the tree are transformed into an offset in the corresponding submatrix, transforming each c_i into $c_i \bmod (n/K^\ell)$

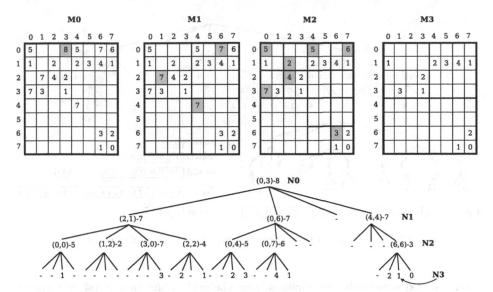

Fig. 1. Example of K^2-treap construction from a matrix

using $\lceil \lg(n) - \ell \lg K \rceil$ bits. For example, in Fig. 2 (top) the coordinates of node $N1$ have been transformed from the global value $(4, 4)$ to a local offset $(0, 0)$. In the bottom of Fig. 2 we highlight the coordinates of nodes $N0$, $N1$ and $N2$ in the corresponding *coord* arrays. In the last level all nodes represent single cells, so there is no *coord* array in this level. With this representation, the worst-case space for storing t points is $\sum_{\ell=0}^{\log_{K^2}(t)} 2K^{2\ell} \lg \frac{n}{K^\ell} = t \lg \frac{n^2}{t}(1 + O(1/K^2))$, that is, the same as if we stored the points using the K^2-tree.

Local maximum values: The maximum value in each node is encoded differentially with respect to the maximum of its parent node. The result of the differential encoding is a new sequence of non-negative values, smaller than the original. Now the K^2-treap is traversed level-wise and the complete sequence of values is stored in a single sequence named *values*. To exploit the small values while allowing efficient direct access to the array, we represent *values* with Direct Access Codes [4]. Following the example in Fig. 2, the value of node $N1$ has been transformed from 7 to $8 - 7 = 1$. In the bottom of the figure the complete sequence *values* is depicted. We also store a small array $first[0, \lg_K n]$ that stores the offset in *values* where each level starts.

Tree structure: We separate the structure of the tree from the values stored in the nodes. The tree structure of the K^2-treap is stored in a K^2-tree. Fig. 2 shows the K^2-tree representation of the example tree, where only cells with value are labeled with a 1. We will consider a K^2-tree stored in a single bitmap T with *rank* support, that contains the sequence of bits from all the levels of the tree. Our representation differs from a classic K^2-tree (that uses two bitmaps T and L and only adds rank support to T) because we will need to perform rank operations also in the last level of the tree. The other difference is that

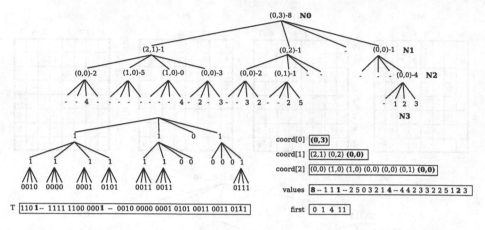

Fig. 2. Storage of the conceptual tree in our data structures

points stored separately are removed from the grid. Thus, in a worst-case space analysis, it turns out that the space used to represent those explicit coordinates is subtracted from the space the K^2-tree would use, therefore storing those explicit coordinates is free in the worst case.

4.2   Query Algorithms

Basic navigation. To access a cell $C = (x, y)$ in the K^2-treap we start by accessing the K^2-tree root. The coordinates and weight of the element stored at the root node are $(x_0, y_0) = coord[0][0]$ and $w_0 = values[0]$. If $(x_0, y_0) = C$, we return w_0 immediately. Otherwise, we find the quadrant where the cell would be located and navigate to that node in the K^2-tree. Let p be the position of the node in T. If $T[p] = 0$ we know that the complete submatrix is empty and return immediately. Otherwise, we need to find the coordinates and weight of the new node. Since only nodes set to 1 in T have coordinates and weights, we compute $r = rank_1(T, p)$. The value of the current node will be at $values[r]$, and its coordinates at $coord[\ell][r - first[\ell]]$, where ℓ is the current level. We rebuild the absolute value and coordinates, w_1 as $w_0 - values[r]$ and (x_1, y_1) adding the current submatrix offset to $coord[\ell][r - first[\ell]]$. If $(x_1, y_1) = C$ we return w_1, otherwise we find again the appropriate quadrant in the current submatrix where C would be located, and so on. The formula to find the children is identical to that of the K^2-tree. The process is repeated recursively until we find a 0 bit in the target submatrix, we find a 1 in the last level of the K^2-tree, or we find the coordinates of the cell in an explicit point.

Top-k queries. The process to answer top-k queries starts at the root of the tree. Given a range $Q = [x_1, x_2] \times [y_1, y_2]$, the process initializes an empty max-priority queue and inserts the root of the K^2-tree. The priority queue stores, in general, K^2-tree nodes sorted by their associated maximum weight. Now, we iteratively extract the first priority queue element (the first time this is the root). If the coordinates of its maximum element fall inside Q, we output it as the next answer. In either case, we insert all the children of the extracted node whose submatrix intersects with Q, and iterate. The process finishes when k results have been found or when the priority queue becomes empty (in which case there are less than k elements in Q).

Other supported queries. The K^2-treap can also answer basic range queries (i.e., report all the points that fall in Q). This is similar to the procedure on a K^2-tree, where the submatrices that intersect Q are explored in a depth-first manner. The only difference is that we must also check whether the explicit points associated to the nodes fall within Q, and in that case report those as well. Finally, we can also answer *interval queries*, which ask for all the points in Q whose weight is in a range $[w_1, w_2]$. To do this, we traverse the tree as in a top-k range query, but we only output weights whose value is in $[w_1, w_2]$. Moreover, we discard submatrices whose maximum weight is below w_1.

5   Experiments and Results

To test the efficiency of our proposal we use several synthetic datasets, as well as some real datasets where top-k queries are of interest. Our synthetic datasets are square matrices where only some of the cells have a value set. We build different matrices varying the following parameters: the *size* $s \times s$ of the matrix ($s = 1024, 2048, 4096, 8192$), the number of different weights d in the matrix (16, 128, 1024) and the *percentage* p of cells that have a point (10, 30, 50, 70, 100%). The distribution of the weights in all the datasets is uniform, and the spatial distribution of the cells with points is random. For example, the synthetic dataset with ($s = 2048, d = 128, p = 30$) has size 2048×2048, 30% of its cells have a value and their values are follow a uniform distribution in $[0, 127]$.

We also test our representation using real datasets. We extracted two different views from a real OLAP database storing information about sales achieved per store/seller each hour over several months: *salesDay* stores the number of sales per seller per day, and *salesHour* the number of sales per hour. Huge historical logs are accumulated over time, and are subject to data mining processes for decision making. In this case, finding the places (at various granularities) with most sales in a time period is clearly relevant. Table 1 shows a summary with basic information about the real datasets. For simplicity, in these datasets we ignore the cost of mapping between real timestamps and seller ids to rows/columns in the table, and assume that the queries are given in terms of rows and columns.

We compare the space requirements of the K^2-treap against the solution based on wavelet trees enhanced with RMQ structures [17] introduced in Section 3 (*wtrmq*). Since our matrices can contain none or multiple values per column, we transform our datasets to store them using wavelet trees. The wavelet tree will store a grid with as many columns as values we have in our matrix, in column-major order. A bitmap is used to map the real columns with virtual

Table 1. Real datasets used, and space required to represent them

Dataset	#Sellers (rows)	Time instants (columns)	Number of diff. values	K^2-treap (bits/cell)	mk2tree (bits/cell)	wtrmq (bits/cell)
SalesDay	1314	471	297	2.48	3.75	9.08
SalesHour	1314	6028	158	1.06	0.99	3.90

ones: we append a 0 per new point and a 1 when the column changes. Hence, range queries in the wtrmq require a mapping from real columns to virtual ones (2 $select_1$ operations per query), and the virtual column of each result must be mapped back to the actual value (a $rank_1$ operation per result).

We also compare our proposal with a representation based on constructing multiple K^2-trees, one per different value in the dataset. In this representation (mk2tree), top-k queries are answered by querying consecutively the K^2-tree representations for the higher values. Each K^2-tree representation in this proposal is enhanced with multiple optimizations over the simple bitmap approach we use, like the compression of the lower levels of the tree using DACs (see [5] for a detailed explanation of this and other enhancements of the K^2-tree).

All bitmaps that are employed use a bitmap representation that supports *rank* and *select* using 5% of extra space. The wtmrq was implemented using a pointer-less version of the wavelet tree [7] with a RMQ implementation that requires 2.38 bits per value. For all experiments we use $K = 2$ for the K^2-treap and mk2tree.

We ran all our experiments on a dedicated server with 4 Intel(R) Xeon(R) E5520 CPU cores at 2.27GHz 8MB cache and 72GB of RAM memory. The machine runs Ubuntu GNU/Linux version 9.10 with kernel 2.6.31-19-server (64 bits) and gcc 4.4.1. All data structures were implemented in C/C++, compiled with full optimizations.

5.1 Space Comparison

We start by comparing the compression achieved by the representations. As shown in Table 1, the K^2-treap overcomes the wtrmq in the real datasets studied by a factor over 3.5. The mk2tree representation is competitive with the K^2-treap and even obtains slightly less space in the dataset salesHour, taking advantage of the relatively small number of different values in the matrix.

The K^2-treap also obtains the best space results in most of the synthetic datasets studied. Only in the datasets with a very small number of different values ($d = 16$) the mk2tree uses less space than the K^2-treap. Notice that, since the distribution of values and cells is uniform, the synthetic datasets are close to a worst-case scenario for the K^2-treap and mk2tree. To provide additional insight on the compression capabilities, Fig. 3 provides a summary of the space results for some of the synthetic datasets used. The left plot shows the evolution of compression with the size of the matrix. The K^2-treap is almost unaffected by

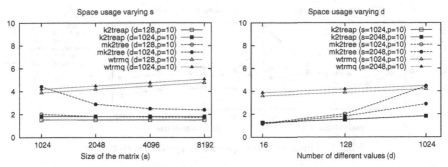

Fig. 3. Evolution of the space usage with s and d in the synthetic datasets, in bits/cell (in the right plot, the two results for the K^2-treap are on top of each other)

the matrix size, as its space is around $t \lg \frac{s^2}{t} = s^2 \frac{p}{100} \lg \frac{100}{p}$ bits, that is, constant per cell as s grows. On the other hand, the *wtrmq* uses $t \lg s = s^2 \frac{p}{100} \lg s$ bits, that is, its space per cell grows logarithmically with s. Finally, the *mk2tree* obtains poor results in the smaller datasets but is more competitive on larger ones (some enhancements in the K^2-tree representations behave worse in smaller matrices). Nevertheless, notice that the improvements in the *mk2tree* compression stall once the matrix reaches a certain size.

The right plot of Fig. 3 shows the space results when varying the number of different weights d. The K^2-treap and the *wtrmq* are affected only logarithmically by d. The *mk2tree*, instead, is sharply affected, since it must build a different K^2-tree for each different value: if d is very small the *mk2tree* representation obtains the best space results also in the synthetic datasets, but for large d its compression degrades significantly.

As the percentage of cells set p increases, the compression in terms of bits/cell (i.e., total bits divided by s^2) will be worse. However, if we measure the compression in bits per point (i.e., total bits divided by t), then the space of the *wtrmq* is independent of p ($\lg s$ bits), whereas the K^2-treap and *mk2tree* use less space as p increases ($\lg \frac{100}{p}$). That is, the space usage of the *wtrmq* increases linearly with p, while that of the K^2-treap and *mk2tree* increases sublinearly. Over all the synthetic datasets, the K^2-treap uses from 1.3 to 13 bits/cell, the *mk2tree* from 1.2 to 19, and the *wtrmq* from 4 to 50 bits/cell.

5.2 Query Times

In this section we analyze the efficiency of top-k queries, comparing our structure with the *mk2tree* and the *wtrmq*. For each dataset, we build multiple sets of top-k queries for different values of k and different spatial ranges (we ensure that the spatial range is at least of size k). All query sets are generated for fixed k and w (side of the spatial window). Each query set contains 1000 queries where the spatial window is placed at a random position within the matrix.

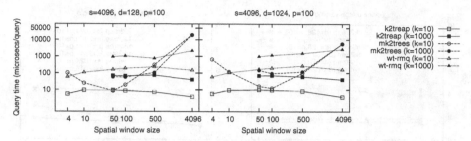

Fig. 4. Times of top-k queries in synthetic datasets

Fig. 4 shows the time required to perform top-k queries in some of our synthetic datasets, for different values of k and w. The K^2-treap obtains better query times than the *wtrmq* in all the queries, and both evolve similarly with the size of the query window. On the other hand, the *mk2tree* representation obtains poor results when the spatial window is small or large, but it is competitive with the K^2-treap for medium-sized ranges. This is due to the procedure to query the multiple K^2-tree representations: for small windows, we may need to query many K^2-trees until we find k results; for very large windows, the K^2-treap starts returning results in the upper levels of the conceptual tree, while the *mk2tree* approach must reach the leaves; for some intermediate values of the spatial window, the K^2-treap still needs to perform several steps to start returning results, and the *mk2tree* representation may find the required results in a single K^2-tree. Notice that the K^2-treap is more efficient when no range limitations are given (that is, when $w = s$), since it can return after exactly K iterations. Fig. 4 only shows the results for two of the datasets, but similar comparison results have been obtained in all the synthetic datasets studied, with the K^2-treap outperforming the alternative approaches in most of the cases, except in some queries with medium-sized query windows, when the *mk2tree* can obtain slightly better query times.

Next we perform a set of queries that would be interesting in our real datasets. We start with the same $w \times w$ queries as before, which filter a range of rows (sellers) and columns (days/hours). Fig. 5 shows the results of these range queries. As we can see, the K^2-treap outperforms both, the *mk2tree* and *wtrmq*, in all cases. Similarly to the previous results, the *mk2tree* approach also obtains poor query times for small ranges but is better in larger ranges.

We run two more specific sets of queries that may be interesting in many datasets, and particularly in our examples: "column-oriented" and "row-oriented" range queries, that only restrict one of the dimensions of the matrix. Row-oriented queries ask for a single row (or a small range of rows) but do not restrict the columns, and column-oriented ask for single columns. We build sets of 10,000 top-k queries for random rows/columns with different values of k. Fig. 6 (left) shows that in column-oriented queries the *wtrmq* is faster than the K^2-treap for small values of k, but our proposal is still faster as k grows. The reason for this difference is that in "square" range queries, the K^2-treap only visits a small

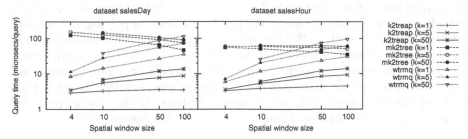

Fig. 5. Query times of top-k queries in real datasets

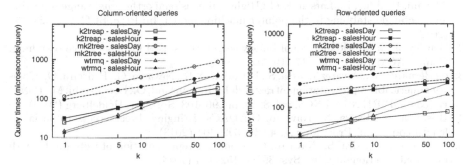

Fig. 6. Query times of row-oriented and column-oriented top-k queries

set of submatrices that overlap the region; in row-oriented or column-oriented queries, the K^2-treap is forced to check many submatrices to find only a few results. The *mk2tree* suffers from the same problem of the K^2-treap, being unable to filter efficiently the matrix, and obtains the worst query times in all cases.

In row-oriented queries (Fig. 6, right) the *wtrmq* is even more competitive, obtaining the best results in many queries. The reason for the differences found between row-oriented and column-oriented queries in the *wtrmq* is the mapping between real and virtual columns: column ranges are expanded to much longer intervals in the wavelet tree, while row ranges are left unchanged. Notice anyway that our proposal is still competitive in the cases where k is relatively large.

6   Conclusions and Future Work

We have introduced a new compact data structure that performs top-k range queries on grids up to 10 times faster than current state-of-the-art solutions and requires as little as 30% of the space, both in synthetic and real OLAP databases, and including uniform distributions, which is the worst scenario for K^2-treaps.

The K^2-treap can be generalized to represent grids in higher dimensions, by simply replacing our underlying K^2-tree with its generalization to d dimensions, the K^d-tree [3] (not to be confused with kd-trees [2]). The algorithms stay identical, but an empirical evaluation is left for future work. In the worst case, a grid of t points on $[n]^d$ will require $O(t \lg \frac{n^d}{t})$ bits, which is of the same order of the data, and much less space will be used on clustered data. Instead, an extension of the wavelet tree will require $O(n \log^d n)$ bits, which quickly becomes impractical. Indeed, any structure able to report the points in a range in polylogarithmic time requires $\Omega(n(\log n/ \log \log n)^{d-1})$ words of space [6], and with polylogarithmic space one needs time at least $\Omega(\log n(\log n/ \log \log n)^{\lfloor d/2 \rfloor - 2})$ [1]. As with top-k queries one can report all the points in a range, there is no hope to obtain good worst-case time and space bounds in high dimensions, and thus heuristics like K^d-treaps are the only practical approaches (kd-trees do offer linear space, but their time guarantee is rather loose, $O(n^{1-1/d})$ for n points on $[n]^d$).

References

1. Afshani, P., Arge, L., Larsen, K.G.: Higher-dimensional orthogonal range reporting and rectangle stabbing in the pointer machine model. In: Proc. SCG, pp. 323–332 (2012)
2. Bentley, J.L.: Multidimensional binary search trees used for associative searching. Comm. ACM 18(9), 509–517 (1975)
3. de Bernardo, G., Álvarez-García, S., Brisaboa, N.R., Navarro, G., Pedreira, O.: Compact querieable representations of raster data. In: Kurland, O., Lewenstein, M., Porat, E. (eds.) SPIRE 2013. LNCS, vol. 8214, pp. 96–108. Springer, Heidelberg (2013)
4. Brisaboa, N., Ladra, S., Navarro, G.: DACs: Bringing direct access to variable-length codes. Inf. Proc. Manag. 49(1), 392–404 (2013)
5. Brisaboa, N., Ladra, S., Navarro, G.: Compact representation of web graphs with extended functionality. Inf. Sys. 39(1), 152–174 (2014)
6. Chazelle, B.: Lower bounds for orthogonal range searching I: The reporting case. J. ACM 37(2), 200–212 (1990)
7. Claude, F., Navarro, G.: Practical rank/select queries over arbitrary sequences. In: Amir, A., Turpin, A., Moffat, A. (eds.) SPIRE 2008. LNCS, vol. 5280, pp. 176–187. Springer, Heidelberg (2008)
8. Fischer, J., Heun, V.: Space-efficient preprocessing schemes for range minimum queries on static arrays. SIAM J. Comp. 40(2), 465–492 (2011)
9. Golin, M., Iacono, J., Krizanc, D., Raman, R., Rao, S.S.: Encoding 2D range maximum queries. In: Asano, T., Nakano, S.-i., Okamoto, Y., Watanabe, O. (eds.) ISAAC 2011. LNCS, vol. 7074, pp. 180–189. Springer, Heidelberg (2011)
10. Grossi, R., Gupta, A., Vitter, J.: High-order entropy-compressed text indexes. In: Proc. 14th SODA, pp. 841–850 (2003)
11. Konow, R., Navarro, G., Clarke, C., López-Ortíz, A.: Faster and smaller inverted indices with treaps. In: Proc. 36th SIGIR, pp. 193–202 (2013)
12. Mäkinen, V., Navarro, G.: Position-restricted substring searching. In: Correa, J.R., Hevia, A., Kiwi, M. (eds.) LATIN 2006. LNCS, vol. 3887, pp. 703–714. Springer, Heidelberg (2006)
13. Martínez, C., Roura, S.: Randomized binary search trees. J. ACM 45(2), 288–323 (1997)
14. McCreight, E.M.: Priority search trees. SIAM J. Comp. 14(2), 257–276 (1985)
15. Munro, J.I.: Tables. In: Chandru, V., Vinay, V. (eds.) FSTTCS 1996. LNCS, vol. 1180, pp. 37–42. Springer, Heidelberg (1996)
16. Navarro, G., Nekrich, Y.: Top-k document retrieval in optimal time and linear space. In: Proc. 23rd SODA, pp. 1066–1078 (2012)
17. Navarro, G., Nekrich, Y., Russo, L.: Space-efficient data-analysis queries on grids. Theor. Comp. Sci. 482, 60–72 (2013)
18. Seidel, R., Aragon, C.: Randomized search trees. Algorithmica 16(4/5), 464–497 (1996)

Performance Improvements for Search Systems Using an Integrated Cache of Lists+Intersections*

Gabriel Tolosa[1,2], Luca Becchetti[3],
Esteban Feuerstein[1], and Alberto Marchetti-Spaccamela[3]

[1] University of Buenos Aires, Argentina
[2] National University of Luján, Argentina
[3] Sapienza University of Rome, Italy

Abstract. Modern information retrieval systems use several levels of caching to speedup computation by exploiting frequent, recent or costly data used in the past. In this study we propose and evaluate a static cache that works simultaneously as list and intersection cache, offering a more efficient way of handling cache space. In addition, we propose effective strategies to select the term pairs that should populate the cache. Simulation using two datasets and a real query log reveal that the proposed approach improves overall performance in terms of total processing time, achieving savings of up to 40% in the best case.

1 Introduction

Modern high scale information retrieval systems such as Web Search Engines (WSE) use sophisticated techniques for efficiency and scalability purposes. In such scenarios, caching is an important and crucial tool to achieve fast response times and to increase query throughput.

It is known that the total cost of a query is the sum of processing time (C_{cpu}) and disk access times (C_{disk}). C_{cpu} involves decompressing the posting lists, computing the query-document similarity scores and determining the top-k documents that form the final answer set. In most cases a conjunctive semantic is considered because intersections produce shorter lists than unions, which leads to smaller query latencies [3] and higher precision levels. On the other hand, C_{disk} involves fetching from hard disk the posting lists of all the query terms.

The main goal of a cache is to speedup computation by storing frequent, recent or costly data. The typical architecture of a search engine involves different cache levels: essentially, caching involves both query result pages (*Result cache*) at the broker level and the posting lists of terms that appear in the queries (*List Cache*) at search node level. The first level tries to minimize recomputation of results for queries that appeared in the past, the latter attempts at reducing the

---

* This work was partially supported by EU-IRSES project EUSACOU 247574, by EU FET project MULTIPLEX 317532 and by UBACyT Project 20020120100058 "Herramientas algorítmicas avanzadas para aplicaciones de búsqueda en Internet - Parte 2".

E. Moura and M. Crochemore (Eds.): SPIRE 2014, LNCS 8799, pp. 227–235, 2014.
© Springer International Publishing Switzerland 2014

amount of disk fetch operations, which are very expensive compared to CPU processing times. A further approach involves caching portions of a query (i.e., pairs of terms), as initially proposed in [13] and extended in [6]. This approach is named *Intersection Caching* and is implemented at search node level as well. The idea in this case is to exploit term co-occurrence patterns, e.g., by keeping the intersection of the postings lists of frequently co-occurring pairs of terms in the memory of the search node, in order to not only save disk access time, but CPU time too.

In the case of industry-scale search engines that store the entire index in main memory [5] the *List Cache* becomes useless, but the intersection cache is still useful [9] because it allows to save CPU time (i.e. the cost of intersecting two posting lists). For more general cases such as medium-scale systems, only a fraction of the index is maintained in cache. Here, lists and intersections caches are both helpful to reduce disk access and processing time.

List and intersection caches are implemented at search node level and, usually, they are independent and exploit different phenomena. While the *List Cache* achieves higher hit rates because the frequency of individual terms is higher than that of term pairs, each hit in the *Intersection Cache* entails a higher benefit because the intersected lists of term pairs are shorter. Based on the observation that many terms co-occur frequently in different queries, our goal is to build a single cache that benefits from both approaches. To do this, we implement a data structure previously proposed by Lam *et al.* [12]. The original idea is to merge the entries of two frequently co-occurring terms to form a single, more compact inverted list. We adapt this structure to a static cache in which the selected term pairs strike a good balance between hit rate and cost benefits, leading to an improvement in the total cost of solving a query. We investigate different ways of choosing and combining terms.

Related Work. There is a large body of work devoted to caching in text search systems, an active research area. Baeza *et al.* [1] analyze the problem of posting list caching. They propose an algorithm that selects terms to put in cache according to their $\frac{frequency(t)}{size(t)}$ ratios. The most important observation is that this static policy has a better hit rate than all dynamic counterparts. In [21] inverted index compression algorithms and list caching policies are explored.

The work in [14] is the first approach on the problem of result caching. The author proposes to consider both frequency and recency into the policy. More recently, Fagni *et al.* [7] proposed SDC (Static/Dynamic Cache) to handle both long term popular queries and shorter query bursts. Gan and Suel [10] study the problem of weighted result caching. In [15], cost aware strategies are extensively evaluated incorporating query costs into the caching policies.

Saraiva *et al.* [18] propose a two-level caching scheme that combines caching of search results with the caching of frequently accessed postings lists. The first proposal on intersection caching appears in [13], where the authors introduce a three-level caching architecture for a web search engine. Further studies on cost aware intersection caching are presented in [8] and [9]. In a more recent work, Ozcan *et al.* [16] introduce a 5-level static caching architecture.

Our Contribution. In this work, we explore the possibility of reserving the whole memory space allocated to caching at search nodes to an integrated cache, in order to reduce query processing time. More precisely, as our main contribution we propose a static cache (named *Integrated Cache*) that replaces both list and intersection caches using a data structure previously used for pairing terms in disk inverted indexes. This data structure already makes an efficient use of memory space, but we design a specific cache management strategy that avoids the duplication of cached terms and we adopt a query resolution strategy (named S4 in [9]) that tries to maximize the hit ratio.

We consider different strategies to populate the cache: a strong baseline, greedy strategies that take into account both frequency of term co-occurrence and postings list size and a new strategy, that relies on casting the problem of selecting term pairs as a maximum weighted matching. We evaluate the proposal against a competitive list caching policy using two real web crawls with pretty different characteristics and a well-known query log over a simulation framework. Rather than hit ratio, the overall time needed by the different strategies to process all queries is our performance metric. Experimental evidence shows that substantial savings are possible using the proposed approach.

Background. A query $q = \{t_1, t_2, t_3, ..., t_n\}$ is a set of terms that represents the user's information need. The inverted index [22] stores the set of all unique terms in the document collection (vocabulary) associated to a set of entries that form a posting list. Each entry represents the occurrence of a term t within a document d and it consists of a a document identifier (DocID) and a payload that is used to store information about the occurrence of t within d. Each posting list is sorted in an order that depends on the specific strategy [2, 20, 22].

In a distributed search system queries are usually answered as follows [3]: the broker machine receives the query and searches the result in cache. If the result is found the answer is returned to the user at no extra cost. Otherwise, the query is sent to the search nodes in the cluster where an inverted index resides.

Each search node fetches the posting lists of the query terms (from disk or cache), reorders these in ascending order of their lengths, executes the intersection of the lists and finally returns the top-k document identifiers in the ranked result set. This requires use of the disks, a time-consuming task that is critical for the scalability of the system.

Term-at-a-time (TAAT) and Document-at-a-time (DAAT) [19] are the two main strategies to solve a query. In the TAAT approach, the posting lists of the query terms are sequentially evaluated, starting from the shortest to the longest one. This basically computes the result as $R = \cap_{i=1}^{n} t_i = (((t_1 \cap t_2) \cap t_3) ... \cap t_n)$. On the other hand, in the DAAT approach the posting lists are traversed in parallel for each document and only the current k-th best candidates are maintained in memory. The Max Successor algorithm [4] is an efficient strategy for DAAT processing. However, the presence of an *Intersection Cache* enables other possibilities such as the S4 strategy introduced in [9], that achieves up to 30% performance improvements when combined with cost aware cache policies.

2   Integrated Cache

Our proposal relies on the paired data representation presented in [12] to build an integrated cache that works as list and intersection cache at the same time. In that work, an index compression technique based on pairing posting lists of frequently co-occurring terms was proposed. The idea is to merge the lists of two frequently co-occurring terms to build a new *paired* list in the inverted index. This is obviously a more compact representation that may reduce query processing time. This data structure is introduced as a compression technique for inverted indexes combined with Gamma Coding and Variable Byte Coding [2] schemes. We extend this approach to an integrated cache of lists+intersections.

In our approach, we use the "Separated Union" [12] representation to maintain an in-memory data structure, that replaces both the list and intersection. Regarding space savings, the main idea is to keep in cache those pairs of terms that maximize the high hit ratio of the *List Cache* and the savings of the most valuable precomputed intersections. We also avoid the repetition of single term lists when these can be reconstructed using information held in previous entries. This leads to extra space savings and a more efficient use of memory, at the expense of some extra computational cost.

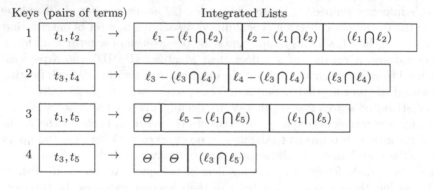

Fig. 1. Data Structure used for the *Integrated Cache*

The idea is best illustrated with an example. In Fig. 1 we show the SU data representation (entries 1 and 2) and the "extra" improvements we propose (lines 3 and 4) to get an even more efficient storage. In line 1, the entry for terms t_1 and t_2 is shown (Lets assume ℓ_i represents the inverted list of t_i). This contains the DocIDs for the first term only $(\ell_1 - (\ell_1 \cap \ell_2))$, then the postings of the second term only $(\ell_2 - (\ell_1 \cap \ell_2))$, and finally the last area with the postings common to both terms (i.e. the intersection $(\ell_1 \cap \ell_2)$). Entry in line 2 is similar to the previous one. Although we incur in an extra cost to reconstruct the posting list, this is cheaper than loading it from disk. In line 3, we show an entry that contains a previously cached term, t_1 (i.e. in the first intersection). To avoid the repetition of part of the postings we propose to reconstruct the full posting list of t_1 from the first entry (with a computational cost overhead) and include in the entry a redirection (Θ). In the case we want to cache a single term, lets say t_1, the list is completely stored in the first area and the remaining two are kept empty ($t_1 \rightarrow |\ell_1|\phi|\phi|$).

3   Selecting Term Pairs

We consider several strategies to select the "best" intersections (bigrams) to keep in cache. To this aim, each postings list is weighted according to the $f(t_i) \times |t_i|$ product, where $f(t_i)$ is the raw frequency of term t_i in a query log training set and $|t_i|$ is the length of the posting list of term t_i in the reference collection. Hereafter, we refer to this metric as FxS. We observed before that $Q_{tf}D_f$ algorithm [1] is one of the best for maximizing hit rate. However, in our setup, experimental evidence shows that FxS outperforms that approach when measuring cost.

Greedy Methods. We start with a naive approach that orders posting lists according to their FxS products and then merges lists by pairing together consecutive term pairs as $(1^{st}, 2^{nd}), (3^{rd}, 4^{th}), ..., ((n-1)^{th}, n^{th})$. We refer to this method as *PfBT-seq*. This approach only groups good terms but doesn't take into account the size of their intersections (while if the size of the intersection is larger, the space saving is larger too). The second approach (*PfBT-cuad*) computes the intersection of each possible bigram (for all term lists) and then selects the pairs that maximize $(t_i \cap t_j)$ without repetitions of terms. This algorithm is time consuming ($O(n^2)$) so we run it considering only sub-groups of lists that we estimate may fit in cache (according to its size). For example, 1GB cache holds roughly 1000 lists for a given collection, so we compute the 500 best pairs and then we fill the remaining space with pairs picked sequentially (as in *PfBT-seq*). The third approach (named *PfBT-win*) is a particular case of the previous one that tries to maximize the space saving among a group of posting lists. It sets a window of w terms (instead of all terms) and computes the intersection of each possible pair. Finally, it selects term pairs using the same criterion as before.

Term Pairing as a Matching Problem. Our last method considers the term pairing as an optimization problem, reducing it to the Maximum Weighted Matching (MWM). In graph theory, a matching is a subset of edges such that none of the selected edges share a common vertex. This is similar to [12] but we apply a different weighting criterion. We formalize the problem as follows: Let $G(T, E)$ be a graph with vertex set the set T of terms and such that, for every $t_i, t_j \in T$, edge $e_{ij} \in E$ exists if and only if $|t_i \cap t_j| \geq 0$. Moreover, we weight each edge e_{ij} by the size of the intersection $|t_i \cap t_j|$. The MWM is a matching in G that maximizes the sum of the weights of the matched edges. In our experiments we refer to this method as *PfBT-mwm*.

All above strategies select *pairs* of terms to fill the cache, so there will not be cases similar to the ones depicted in lines 3 and 4 of Fig. 1. We plan to consider other strategies in the future that will benefit from this idea.

4   Experimental Setup and Results

We select two completely different document collections to evaluate our *Integrated Cache*. Our goal is to simulate two scenarios whose behaviors may be rather distinct. The first document corpus is a subset of a large crawl of the UK web obtained by Yahoo! in 2005. It includes 1.479.139 documents, with 6.493.453 distinct index terms and requires 29 GB of disk space in HTML format (uncompressed). We refer to this corpus as UK. The second collection is a crawl derived from the Stanford WebBase Project[1] [11]. We select a recent sample (march, 2013) that contains about 7.774.632 documents (241 GB of disk space).

To evaluate our proposal we use the AOL Query Log [17] that contains around 20 million queries. We select a subset of 6M queries to compute statistics and around 2.7M queries as the test set (AOL-1). Then, we filter the file keeping only unique queries. This allows to isolate the effect of the *Result Cache* simulating that it captures all query repetitions (in the case of having a cache of infinite size), thus giving a lower bound on the performance improvement due to our cache. This second test file is about 800K queries (AOL-2).

We use Zettair[2] to index the collections and to obtain real fetching times of the posting lists. The size of the (compressed) index for the UK collection is about 1.8 GB, which grows up to 23 GB for WB. Zettair compresses posting lists using a variable-byte scheme with a B+Tree structure to handle the vocabulary. Our *Integrated Cache* implementation reserves eight bytes for each posting in the pure terms area (DocID and frequency uses four bytes each) while the intersection area requires twelve bytes because it stores the frequencies of both terms.

Cost Model. We use the cost estimation methodology introduced in [8]. The cost of processing a query in a node is modeled in terms of disk fetch and CPU times: $C_q = C_{disk} + C_{cpu}$. C_{disk} is calculated fetching all the terms from disk using Zettair (we retrieve the whole posting list and measure the corresponding fetching time). To calculate C_{cpu} we run a list intersection benchmark.

We provide a simulation-based evaluation of the proposal using both document collections. The total amount of memory reserved for the cache ranges from 100MB to 1GB for the UK collection, while we increase the size up to 16GB for the WB collection. This sizes allow to store about 60% and 70% of the indexes respectively. For each query we log the total cost incurred using a static version of the *List Cache* filled with the top-k most valuable posting lists according to the FxS metric (*baseline*). Then, we evaluate the *Integrated Cache* filling it with data from the proposed four approaches that are also based on the FxS metric (to allow a fair comparison against the list cache). We set $w = 10$ for the *PfBT-win* method. A deeper analysis of the optimal value of w will be part of future work. Finally, we normalize the final costs to get a clearer comparison.

Results. For the sake of space, we show results for the second experiment only (see Figure 2), using the dataset of unique queries (AOL-2). All evaluated strate-

---

[1] http://dbpubs.stanford.edu:8091/testbed/doc2/WebBase/
[2] http://www.seg.rmit.edu.au/zettair/

gies outperform the baseline and the best strategy is PfBT-mwm. Improvements range from 7% up to 22% for the UK collection. The behavior is again different for the WB collection. For smaller cache sizes, the performance is worse (or just slightly better) up to 1GB cache and it increases up to 30% in the best case (16GB). This is because this collection has longer posting lists and only a few are loaded in smaller caches. As expected, performance is even better if we consider the first dataset (AOL-1), since it contains repeated queries.

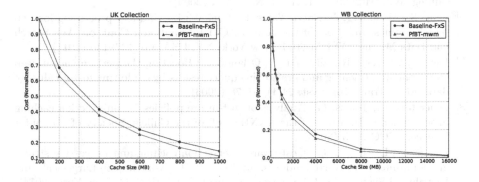

Fig. 2. Performance of the PfBT-mwm approach using the the AOL-2 query set

5   Conclusions and Future Work

We proposed an integrated cache for lists+intersections and considered several heuristics to populate the cache, including one based on casting the problem as a maximum weighted matching one. We provided an evaluation using two document collections and subsets of a real query log. We showed that the proposed *Integrated Cache* outperforms the solely posting lists cache up to a 40%.

Several interesting open problems remain. First, we plan to extend this proposal to consider trigrams or more complex combinations of terms. Another interesting open question concerns the design and implementation of a dynamic version of this cache. Here, the access and eviction policies should contemplate not only the terms but also the pairs. It is not clear how to best apply standard replacement algorithms in an online fashion.

References

[1] Baeza-Yates, R., Gionis, A., Junqueira, F., Murdock, V., Plachouras, V., Silvestri, F.: The impact of caching on search engines. In: Proc. of the 30th Annual Int. Conf. on Research and Development in Information Retrieval (2007)
[2] Baeza-Yates, R., Ribeiro-Neto, B.: Modern Information Retrieval: The Concepts and Technology behind Search, 2nd edn. Addison-Wesley Prof., Inc. (2011)

[3] Cambazoglu, B.B., Zaragoza, H., Chapelle, O., Chen, J., Liao, C., Zheng, Z., Degenhardt, J.: Early exit optimizations for additive machine learned ranking systems. In: Proc. of the Third ACM Int. Conf. on Web Search and Data Mining (2010)

[4] Culpepper, J.S., Moffat, A.: Compact set representation for information retrieval. In: Ziviani, N., Baeza-Yates, R. (eds.) SPIRE 2007. LNCS, vol. 4726, pp. 137–148. Springer, Heidelberg (2007)

[5] Dean, J.: Challenges in building large-scale information retrieval systems: Invited talk. In: Proc. of the Second ACM International Conf. on Web Search and Data Mining, WSDM 2009, p. 1. ACM, New York (2009)

[6] Ding, S., Attenberg, J., Baeza-Yates, R., Suel, T.: Batch query processing for web search engines. In: Proc. of the Fourth ACM International Conf. on Web Search and Data Mining, WSDM 2011, New York, NY, USA, pp. 137–146 (2011)

[7] Fagni, T., Perego, R., Silvestri, F., Orlando, S.: Boosting the performance of web search engines: Caching and prefetching query results by exploiting historicalusage data. ACM Trans. Inf. Syst. 24(1), 51–78 (2006)

[8] Feuerstein, E., Tolosa, G.: Analysis of cost-aware policies for intersection caching in search nodes. In: Proc. of the XXXII Conf. of the Chilean Society of Computer Science, SCCC 2013 (2013)

[9] Feuerstein, E., Tolosa, G.: Cost-aware intersection caching and processing strategies for in-memory inverted indexes. In: Proc. of 11th Workshop on Large-scale and Distributed Systems for Information Retrieval, LSDS-IR 2014, New York (2014)

[10] Gan, Q., Suel, T.: Improved techniques for result caching in web search engines. In: Proc. of the 18th Int. Conf. on World Wide Web, WWW 2009, pp. 431–440 (2009)

[11] Hirai, J., Raghavan, S., Garcia-Molina, H., Paepcke, A.: Webbase: A repository of web pages. In: Proc. of the 9th International World Wide Web Conf. on Computer Networks. North-Holland Publishing Co. (2000)

[12] Lam, H.T., Perego, R., Quan, N.T.M., Silvestri, F.: Entry pairing in inverted file. In: Vossen, G., Long, D.D.E., Yu, J.X. (eds.) WISE 2009. LNCS, vol. 5802, pp. 511–522. Springer, Heidelberg (2009)

[13] Long, X., Suel, T.: Three-level caching for efficient query processing in large web search engines. In: Proc. of the 14th Int. Conf. on World Wide Web, WWW 2005, USA, pp. 257–266 (2005)

[14] Markatos, E.: On caching search engine query results. Comput. Commun. 24(2), 137–143 (2001)

[15] Ozcan, R., Altingovde, I.S., Ulusoy, O.: Cost-aware strategies for query result caching in web search engines. ACM Trans. Web 5(2), 9:1–9:25 (2011)

[16] Ozcan, R., Sengor Altingovde, I., Barla Cambazoglu, B., Junqueira, F.P., Ulusoy, O.: A five-level static cache architecture for web search engines. Information Processing & Management 48(5), 828–840 (2012)

[17] Pass, G., Chowdhury, A., Torgeson, C.: A picture of search. In: Proc. of the 1st International Conf. on Scalable Information Systems, InfoScale 2006. ACM (2006)

[18] Saraiva, P.C., Silva de Moura, E., Ziviani, N., Meira, W., Fonseca, R., Riberio-Neto, B.: Rank-preserving two-level caching for scalable search engines. In: Proc. of the 24th Annual Int. Conf. on Research and Development in Information Retrieval, SIGIR 2001, USA, pp. 51–58 (2001)

[19] Turtle, H., Flood, J.: Query evaluation: Strategies and optimizations. Information Processing and Management 31(6), 831–850 (1995)

[20] Witten, I.H., Moffat, A., Bell, T.C.: Managing Gigabytes: Compressing and Indexing Documents and Images, 2nd edn. Morgan Kaufmann Publishers Inc., San Francisco (1999)

[21] Zhang, J., Long, X., Suel, T.: Performance of compressed inverted list caching in search engines. In: Proc. of the 17th Int. Conf. on World Wide Web, WWW 2008, USA, pp. 387–396 (2008)

[22] Zobel, J., Moffat, A.: Inverted files for text search engines. ACM Comput. Surv. 38(2) (July 2006)

Information-Theoretic Term Selection
for New Item Recommendation

Thales F. Costa[1], Anisio Lacerda[1], Rodrygo L.T. Santos[1], and Nivio Ziviani[1,2]

[1] Department of Computer Science
Universidade Federal de Minas Gerais
Belo Horizonte, MG, Brazil
{thalesfc,anisio,rodrygo,nivio}@dcc.ufmg.br
[2] Zunnit Technologies
Belo Horizonte, MG, Brazil
nivio@zunnit.com

Abstract. Recommender systems aim at predicting the preference of a user towards a given item (e.g., a movie, a song). For systems that must cope with continuously evolving item catalogs, there will be a considerable rate of new items for which no past preference is known that could otherwise inform preference-based recommendations. In contrast, pure content-based recommendations may suffer from noisy item descriptions. To overcome these problems, we propose an information-theoretic approach that exploits a taxonomy of categories associated with the cataloged items in order to select informative terms for an improved recommendation. Our experiments using two publicly available datasets attest the effectiveness of the proposed approach, which significantly outperforms state-of-the-art content-based recommenders from the literature.

1 Introduction

Recommender systems are information systems designed to recommend items potentially relevant to users. With the continuous evolution of their item catalog, recommender systems are often faced with the problem of how to provide recommendations of new items. This so-called *cold-start problem*, or *new item recommendation problem* [10], may hamper the performance of such systems, as they are unable to draw any inference of the preferences of a given user for a particular item. Effectively tackling this problem is critical because it reduces the *item latency*, which is the time between the release of a new item and its first appearance within a recommendation list. At the same time, to sustain *customer loyalty*, the new items included in a recommendation must also be relevant [4].

Collaborative filtering algorithms are generally reported to have the best accuracy in traditional recommendation scenarios [3]. However, these algorithms cannot cope effectively with the new item recommendation problem [13], when there are not enough ratings to model the users' preferences towards new items. As an alternative, existing content-based recommendation approaches [7,10] typically leverage domain-specific features such as cast and director for movies, or author

E. Moura and M. Crochemore (Eds.): SPIRE 2014, LNCS 8799, pp. 236–243, 2014.
© Springer International Publishing Switzerland 2014

and publisher for books. Nonetheless, these approaches have a clear limitation when generalizing to different domains. In order to overcome this limitation, we propose an information-theoretic content-based recommendation approach, by relying on content features generally available on any domain.

In this paper, we introduce *Information-aware Content-based Recommender* (ICBR), a novel supervised recommendation approach, which exploits the taxonomy categories associated with each item to improve the matching between new items and potentially interested users. To this end, we perform a systematic exploration of information-theoretic metrics for selecting effective item descriptors, and of topic models as an alternative to an explicit taxonomy. Our experimental results attest the effectiveness of our approach across two publicly available datasets for movie and book recommendations, with consistent and significant gains compared to state-of-the-art baselines from the literature, for items with various levels of difficulty, and even when no explicit taxonomy is available.

In the remainder of this paper, Section 2 describes related approaches for the cold-start problem. Section 3 presents the ICBR model. Section 4 details the experimental setup and the main research questions addressed in this paper. Section 5 presents the results of our thorough experiments. Finally, Section 6 provides our concluding remarks and directions for future work.

2 Related Work

Several approaches have been proposed in recent years to tackle the new item recommendation problem. A traditional solution to this problem relies on the identification of users who have previously manifested an interest towards cataloged items with content similar to that of the new item. While such content-based approaches are generally effective in a cold-start scenario, they also have shortcomings [9]. In particular, word-level features may not capture the preferences of a user towards an item as well as explicit ratings would. In addition, domain-specific features (e.g., author and publisher for books) may not generalize well across different domains. To overcome these limitations, an alternative approach is latent semantic analysis (LSA), which represents the cataloged items in a lower dimensional space of latent concepts [5]. In contrast to these approaches, we exploit features derived from the categories underlying a taxonomy of items under an information-theoretic recommendation model, as we will describe in Section 3. In our investigations in Section 5, LSA is used as a baseline.

As an alternative to content-based approaches, traditional collaborative filtering (CF) approaches address a slightly relaxed version of the problem, in which a few ratings (as opposed to none) are available for a new item. In this context, several CF approaches have been proposed to weigh content-based features: aspect models [13], Boltzmann machines [7], association rules [8], item-based CF [3] and linear transformation [11]. In contrast to these approaches, we tackle the strict version of the problem, in which no ratings are available for the new item, a crucial scenario where CF approaches cannot be applied [10].

3   Selecting Informative Item Descriptors

Content-based recommenders based on raw textual features typically suffer with non-informative item descriptors. For instance, consider the movie "Titanic", whose description includes the terms "crash" and "freezing". While these terms describe important elements of this movie, a user who likes the movie is arguably more interested in love stories than in maritime collisions on freezing waters. To achieve an improved content-based recommendation for new items, we introduce Information-aware Content-based Recommender (ICBR). In particular, ICBR represents a new item as a "*query*", and each user as a virtual "*document*" that is potentially "*relevant*" for this query. Formally, given a new item i and a user u, ICBR estimates the relevance of u given i according to:

$$score(i,u) = \sum_{t \in \hat{i}} \left(1 + \log(tf_{t,\hat{u}})\right) \times \log\left(\frac{n}{n_t} + 1\right), \tag{1}$$

where $\hat{i}$ and $\hat{u}$ are term-based representations of the item i and the user u, respectively, $tf_{t,\hat{u}}$ is the frequency of the term t in $\hat{u}$, n_t is the number of users whose representation include t, and n is the total number of users in the system.

To ensure we have meaningful representations for the new item i and each user u, we propose an information-theoretic term selection approach by exploiting a taxonomy associated with the cataloged items. In particular, taxonomy-oriented terms could better explain the interests of a user for a particular item. For instance, the interests of a user who likes the movie "Titanic" are arguably better represented by terms related to the category *Drama* than to ordinary terms such as "crash". Given an item i whose description comprises a set of terms I, and a taxonomy C of classes defined over the entire item catalog, the term-based representation $\hat{i}$ of the item i is defined according to:

$$\hat{i} = \arg\max_{T \subseteq I} \sum_{t \in T} \sum_{c \in C} w(t,c), \text{ s.t. } |T| \leq m, \tag{6}$$

where $T \subseteq I$ is a subset of the terms in I, m is the maximum number of terms to be selected, and $w(t,c)$ is the weight of the term t for each category $c \in C$.

Table 1 describes several information-theoretic metrics that are used as alternative term-category weighting schemes in our experiments. In particular,

Table 1. Term weighting functions $w(t,c)$ for a term t and category c

CHI2:	$\dfrac{\left(p(t\|c) - p(t)\right)^2}{p(t)}$	(2)	DICE:	$2 \times \dfrac{\|E_{t,c}\|}{\|E_t\| + \|E_c\|}$	(3)
KLD:	$p(t\|c) \times \log\left(\dfrac{p(t\|c)}{p(t)}\right)$	(4)	MI:	$\log\left(\dfrac{\|E_{t,c}\|}{\|E_t\| \times \|E_c\|}\right)$	(5)

Table 2. Statistics of Book-Crossing (BX) and MovieLens-1M (ML). The ⋆ symbol denotes statistics affected by the augmentation step described in Section 4.

Elements	BX	ML	Ratings	BX	ML
# Items	5712*	3706	Full range	1 ∼ 10	1 ∼ 5
# Users	3786*	6040	Non-relevant (0)	0 ∼ 6	0 ∼ 3
# Ratings	≈ 206k*	≈ 1M	Relevant (1)	7 ∼ 8	4
# Categories	855*	18*	Highly relevant (2)	9 ∼ 10	5

we hypothesize that these metrics provide a simple and sound mechanism for selecting terms that better convey a user's interest for a particular item. Lastly, to represent a user u, we concatenate the representation of all the items $i \in R_u^+$ that the user has positively rated in the past, according to $\hat{u} = \bigcup_{i \in R_u^+} \hat{i}$.

4   Experimental Setup

In this section, we detail the experimental setup that supports our investigations in Section 5. In particular, we aim to answer the following research questions:

Q1. How effective is our approach for recommending new items to users?
Q2. How does our approach perform for items with various levels of difficulty?
Q3. How does our approach perform without a purposely built taxonomy?

In the following, we describe the datasets, the recommendation baselines, and the training and evaluation procedures used in our investigations.

Recommendation datasets. We report our experimental results on two publicly available datasets: Book-Crossing (BX), a book recommendation dataset, and MovieLens-1M (ML), a movie recommendation dataset. Due to BX's extreme sparsity [6], we discard items with less than five ratings, as well as users who have rated less than five items. In addition, we complement the BX dataset with the description and category of each book, further discarding books with no associated description. Likewise, we complement ML with the synopsis and genre of each movie. Salient statistics from both datasets are presented in Table 2.

Recommendation baselines. We evaluate our ICBR model in comparison to four effective baseline recommenders. Our first baseline is top popular user (TPU), which scores users proportionally to their number of ratings. Formally, $score(i, u) = |R_u^+|$, where R_u^+ is the set of training items positively rated by the user u. TPU is a baseline since previous results suggest that non-personalized algorithms may perform well in extremely sparse scenarios [10].

Our second baseline is latent semantic analysis (LSA), a state-of-the-art content-based approach, which projects both items as well as users into a lower dimensional space obtained via singular value decomposition [1,4]. Using LSA, the score of a user u for an item i is computed as $score(i, u) = sim(\tilde{i}, \tilde{u})$, where $\tilde{i}$

and $\tilde{u}$ are the vector representations of the item i and the user u in the resulting space of latent factors, and $sim(\tilde{i}, \tilde{u})$ is the cosine similarity between $\tilde{i}$ and $\tilde{u}$. We set the number of latent factors to 2,000 through cross-validation.

Our third baseline extends LSA to leverage taxonomy features [1]. In this extension, named LSA_{tax}, we augment the item-term matrix such that $M = \omega B$, where B is a binary matrix, indicating the membership of each cataloged item to each taxonomy category, and ω is a weight assigned uniformly to all categories. Through cross-validation, we set $\omega = 3$ in our experiments.

Lastly, as a fourth baseline, we consider $ICBR_{all}$, a variant of our model that represents an item using its entire description, replacing Equation (6) with $i = I$.

Training and evaluation procedures. To tune the parameter m of ICBR as well as the parameters of our baselines recommenders, we perform a k-fold cross-validation, with $k = 5$ for BX [12] and $k = 10$ for ML [14]. Since we are simulating an extreme occurrence of the new item recommendation problem, if an item is in the test set, none of its ratings is used for training. In our investigations, each considered recommendation approach is assessed regarding its ability to rank users according to these users' interest for a new item. Arguably, the interest of a user for a particular item can be approximated by the rating that the user gives to the item. Accordingly, for each item i, we label each user u as either non-relevant (0), relevant (1), or highly relevant (2). Table 2 defines the mapping between rating ranges and the aforementioned relevance levels in each of our considered datasets. Based upon this definition, we report normalized-discounted cumulative gain (nDCG) figures at different rank cutoffs, as an average across all test folds, accompanied by a 95% confidence interval.

5   Experimental Evaluation

In this section, we assess the effectiveness of our approach for recommending new items, in order to answer the research questions stated in Section 4.

Recommendation effectiveness (Q1). To assess the usefulness of a taxonomy within our information-theoretic model, we compare multiple variants of ICBR, each of which leveraging a different information-theoretic term selection scheme among those described in Table 1, to $ICBR_{all}$, a baseline ICBR variant that does not perform any term selection. Table 3 shows the results of this assessment in terms of nDCG at two rank cutoffs for both the BX and ML datasets.

Compared to the baseline variant $ICBR_{all}$, we note that most variants of ICBR improve, often significantly. Since CHI2 and KLD are consistently effective across the two datasets, we combine them into $ICBR_{comb}$, a variant of ICBR that estimates the relevance of a given new item to a user by uniformly interpolating the KLD and CHI2 scores. This combination outperforms all individual metrics.

From Table 3, we can also contrast the effectiveness of our model to state-of-the-art baselines from the literature, namely, TPU, LSA, and LSA_{tax}. In particular, for the BX dataset, $ICBR_{comb}$ significantly outperforms all baselines, with

Table 3. Recommendation performance with various term selection schemes

Model	nDCG@20	nDCG@100	nDCG@20	nDCG@100
TPU	0.1020 ± 0.003	0.1373 ± 0.002	0.2003 ± 0.010	0.1902 ± 0.007
LSA	0.1475 ± 0.009	0.2025 ± 0.008	0.1375 ± 0.008	0.1592 ± 0.007
LSA_{tax}	0.1559 ± 0.008	0.2151 ± 0.007	0.1578 ± 0.008	0.1801 ± 0.007
$ICBR_{all}$	0.1301 ± 0.003	0.1777 ± 0.004	0.2195 ± 0.008	0.2116 ± 0.006
$ICBR_{mi}$	0.1383 ± 0.004	0.1889 ± 0.003	0.2215 ± 0.011	0.2185 ± 0.008
$ICBR_{dice}$	0.1410 ± 0.004	0.1888 ± 0.004	0.2173 ± 0.011	0.2150 ± 0.007
$ICBR_{kld}$	0.1672 ± 0.004	0.2213 ± 0.004	0.2372 ± 0.012	0.2376 ± 0.009
$ICBR_{chi2}$	0.1725 ± 0.005	0.2259 ± 0.005	0.2395 ± 0.014	0.2394 ± 0.010
$ICBR_{comb}$	$\underline{0.1733} \pm 0.005$	$\underline{0.2286} \pm 0.005$	$\underline{0.2438} \pm 0.013$	$\underline{0.2448} \pm 0.010$

(BX = first two columns, ML = last two columns)

gains in nDCG as high as 69.90% over TPU, and 17.49% over LSA. For the ML dataset, the gains over TPU are as high as 28.70%, while LSA is outperformed by 77.30%. Finally, compared with LSA_{tax}, which also exploits taxonomies to improve upon the pure content-based LSA approach, $ICBR_{comb}$ attains significant improvements of up to 11.16% for the BX dataset, and of up to 54.49% for the ML dataset. Overall, these results answer question Q1, by attesting the effectiveness of our ICBR model in contrast to state-of-the-art approaches from the literature for the new item recommendation problem.

Recommendation difficulty (Q2). In our training and evaluation procedure, the actual number of relevant users per item varies considerably. As a result, the recommendation for items with fewer relevant users can be regarded as more difficult than for those with many relevant users. Hence, we assess the effectiveness of ICBR in contrast to the aforementioned baselines (TPU, LSA, LSA_{tax}, and $ICBR_{all}$) for input items with various levels of difficulty. Figures 1a and 1b show the results of this investigation. In both figures, the available items are grouped into five bins with roughly the same number of items, organized according to the number of relevant users per item, as indicated in the x axis.

From Figures 1a and 1b, we first observe that all approaches generally improve as the level of difficulty is reduced, i.e., when more relevant users can be potentially recommended for a given new item. More importantly, for both the BX and the ML datasets, we note that $ICBR_{comb}$ outperforms all baseline recommenders across all difficulty levels, with gains ranging between 3.59% and 9.72% for BX, and 8.62% and 28.58% for ML. These results answer question Q2, by further attesting the effectiveness of our proposed approach as well as its robustness for recommending new items with different levels of difficulty.

Effectiveness with latent topics (Q3). To assess the effectiveness of ICBR for domains where an explicit taxonomy is not available, we deploy its best variant, $ICBR_{comb}$, using either an explicit taxonomy or a latent one, with categories represented as topics identified using latent Dirichlet allocation (LDA) [2]. For the sake of clarity, we refer to the latter as $ICBR_{lda}$. In order to assess the effectiveness of our approach, we compare our results with LSA_{lda}, an extended version

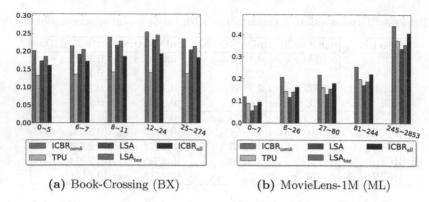

(a) Book-Crossing (BX)   **(b)** MovieLens-1M (ML)

Fig. 1. Recommendation performance (nDCG@100) for various difficulty levels

Table 4. Recommendation performance using latent topics

	Model	nDCG@20	nDCG@100	nDCG@20	nDCG@100
BX	$ICBR_{all}$	0.1301 ± 0.003	0.1777 ± 0.004	0.2195 ± 0.008	0.2116 ± 0.006
	$ICBR_{comb}$	0.1733 ± 0.005	0.2286 ± 0.005	0.2438 ± 0.013	0.2448 ± 0.010
	LSA_{lda}	0.1477 ± 0.009	0.2041 ± 0.008	0.1396 ± 0.008	0.1625 ± 0.007
	$ICBR_{lda}$	0.1608 ± 0.004	0.2121 ± 0.005	0.2227 ± 0.012	0.2209 ± 0.008

of the LSA algorithm which leverages latent topics in the same manner as how LSA_{tax} leverages explicit categories. In addition, we once again include $ICBR_{all}$, the baseline variant of our model, which performs no term selection. Table 4 presents nDCG figures for the aforementioned recommendation approaches.

Compared to LSA_{lda}, $ICBR_{lda}$ is superior in terms of nDCG. For BX, $ICBR_{lda}$ improves by up to 8.86%. For ML, gains are as high as 59.52%. These results further attest the effectiveness of ICBR even when no explicit taxonomy is available. Nonetheless, while categories automatically derived using LDA can be used effectively by our model, the results in this section also show that the availability of a manually curated taxonomy can provide further gains.

6   Conclusions and Future Work

In this paper, we introduced Information-aware Content-based Recommender (ICBR), a novel supervised approach for new item recommendation, which models the terms that describe the new item as a "query", and each candidate user who could be recommended the item as a virtual "document", comprising the terms in the description of the items that the user has positively rated in the past. In order to improve this content-based representation, we proposed a term selection mechanism aimed to weigh the informativeness of each term with respect to the taxonomy categories covered by each item.

By contrasting our model with a variant that performs no term selection, we demonstrated the usefulness of our information-theoretic term selection schemes for improving the underlying content-based representation of items and users. This improved representation outperformed state-of-the-art content-based recommenders from the literature. We also demonstrated the effectiveness of our model across new items with different levels of difficulty and for domains where an explicit taxonomy is not available. In the future, we plan to investigate supervised approaches to combine multiple alternative representations for items.

Acknowledgements. We thank the partial support given by the Brazilian National Institute of Science and Technology for the Web (grant MCT-CNPq 573871/2008-6) and authors' individual grants and scholarships from CNPq and CAPES.

References

1. Bambini, R., Cremonesi, P., Turrin, R.: A recommender system for an IPTV service provider: A real large-scale production environment. In: Recommender Systems Handbook, pp. 299–331. Springer (2011)
2. Blei, D.M., Ng, A.Y., Jordan, M.I.: Latent dirichlet allocation. J. Mach. Learn. Res. 3, 993–1022 (2003)
3. Cremonesi, P., Koren, Y., Turrin, R.: Performance of recommender algorithms on top-n recommendation tasks. In: RecSys, pp. 39–46 (2010)
4. Cremonesi, P., Turrin, R., Airoldi, F.: Hybrid algorithms for recommending new items. In: HetRec, pp. 33–40 (2011)
5. Furnas, G.W., Deerwester, S., Dumais, S.T., Landauer, T.K., Harshman, R.A., Streeter, L.A., Lochbaum, K.E.: Information retrieval using a singular value decomposition model of latent semantic structure. In: SIGIR, pp. 465–480 (1988)
6. Gedikli, F., Jannach, D.: Recommending based on rating frequencies. In: RecSys, pp. 233–236 (2010)
7. Gunawardana, A., Meek, C.: Tied boltzmann machines for cold start recommendations. In: RecSys, pp. 19–26 (2008)
8. Leung, C.W.-K., Chan, S.C.-F., Chung, F.-l.: An empirical study of a cross-level association rule mining approach to cold-start recommendations. Know.-Based Syst., 515–529 (2008)
9. Lops, P., Gemmis, M., Semeraro, G.: Content-based recommender systems: State of the art and trends. In: Recommender Systems Handbook. Springer (2011)
10. Park, S.-T., Chu, W.: Pairwise preference regression for cold-start recommendation. In: RecSys, pp. 21–28 (2009)
11. Pilászy, I., Tikk, D.: Recommending new movies: even a few ratings are more valuable than metadata. In: RecSys, pp. 93–100 (2009)
12. Qumsiyeh, R., Ng, Y.-K.: Predicting the ratings of multimedia items for making personalized recommendations. In: SIGIR, pp. 475–484 (2012)
13. Schein, A.I., Popescul, A., Ungar, L.H., Pennock, D.M.: Methods and metrics for cold-start recommendations. In: SIGIR, pp. 253–260 (2002)
14. Schifanella, R., Panisson, A., Gena, C., Ruffo, G.: Mobhinter: epidemic collaborative filtering and self-organization in mobile ad-hoc networks. In: RecSys, pp. 27–34 (2008)

On the String Consensus Problem
and the Manhattan Sequence Consensus Problem

Tomasz Kociumaka[1,*], Jakub W. Pachocki[2], Jakub Radoszewski[1,**],
Wojciech Rytter[1,3], and Tomasz Waleń[1]

[1] Faculty of Mathematics, Informatics and Mechanics,
University of Warsaw, Warsaw, Poland
{kociumaka,jrad,rytter,walen}@mimuw.edu.pl
[2] Carnegie Mellon University
pachocki@cs.cmu.edu
[3] Faculty of Mathematics and Computer Science,
Copernicus University, Toruń, Poland

Abstract. In the MANHATTAN SEQUENCE CONSENSUS problem (MSC
problem) we are given k integer sequences, each of length ℓ, and we are
to find an integer sequence $\mathbf{x}$ of length ℓ (called a consensus sequence),
such that the maximum Manhattan distance of $\mathbf{x}$ from each of the input
sequences is minimized. For binary sequences Manhattan distance coincides with Hamming distance, hence in this case the string consensus
problem (also called string center problem or closest string problem) is a
special case of MSC. Our main result is a practically efficient $\mathcal{O}(\ell)$-time
algorithm solving MSC for $k \leq 5$ sequences. Practicality of our algorithms has been verified experimentally. It improves upon the quadratic
algorithm by Amir et al. (SPIRE 2012) for string consensus problem
for $k = 5$ binary strings. Similarly as in Amir's algorithm we use a
column-based framework. We replace the implied general integer linear
programming by its easy special cases, due to combinatorial properties
of the MSC for $k \leq 5$. We also show that for a general parameter k
any instance can be reduced in linear time to a kernel of size $k!$, so the
problem is fixed-parameter tractable. Nevertheless, for $k \geq 4$ this is still
too much for any naive solution to be feasible in practice.

1 Introduction

In the sequence consensus problems, given a set of sequences of length ℓ we are
searching for a new sequence of length ℓ which minimizes the maximum distance
to all the given sequences in some particular metric. Finding the consensus sequence is a tool for many clustering algorithms and as such has applications
in unsupervised learning, classification, databases, spatial range searching, data
mining etc [4]. It is also one of popular methods for detecting data common-
alities of many strings (see [1]) and has a considerable number of applications

---

* Supported by Polish budget funds for science in 2013-2017 as a research project
under the 'Diamond Grant' program.
** The author receives financial support of Foundation for Polish Science.

E. Moura and M. Crochemore (Eds.): SPIRE 2014, LNCS 8799, pp. 244–255, 2014.
© Springer International Publishing Switzerland 2014

in coding theory [6,8], data compression [11] and bioinformatics [12,16]. The consensus problem has previously been studied mainly in $\mathbb{R}^\ell$ space with the Euclidean distance and in Σ^ℓ (that is, the space of sequences over a finite alphabet Σ) with the Hamming distance. Other metrics were considered in [2]. We study the sequence consensus problem for Manhattan metric (ℓ_1 norm) in correlation with the Hamming-metric variant of the problem.

The Euclidean variant of the sequence consensus problem is also known as the bounding sphere, enclosing sphere or enclosing ball problem. It was initially introduced in 2 dimensions (i.e., the smallest circle problem) by Sylvester in 1857 [22]. For an arbitrary number of dimensions, several approximation algorithms [4,15,21] and practical exact algorithms [7,10] have been proposed.

The Hamming-distance variant of the sequence consensus problem is known under the names of string consensus, center string or closest string problem. The problem is known to be NP-complete even for binary alphabet [8]. The algorithmic study of Hamming string consensus (HSC) problem started in 1999 with the first approximation algorithms [16]. Afterwards polynomial-time approximation schemes (PTAS) with different running times were presented [3,19,20]. A number of exact algorithms have also been proposed. Many of these consider a decision version of the problem, in which we are to check if there is a solution to HSC problem with distance at most d to the input sequences. Thus FPT algorithms with time complexities $\mathcal{O}(k\ell + kd^{d+1})$ and $\mathcal{O}(k\ell + kd(16|\Sigma|)^d)$ were presented in [12] and [19], respectively.

An FPT algorithm parameterized only by k was given in [12]. It uses Lenstra's algorithm [17] for a solution of an integer linear program of size exponential in k (which requires $\mathcal{O}(k!^{4.5k!}\ell)$ operations on integers of magnitude $\mathcal{O}(k!^{2k!}\ell)$, see [1]) and due to extremely large constants is not feasible for $k \geq 4$. This opened a line of research with efficient algorithms for small constant k. A linear-time algorithm for $k = 3$ was presented in [12], a linear-time algorithm for $k = 4$ and binary alphabet was given in [5], and recently an $\mathcal{O}(\ell^2)$-time algorithm for $k = 5$ and also binary alphabet was developed in [1].

For two sequences $\mathbf{x} = (x_1, \ldots, x_\ell)$ and $\mathbf{y} = (y_1, \ldots, y_\ell)$ the Manhattan distance (also known as rectilinear or taxicab distance) between $\mathbf{x}$ and $\mathbf{y}$ is defined as follows:

$$dist(\mathbf{x}, \mathbf{y}) = \sum_{j=1}^{\ell} |x_j - y_j|.$$

The Manhattan version of the consensus problem is formally defined as follows:

MANHATTAN SEQUENCE CONSENSUS **problem**

Input: A collection $\mathcal{A}$ of k integer sequences $\mathbf{a}_i$, each of length ℓ;

Output: $\mathrm{OPT}(\mathcal{A}) = \min_{\mathbf{x}} \max \{dist(\mathbf{x}, \mathbf{a}_i) : 1 \leq i \leq k\}$,
and the corresponding integer consensus sequence $\mathbf{x}$.

We assume that integers $a_{i,j}$ satisfy $|a_{i,j}| \leq M$, and all ℓ, k, M fit in a machine word, so that arithmetics on integers of magnitude $\mathcal{O}(\ell M)$ take constant time.

For simplicity in this version of the paper we concentrate on computing OPT($\mathcal{A}$) and omit the details of recovering the corresponding consensus sequence **x**. Nevertheless, this step is included in the implementation provided.

Example 1. Let $\mathcal{A} = ((120, 0, 80), (20, 40, 130), (0, 100, 0))$. Then OPT($\mathcal{A}$) = 150 and a consensus sequence is **x** = $(30, 40, 60)$, see also Fig. 1.

Our results are the following:

- We show that MANHATTAN SEQUENCE CONSENSUS problem has a kernel with $\ell \leq k!$ and give an algorithm which works in linear time for any fixed k.
- We present a practical linear-time algorithm for the MANHATTAN SEQUENCE CONSENSUS problem for $k = 5$ (which obviously can be used for any $k \leq 5$).

Note that binary HSC problem is a special case of MSC problem. Hence, the latter problem is NP-complete. Moreover, the efficient linear-time algorithm presented here for MSC problem for $k = 5$ yields an equally efficient linear-time algorithm for the binary HSC problem and thus improves the result of [1].

Organization of the Paper. Our approach is based on a reduction of the MSC problem to instances of integer linear programming (ILP). For general constant k we obtain a constant, though a very large, number of instances with a constant number of variables that we solve using Lenstra's algorithm [17] which works in constant time (the constant coefficient of this algorithm is also very large). This idea is similar to the one used in the FPT algorithm for HSC problem [12], however for MSC it requires an additional combinatorial observation. For $k \leq 5$ we obtain a more efficient reduction of MSC to at most 20 instances of very special ILP which we solve efficiently without applying a general ILP solver.

In Section 2 we show the first steps of the reduction of MSC to ILP. In Section 3 we show a kernel for the problem of $\mathcal{O}(k!)$ size. In Section 4 we perform a combinatorial analysis of the case $k = 5$ which leaves 20 simple types of the sequence **x** to be considered. This analysis is used in Section 5 to obtain 20 special ILP instances with only 4 variables. They could be solved using Lenstra's ILP solver. However, there exists an efficient algorithm tailored for this type of special instances. Due to space constraints, it is omitted in this version; it can be found in [14]. Finally we analyze the performance of a C++ implementation of our algorithm in the Conclusions (Section 6).

2   From MSC Problem to ILP

Let us fix a collection $\mathcal{A} = (\mathbf{a}_1, \ldots, \mathbf{a}_k)$ of the input sequences. The elements of $\mathbf{a}_i$ are denoted by $a_{i,j}$ (for $1 \leq j \leq \ell$). We also denote $dist(\mathbf{x}, \mathcal{A}) = \max \{ dist(\mathbf{x}, \mathbf{a}_i) : 1 \leq i \leq k \}$.

For $j \in \{1, \ldots, n\}$ let π_j be a permutation of $\{1, \ldots, k\}$ such that $a_{\pi_j(1),j} \leq \ldots \leq a_{\pi_j(k),j}$, i.e. π_j is the ordering permutation of elements $a_{1,j}, \ldots, a_{k,j}$. We also set $s_{i,j} = a_{\pi_j(i),j}$, see Example 2. For some j there might be several possibilities for π_j (if $a_{i,j} = a_{i',j}$ for some $i \neq i'$), we fix a single choice for each j.

Example 2. Consider the following three sequences $\mathbf{a}_i$ and sequences $\mathbf{s}_i$ obtained by sorting columns:

$$[a_{i,j}] = \begin{bmatrix} 120 & 0 & 80 \\ 20 & 40 & 130 \\ 0 & 100 & 0 \end{bmatrix}, \quad [s_{i,j}] = \begin{bmatrix} 0 & 0 & 0 \\ 20 & 40 & 80 \\ 120 & 100 & 130 \end{bmatrix}.$$

The Manhattan consensus sequence is $\mathbf{x} = (30, 40, 60)$, see Fig. 1. In the figure, the circled numbers in j-th column are $\pi_j(1), \pi_j(2), \ldots, \pi_j(k)$ (top-down).

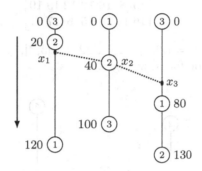

Fig. 1. Illustration of Example 2; $\pi_1 = (3, 2, 1)$, $\pi_2 = (1, 2, 3)$, $\pi_3 = (3, 1, 2)$

Definition 3. *A basic interval is an interval of the form $[i, i + 1]$ (for $i = 1, \ldots, k - 1$) or $[i, i]$ (for $i = 1, \ldots, k$). The former is called* proper, *and the latter* degenerate. *An interval system is a sequence $\mathcal{I} = (I_1, \ldots, I_\ell)$ of basic intervals I_j.*

For a basic interval I_j we say that a value x_j is consistent with I_j if $x_j \in \{s_{i,j}, \ldots, s_{i+1,j}\}$ when $I_j = [i, i+1]$ is proper, and if $x_j = s_{i,j}$ when $I_j = [i, i]$ is degenerate. A sequence $\mathbf{x}$ is called consistent with an interval system $\mathcal{I} = (I_j)_{j=1}^\ell$ if for each j the value x_j is consistent with I_j.

For an interval system $\mathcal{I}$ we define $\mathrm{OPT}(\mathcal{A}, \mathcal{I})$ as the minimum $dist(\mathbf{x}, \mathcal{A})$ among all integer sequences $\mathbf{x}$ consistent with $\mathcal{I}$. Due to the following trivial observation, for every $\mathcal{A}$ there exists an interval system $\mathcal{I}$ such that $\mathrm{OPT}(\mathcal{A}) = \mathrm{OPT}(\mathcal{A}, \mathcal{I})$.

Observation 4. *If $\mathbf{x}$ is a Manhattan consensus sequence then for each j, $s_{1,j} \leq x_j \leq s_{k,j}$.*

Transformation of the Input to an ILP. Note that for all sequences $\mathbf{x}$ consistent with a fixed $\mathcal{I}$, the Manhattan distances $dist(\mathbf{x}, \mathbf{a}_i)$ can be expressed as $d_i + \sum_{j=1}^\ell e_{i,j} x_j$ with $e_{i,j} = \pm 1$. Thus, the problem of finding $\mathrm{OPT}(\mathcal{A}, \mathcal{I})$ can be formulated as an ILP, which we denote $\mathrm{ILP}(\mathcal{I})$. If I_j is a proper interval

$[i, i+1]$, we introduce a variable $x_j \in \{s_{i,j}, \ldots, s_{i+1,j}\}$. Otherwise we do not need a variable x_j. The i-th constraint of ILP($\mathcal{I}$) algebraically represents $dist(\mathbf{x}, \mathbf{a}_i)$, see Example 6.

Observation 5. *The optimal value of* ILP($\mathcal{I}$) *is equal to* OPT($\mathcal{A}, \mathcal{I}$).

Example 6. Consider the following 5 sequences of length 7:

$$[a_{i,j}] = \begin{bmatrix} 20 & 18 & 20 & 10 & 16 & 8 & 10 \\ 11 & 6 & 7 & 17 & 14 & 14 & 17 \\ 14 & 12 & 18 & 13 & 11 & 6 & 12 \\ 19 & 8 & 16 & 18 & 12 & 19 & 19 \\ 16 & 15 & 11 & 15 & 6 & 17 & 11 \end{bmatrix}$$

and an interval system $\mathcal{I} = ([2,3], [2,3], [2,3], [3,4], [3,4], [3,3], [3,4])$. An illustration of both can be found in Fig. 2.

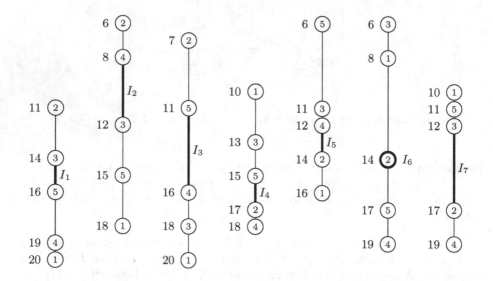

Fig. 2. Illustration of Example 6: 5 sequences of length 7 together with an interval system. Notice that I_6 is a degenerate interval.

We obtain the following ILP($\mathcal{I}$), where $x_1 \in [14, 16]$, $x_2 \in [8, 12]$, $x_3 \in [11, 16]$, $x_4 \in [15, 17]$, $x_5 \in [12, 14]$, $x_7 \in [12, 17]$ and the sequence $\mathbf{x}$ can be retrieved as $\mathbf{x} = (x_1, x_2, x_3, x_4, x_5, 14, x_7)$:

$$\min z$$

$$\begin{array}{llllllll}
20 - x_1 & + 18 - x_2 & + 20 - x_3 & + x_4 - 10 & + 16 - x_5 & + 6 & + x_7 - 10 & \leq z \\
x_1 - 11 & + x_2 - 6 & + x_3 - 7 & + 17 - x_4 & + 14 - x_5 & + 0 & + 17 - x_7 & \leq z \\
x_1 - 14 & + 12 - x_2 & + 18 - x_3 & + x_4 - 13 & + x_5 - 11 & + 8 & + x_7 - 12 & \leq z \\
19 - x_1 & + x_2 - 8 & + 16 - x_3 & + 18 - x_4 & + x_5 - 12 & + 5 & + 19 - x_7 & \leq z \\
16 - x_1 & + 15 - x_2 & + x_3 - 11 & + x_4 - 15 & + x_5 - 6 & + 3 & + x_7 - 11 & \leq z
\end{array}$$

Note that $P = \mathrm{ILP}(\mathcal{I})$ has the following special form, which we call $(\pm)$ILP:

$$\min z$$

$$d_i + \sum_j x_j e_{i,j} \leq z$$

$$x_j \in R_P(x_j)$$

where $e_{i,j} = \pm 1$ and $R_P(x_j) = \{\ell_j, \ldots, r_j\}$ for integers $\ell_j \leq r_j$. Whenever we refer to variables, it does not apply to z, which is of auxiliary character. Also, "$x_j \in R_P(x_j)$" are called variable ranges rather than constraints. We say that $(e_{1,j}, \ldots, e_{k,j})$ is a *coefficient vector* of x_j and denote it as $E_P(x_j)$. If the program P is apparent from the context, we omit the subscript.

Simplification of ILP. The following two facts are used to reduce the number of variables of a $(\pm)$ILP. For $A, B \subseteq \mathbb{Z}$ we define $-A = \{-a : a \in A\}$ and $A + B = \{a + b : a \in A, b \in B\}$.

Fact 7. *Let P be a $(\pm)$ILP. Let P' be a program obtained from P by replacing a variable x_j with $-x_j$, i.e. setting $E_{P'}(x_j) = -E_P(x_j)$ and $R_{P'}(x_j) = -R_P(x_j)$. Then $\mathrm{OPT}(P) = \mathrm{OPT}(P')$.*

Fact 8. *Let P be a $(\pm)$ILP. Assume $E_P(x_j) = E_P(x_{j'})$ for $j \neq j'$. Let P' be a program obtained from P by removing the variable $x_{j'}$ and replacing x_j with $x_j + x_{j'}$, i.e. setting $R_{P'}(x_j) = R_P(x_j) + R_P(x_{j'})$. Then $\mathrm{OPT}(P) = \mathrm{OPT}(P')$.*

Proof. Let $(z, x_1, \ldots, x_n)$ be a feasible solution of P. Then setting $x_j := x_j + x_{j'}$ and removing the variable $x_{j'}$ we obtain a feasible solution of P'. Therefore $\mathrm{OPT}(P') \leq \mathrm{OPT}(P)$. For the proof of the other inequality, take a feasible solution $(z, x_1, \ldots, x_n)$ (with $x_{j'}$ missing) of P'. Note that $x_j \in R_{P'}(x_j) = R_P(x_j) + R_P(x_{j'})$. Therefore one can split x_j into $x_j + x_{j'}$ so that $x_j \in R_P(x_j)$ and $x_{j'} \in R_P(x_{j'})$. This way we obtain a feasible solution of P and thus prove that $\mathrm{OPT}(P) \leq \mathrm{OPT}(P')$. $\qquad\square$

Corollary 9. *For a $(\pm)$ILP with k constraints one can compute in linear time an equivalent $(\pm)$ILP with k constraints and up to 2^{k-1} variables.*

Proof. We apply Fact 7 to obtain $e_{1,1} = e_{1,2} = \ldots = e_{1,\ell}$, this leaves at most 2^{k-1} different coefficient vectors. Afterwards we apply Fact 8 as many times as possible until there is exactly one variable with each coefficient vector. $\qquad\square$

Example 10. Consider the $(\pm)$ILP P from Example 6. Observe that $E_P(x_4) = E_P(x_7) = -E_P(x_2)$ and thus Facts 7 and 8 let us merge x_2 and x_7 into x_4 with

$$R_{P'}(x_4) = R_P(x_4) + R_P(x_7) - R_P(x_2) = [15, 17] + [12, 17] + [-12, -8] = [15, 26].$$

Simplifying the constant terms we obtain the following $(\pm)$ILP P':

$$
\begin{aligned}
&\min z\\
-x_1 - x_3 + x_4 - x_5 + 60 &\le z\\
+x_1 + x_3 - x_4 - x_5 + 24 &\le z\\
+x_1 - x_3 + x_4 + x_5 - 12 &\le z\\
-x_1 - x_3 - x_4 + x_5 + 57 &\le z\\
-x_1 + x_3 + x_4 + x_5 - 9 &\le z
\end{aligned}
$$

3   Kernel of MSC for Arbitrary k

In this section we give a kernel for the MSC problem parameterized with k, which we then apply to develop a linear-time FPT algorithm. To obtain the kernel we need a combinatorial observation that if $\pi_j = \pi_{j'}$ then the j-th and the j'-th column in $\mathcal{A}$ can be merged. This is stated formally in the following lemma.

Lemma 11. *Let $\mathcal{A} = (\mathbf{a}_1, \ldots, \mathbf{a}_k)$ be a collection of sequences of length ℓ. Assume that $\pi_j = \pi_{j'}$ for some $1 \le j < j' \le \ell$. Let $\mathcal{A}' = (\mathbf{a}'_1, \ldots, \mathbf{a}'_k)$ be a collection of sequences of length $\ell - 1$ obtained from $\mathcal{A}$ by removing the j'-th column and setting $a'_{i,j} = a_{i,j} + a_{i,j'}$. Then $\mathrm{OPT}(\mathcal{A}) = \mathrm{OPT}(\mathcal{A}')$.*

Proof. First, let us show that $\mathrm{OPT}(\mathcal{A}') \le \mathrm{OPT}(\mathcal{A})$. Let $\mathbf{x}$ be a Manhattan consensus sequence for $\mathcal{A}$ and let $\mathbf{x}'$ be obtained from $\mathbf{x}$ by removing the j'-th entry and setting $x'_j = x_j + x_{j'}$. We claim that $dist(\mathbf{x}', \mathcal{A}') \le dist(\mathbf{x}, \mathcal{A})$. Note that it suffices to show that $|x'_j - a'_{i,j}| \le |x_j - a_{i,j}| + |x_{j'} - a_{i,j'}|$ for all i. However, with $x'_j = x_j + x_{j'}$ and $a'_{i,j} = a_{i,j} + a_{i,j'}$, this is a direct consequence of the triangle inequality.

It remains to prove that $\mathrm{OPT}(\mathcal{A}) \le \mathrm{OPT}(\mathcal{A}')$. Let $\mathbf{x}'$ be a Manhattan consensus sequence for $\mathcal{A}'$. By Observation 4, x'_j is consistent with some proper basic interval $[i, i + 1]$. Let $d'_{i,j} = x'_j - s'_{i,j}$ and $D'_{i,j} = s'_{i+1,j} - s'_{i,j}$. Also, let $D_{i,j} = s_{i+1,j} - s_{i,j}$ and $D_{i,j'} = s_{i+1,j'} - s_{i,j'}$. Note that, since $\pi_j = \pi_{j'}$, $D'_{i,j} = D_{i,j} + D_{i,j'}$. Thus, one can partition $d'_{i,j} = d_{i,j} + d_{i,j'}$ so that both $d_{i,j}$ and $d_{i,j'}$ are non-negative integers not exceeding $D_{i,j}$ and $D_{i,j'}$ respectively. We set $x_j = s_{i,j} + d_{i,j}$ and $x_{j'} = s_{i,j'} + d_{i,j'}$. The remaining components of $\mathbf{x}$ correspond to components of $\mathbf{x}'$. Note that $x'_j = x_j + x_{j'}$ and that both x_j and $x_{j'}$ are consistent with $[i, i+1]$. Consequently for any sequence $\mathbf{a}_m$ it holds that $dist(\mathbf{x}, \mathbf{a}_m) = dist(\mathbf{x}', \mathbf{a}_m)$ and therefore $dist(\mathbf{x}, \mathcal{A}) = dist(\mathbf{x}', \mathcal{A}')$, which concludes the proof.

Finally, note that the procedures given above can be used to efficiently convert between the optimum solutions $\mathbf{x}$ and $\mathbf{x}'$. $\qquad\square$

By Lemma 11, to obtain the desired kernel we need to sort the elements in columns of $\mathcal{A}$ and afterwards group by the resulting permutations π_j.

Theorem 12. *In $\mathcal{O}(\ell k \log k)$ time one can reduce any instance of MSC to an instance with k sequences of length ℓ', with $\ell' \le k!$.*

Remark 13. For binary instances, if permutations π_j are chosen appropriately, we can achieve $\ell' \leq 2^k$.

Theorem 14. *For any integer k, the* MANHATTAN SEQUENCE CONSENSUS *problem can be solved in $\mathcal{O}(\ell k \log k + 2^{k! \log k + O(k2^k)} \log M)$ time.*

Proof. We solve the kernel from Theorem 12 by considering all possible interval systems $\mathcal{I}$ composed of proper intervals. The sequences in the kernel have length at most $k!$, which gives $(k-1)^{k!}$ $(\pm)$ILPs of the form ILP$(\mathcal{I})$ to solve.

Each of the $(\pm)$ILPs initially has k constraints on $k!$ variables but, due to Corollary 9, the number of variables can be reduced to 2^{k-1}. Lenstra's algorithm with further improvements [13,9,18] solves ILP with p variables in $\mathcal{O}(p^{2.5p+o(p)} \log L)$ time, where L is the bound on the scope of variables. In our case $L = \mathcal{O}(\ell M)$, which gives the time complexity of:

$$\mathcal{O}\left((k-1)^{k!} \cdot 2^{(k-1)(2.5 \cdot 2^{k-1} + o(2^{k-1}))} \log M\right) = \mathcal{O}\left(2^{k! \log k + O(k2^k)} \log M\right).$$

This concludes the proof of the theorem.   □

4   Combinatorial Characterization of Solutions for $k = 5$

In this section we characterize those Manhattan consensus sequences **x** which additionally minimize $\sum_i dist(\mathbf{x}, \mathbf{a}_i)$ among all Manhattan consensus sequences. Such sequences are called here *sum-MSC sequences*. We show that one can determine a collection of 20 interval systems, so that any sum-MSC sequence is guaranteed to be consistent with one of them. We also prove some structural properties of these systems, which are then useful to efficiently solve the corresponding $(\pm)$ILPs.

We say that x_j is in the *center* if $x_j = s_{3,j}$, i.e. x_j is equal to the column median. Note that if $x_j \neq s_{3,j}$, then moving x_j by one towards the center decreases by one $dist(\mathbf{x}, \mathbf{a}_i)$ for at least three sequences $\mathbf{a}_i$.

Definition 15. *We say that $\mathbf{a}_i$ governs x_j if x_j is in the center or moving x_j towards the center increases $dist(\mathbf{x}, \mathbf{a}_i)$. The set of indices i such that $\mathbf{a}_i$ governs x_j is denoted as $G_j(\mathbf{x})$.*

Observe that if x_j is in the center, then $|G_j(\mathbf{x})| = 5$, and otherwise $|G_j(\mathbf{x})| \leq 2$; see Fig. 3. For $k = 5$ we have 4 proper basic intervals: $[1, 2]$, $[2, 3]$, $[3, 4]$ and $[4, 5]$. We call $[1, 2]$ and $[4, 5]$ *border* intervals, and the other two *middle intervals*. We define $G_j([1, 2]) = \{\pi_j(1)\}$, $G_j([2, 3]) = \{\pi_j(1), \pi_j(2)\}$, $G_j([3, 4]) = \{\pi_j(4), \pi_j(5)\}$ and $G_j([4, 5]) = \{\pi_j(5)\}$. Note that if we know $G_j(\mathbf{x})$ and $|G_j(\mathbf{x})| \leq 2$, then we are guaranteed that x_j is consistent with the basic interval I_j for which $G_j(I_j) = G_j(\mathbf{x})$.

Observe that if **x** is a Manhattan consensus sequence, then $G_j(\mathbf{x}) \neq \emptyset$ for any j. If we additionally assume that **x** is a sum-MSC sequence, we obtain a stronger property.

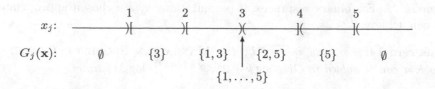

Fig. 3. Assume $a_{1,j} = 2, a_{2,j} = 4, a_{3,j} = 1, a_{4,j} = 3, a_{5,j} = 5$. Then $G_j(\mathbf{x})$ depends on the interval of x_j as shown in the figure; e.g., if $1 \le x_j < 2$ then $G_j(\mathbf{x}) = \{3\}$.

Lemma 16. *Let $\mathbf{x}$ be a sum-MSC sequence. Then $G_j(\mathbf{x}) \cap G_{j'}(\mathbf{x}) \neq \emptyset$ for any j, j'.*

Proof. For a proof by contradiction assume $G_j(\mathbf{x})$ and $G_{j'}(\mathbf{x})$ are disjoint. This implies that neither x_j nor $x_{j'}$ is in the center and thus $|G_j(\mathbf{x})|, |G_{j'}(\mathbf{x})| \le 2$. Let us move both x_j and $x_{j'}$ by one towards the center. Then $dist(\mathbf{x}, \mathbf{a}_i)$ remains unchanged for $i \in G_j(\mathbf{x}) \cup G_{j'}(\mathbf{x})$ (by disjointness), and decreases by two for the remaining sequences $\mathbf{a}_i$. There must be at least one such remaining sequence, which contradicts our choice of $\mathbf{x}$. $\qquad\square$

Additionally, if a sum-MSC sequence $\mathbf{x}$ has a position j with $|G_j(\mathbf{x})| = 1$, the structure of $\mathbf{x}$ needs to be even more regular.

Definition 17. *A sequence $\mathbf{x}$ is called an i-border sequence if for each j it holds that $x_j = a_{i,j}$ or $G_j(\mathbf{x}) = \{i\}$.*

Lemma 18. *Let $\mathbf{x}$ be a sum-MSC sequence. If $G_j(\mathbf{x}) = \{i\}$ for some j, then $\mathbf{x}$ is an i-border sequence.*

Proof. For a proof by contradiction assume $G_{j'}(\mathbf{x}) \neq \{i\}$ and $x_{j'} \neq a_{i,j'}$ for some j'. Let us move x_j towards the center and $x_{j'}$ towards $a_{i,j'}$ both by one. Then for any i' it holds that $dist(\mathbf{x}, \mathbf{a}_{i'})$ does not increase. By Lemma 16 $i \in G_{j'}(\mathbf{x})$, so $x_{j'}$ is moved away from the center. Moreover, x_j is moved towards some $a_{i',j}$ with $i' \neq i$, since $G_{j'}(\mathbf{x}) \neq \{i\}$. Consequently, $dist(\mathbf{x}, \mathbf{a}_{i'})$ decreases by two, which contradicts our choice of $\mathbf{x}$. $\qquad\square$

Definition 19. *A sequence $\mathbf{x}$ is called an i-middle sequence if for each j it holds that $i \in G_j(\mathbf{x})$ and $|G_j(\mathbf{x})| \ge 2$.*

Definition 20. *For a 3-element set $\Delta \subseteq \{1, \ldots, 5\}$ a sequence $\mathbf{x}$ is called a Δ-triangle sequence if for each j it holds that $|\Delta \cap G_j(\mathbf{x})| \ge 2$.*

Lemma 21. *Let $\mathbf{x}$ be a sum-MSC sequence. Then $\mathbf{x}$ is a border sequence, a middle sequence or a triangle sequence.*

Proof. Recall that $G_j(\mathbf{x}) \neq \emptyset$ for each j. By Lemma 18, if $G_j(\mathbf{x}) = \{i\}$ for some j, then $\mathbf{x}$ is an i-border sequence. This lets us assume $|G_j(x)| \ge 2$, i.e. $|G_j(\mathbf{x})| \in \{2, 5\}$, for each j. Let $\mathcal{F}$ be the family of 2-element sets among $G_j(\mathbf{x})$. By Lemma 16 every two of them intersect, so we can apply the following easy set-theoretical claim.

Claim. Let $\mathcal{G}$ be a family of 2-element sets such that every two sets in $\mathcal{G}$ intersect. Then sets in $\mathcal{G}$ share a common element or $\mathcal{G}$ contains exactly three sets with three elements in total.

If all sets in $\mathcal{F}$ share an element i, then $\mathbf{x}$ is clearly an i-middle sequence. Otherwise $\mathbf{x}$ is a Δ-triangle sequence for $\Delta = \bigcup \mathcal{F}$.   □

Fact 22. *There exist 20 interval systems $\mathcal{B}_i, \mathcal{M}_i$ (for $i \in \{1, \ldots, 5\}$) and $\mathcal{T}_\Delta$ (for 3-element sets $\Delta \subseteq \{1, \ldots, 5\}$) such that:*

(a) $\mathcal{B}_i$ is consistent with all i-border sequences, $G_j(\mathcal{B}_{i,j}) = \{i\}$ for proper $\mathcal{B}_{i,j}$;
(b) $\mathcal{M}_i$ is consistent with all i-middle sequences and if $\mathcal{M}_{i,j}$ is proper then $|G_j(\mathcal{M}_{i,j})| = 2$ and $i \in G_j(\mathcal{M}_{i,j})$;
(c) $\mathcal{T}_\Delta$ is consistent with all Δ-triangle sequences and if $\mathcal{T}_{\Delta,j}$ is proper then $|G_j(\mathcal{T}_{\Delta,j})| = 2$ and $G_j(\mathcal{T}_{\Delta,j}) \subseteq \Delta$.

Proof. (a) Let us fix a position j. For any i-border sequence $\mathbf{x}$, we know that $x_j = a_{i,j}$ or $G_j(\mathbf{x}) = \{i\}$. If either of the border intervals I satisfies $G_j(I) = \{i\}$, we set $\mathcal{B}_{i,j} := I$ (observe that $a_{i,j}$ is then consistent with I). Otherwise we choose $\mathcal{B}_{i,j}$ so that it is degenerate and corresponds to $x_j = a_{i,j}$.

(b) Again fix j. For any i-middle sequence $\mathbf{x}$, we know that x_j is consistent with at least one of the two middle intervals (both if x_j is in the center). If either of the middle intervals I satisfies $i \in G_j(I)$, we choose $\mathcal{M}_{i,j} := I$. (Note that this condition cannot hold for both middle intervals). Otherwise we know that x_j is in the center and set $\mathcal{M}_{i,j}$ so that it is degenerate and corresponds to x_j in the center, i.e. $\mathcal{M}_{i,j} := [3, 3]$.

(c) We act as in (b), i.e. if either of the middle intervals I satisfies $|G_j(I) \cap \Delta| = 2$, we choose $\mathcal{T}_{\Delta,j} := I$ (because sets $G_j(I)$ are disjoint for both middle intervals, this condition cannot hold for both of them). Otherwise, we set $\mathcal{T}_{\Delta,j} := [3, 3]$, since x_j is guaranteed to be in the center for any Δ-triangle sequence $\mathbf{x}$.   □

5   Practical Algorithm for $k \leq 5$

It suffices to consider $k = 5$. Using Fact 22 we reduce the number of interval systems from $(k-1)^{k!} = 4^{5!} > 10^{72}$ to 20 compared to the algorithm of Section 3. Moreover, for each of them ILP($\mathcal{I}$) admits structural properties, which lets us compute OPT($\mathcal{A}, \mathcal{I}$) much more efficiently than using a general ILP solver.

Definition 23. *A $(\pm)$ILP is called easy if for each constraint the number of $+1$ coefficients is 0, 1 or n, where n is the number of variables.*

Lemma 24. *For each $\mathcal{I}$ being one of the 20 interval systems $\mathcal{B}_i, \mathcal{M}_i$ and $\mathcal{T}_\Delta$, ILP($\mathcal{I}$) can be reduced to an equivalent easy $(\pm)$ILP with up to 4 variables.*

Proof. Recall that for degenerate intervals I_j, we do not introduce variables. On the other hand, if I_j is proper, possibly negating the variable x_j (Fact 7), we can make sure that the coefficient vector $E(x_j)$ has $+1$ entries corresponding to $i \in G_j(I_j)$ and -1 entries for the remaining i. Moreover, merging the variables

(Fact 8), we end up with a single variable per possible value $G_j(I_j)$. Now we use structural properties stated in Fact 22 to claim that the $(\pm)$ILP we obtain this way, possibly after further variable negations, becomes easy.

Border Sequences. By Fact 22(a), if $\mathcal{B}_{i,j}$ is proper, then $G_j(\mathcal{B}_{i,j}) = \{i\}$ and thus the $(\pm)$ILP has at most 1 variable and consequently is easy.

Middle Sequences. By Fact 22(b), if $\mathcal{M}_{i,j}$ is proper, then $G_j(\mathcal{M}_{i,j}) = \{i, i'\}$ for some $i' \neq i$. Thus there are up to 4 variables, the constraint corresponding to i has only $+1$ coefficients, and the remaining constraints have at most one $+1$.

Triangle Sequences. By Fact 22(c), if $\mathcal{T}_{\Delta,j}$ is proper, then $G_j(\mathcal{T}_{\Delta,j})$ is a 2-element subset of Δ, and thus there are up to three variables. Any $(\pm)$ILP with up to two variables is easy, and if we obtain three variables, then the constraints corresponding to $i \in \Delta$ have exactly two $+1$ coefficients, while the constraints corresponding to $i \notin \Delta$ have just -1 coefficients. Now, negating each variable (Fact 7), we get one $+1$ coefficient in constraints corresponding to $i \in \Delta$ and all $+1$ coefficients for $i \notin \Delta$.   □

The algorithm of Lenstra [17] with further improvements [13,9,18], which runs in roughly $n^{2.5n+o(n)}$ time, could perform reasonably well for $n = 4$. However, there is a simple $\mathcal{O}(n^2)$-time algorithm designed for easy $(\pm)$ILP. Due to space constraints, it is omitted in this paper. It can be found in the full version [14].

In conclusion, the algorithm for MSC problem first proceeds as described in Fact 22 to obtain the interval systems $\mathcal{B}_i, \mathcal{M}_i$ and $\mathcal{T}_\Delta$. For each of them it computes $\mathrm{ILP}(\mathcal{I})$, as described in Section 2, and converts it to an equivalent easy $(\pm)$ILP following Lemma 24. Finally, it uses the efficient algorithm to solve each of these 20 $(\pm)$ILPs. The final result is the minimum of the optima obtained.

6   Conclusions

We have presented an $\mathcal{O}(\ell k \log k)$-time kernelization algorithm, which for any instance of the MSC problem computes an equivalent instance with $\ell' \leq k!$. Although for $k \leq 5$ this gives an instance with $\ell' \leq 120$, i.e. the kernel size is constant, solving it in a practically feasible time remains challenging. Therefore for $k \leq 5$ we have designed an efficient linear-time algorithm.

We have implemented the algorithm,[1] including retrieving the optimum consensus sequence (omitted in the description above). For random input data with $\ell = 10^6$ and $k = 5$, the algorithm without kernelization achieved the running time of 1.48015s, which is roughly twice the time required to read the input file (0.73443s, not included in the former). The algorithm pipelined with the kernelization achieved 0.33415s. The experiments were conducted on a MacBook Pro notebook (2.3 Ghz Intel Core i7, 8 GB RAM).

References

1. Amir, A., Paryenty, H., Roditty, L.: Configurations and minority in the string consensus problem. In: Calderón-Benavides, L., González-Caro, C., Chávez, E., Ziviani, N. (eds.) SPIRE 2012. LNCS, vol. 7608, pp. 42–53. Springer, Heidelberg (2012)

---

[1] Source code is available at http://www.mimuw.edu.pl/~kociumaka/files/msc.cpp.

2. Amir, A., Paryenty, H., Roditty, L.: On the hardness of the consensus string problem. Inf. Process. Lett. 113(10-11), 371–374 (2013)
3. Andoni, A., Indyk, P., Patrascu, M.: On the optimality of the dimensionality reduction method. In: FOCS, pp. 449–458. IEEE Computer Society (2006)
4. Badoiu, M., Har-Peled, S., Indyk, P.: Approximate clustering via core-sets. In: Reif, J.H. (ed.) STOC, pp. 250–257. ACM (2002)
5. Boucher, C., Brown, D.G., Durocher, S.: On the structure of small motif recognition instances. In: Amir, A., Turpin, A., Moffat, A. (eds.) SPIRE 2008. LNCS, vol. 5280, pp. 269–281. Springer, Heidelberg (2008)
6. Cohen, G.D., Honkala, I.S., Litsyn, S., Solé, P.: Long packing and covering codes. IEEE Transactions on Information Theory 43(5), 1617–1619 (1997)
7. Fischer, K., Gärtner, B., Kutz, M.: Fast smallest-enclosing-ball computation in high dimensions. In: Di Battista, G., Zwick, U. (eds.) ESA 2003. LNCS, vol. 2832, pp. 630–641. Springer, Heidelberg (2003)
8. Frances, M., Litman, A.: On covering problems of codes. Theory Comput. Syst. 30(2), 113–119 (1997)
9. Frank, A., Tardos, É.: An application of simultaneous diophantine approximation in combinatorial optimization. Combinatorica 7(1), 49–65 (1987)
10. Gärtner, B., Schönherr, S.: An efficient, exact, and generic quadratic programming solver for geometric optimization. In: Symposium on Computational Geometry, pp. 110–118 (2000)
11. Graham, R.L., Sloane, N.J.A.: On the covering radius of codes. IEEE Transactions on Information Theory 31(3), 385–401 (1985)
12. Gramm, J., Niedermeier, R., Rossmanith, P.: Fixed-parameter algorithms for closest string and related problems. Algorithmica 37(1), 25–42 (2003)
13. Kannan, R.: Minkowski's convex body theorem and integer programming. Mathematics of Operations Reasearch 12, 415–440 (1987)
14. Kociumaka, T., Pachocki, J.W., Radoszewski, J., Rytter, W., Waleń, T.: On the string consensus problem and the Manhattan sequence consensus problem (full version). CoRR, abs/1407.6144 (2014)
15. Kumar, P., Mitchell, J.S.B., Yildirim, E.A.: Computing core-sets and approximate smallest enclosing hyperspheres in high dimensions. In: 5th Workshop on Algorithm Engineering and Experiments (2003)
16. Lanctôt, J.K., Li, M., Ma, B., Wang, S., Zhang, L.: Distinguishing string selection problems. In: Tarjan, R.E., Warnow, T. (eds.) SODA, pp. 633–642. ACM/SIAM (1999)
17. Lenstra Jr., H.W.: Integer programming with a fixed number of variables. Mathematics of Operations Research 8, 538–548 (1983)
18. Lokshtanov, D.: New Methods in Parameterized Algorithms and Complexity. PhD thesis, University of Bergen (2009)
19. Ma, B., Sun, X.: More efficient algorithms for closest string and substring problems. SIAM J. Comput. 39(4), 1432–1443 (2009)
20. Mazumdar, A., Polyanskiy, Y., Saha, B.: On Chebyshev radius of a set in Hamming space and the closest string problem. In: ISIT, pp. 1401–1405. IEEE (2013)
21. Ritter, J.: An efficient bounding sphere. In: Glassner, A.S. (ed.) Gems. Academic Press, Boston (1990)
22. Sylvester, J.J.: A question in the geometry of situation. Quarterly Journal of Pure and Applied Mathematics 1, 79 (1857)

Context-Aware Deal Size Prediction

Anisio Lacerda[1], Adriano Veloso[1], Rodrygo L.T. Santos[1], and Nivio Ziviani[1,2]

[1] Department of Computer Science
Universidade Federal de Minas Gerais
Belo Horizonte, MG, Brazil
{anisio,adrianov,rodrygo,nivio}@dcc.ufmg.br
[2] Zunnit Technologies
Belo Horizonte, MG, Brazil
nivio@zunnit.com

Abstract. Daily deals sites, such as Groupon and LivingSocial, attract millions of customers in the hunt for products and services at substantially reduced prices (i.e., deals). An important aspect for the profitability of these sites is the correct prediction of how many coupons will be sold for each deal in their catalog—a task commonly referred to as deal size prediction. Existing solutions for the deal size prediction problem focus on one deal at a time, neglecting the existence of similar deals in the catalog. In this paper, we propose to improve deal size prediction by taking into account the context in which a given deal is offered. In particular, we propose a topic modeling approach to identify markets with similar deals and an expectation-maximization approach to model intra-market competition while minimizing the prediction error. A systematic set of experiments shows that our approach offers gains in precision ranging from 8.18% to 17.67% when compared against existing solutions.

1 Introduction

In recent years, daily deals sites (or simply DDSs) such as Groupon[1] and Living-Social[2] became an important group-buying alternative for both local merchants and consumers. While local merchants are mainly interested in disseminating their brand to increase revenue, potential consumers seek discounted prices for products and services as diverse as restaurant meals, theater tickets, etc. In this business model, the DDS operates as a mediator for local merchants (sellers) to negotiate a deal (product or service) that is sold to consumers (buyers). In this case, the profitability of the DDS depends directly on two factors: (i) the commission associated with a deal, and (ii) the number of coupons that are actually sold for the deal, a quantity commonly referred to as the deal size. Different from commissions, which are governed by business decisions, the task of predicting the deal size can be modeled as a machine learning regression problem [8].

While being of paramount importance for the success of DDSs, accurately predicting deal sizes is surrounded by challenges. First, the catalog of deals is

---

[1] http://www.groupon.com
[2] http://www.livingsocial.com

E. Moura and M. Crochemore (Eds.): SPIRE 2014, LNCS 8799, pp. 256–267, 2014.
© Springer International Publishing Switzerland 2014

Fig. 1. The deal size depends on the deal and the presented alternatives

usually available for only a limited time frame, which varies from 4 to 5 days on average [8]. This may compromise the amount of historical data that is available for learning predictors, harming the effectiveness of algorithms such as support vector regression (SVR [3,11]). Second, deals may compete among themselves for consumer preference. For instance, consider the example in Fig. 1, which illustrates a commonly observed case in which a consumer is looking for discounts in restaurants and has a limited budget of $50. In this case, she would arguably prefer an Italian dinner over a Kebab, or vice-versa, but she is unlikely to increase her budget in order to buy both deals. An even worse outcome may happen when the similarity between the two services is so high that causes hesitation, leading the consumer to abandon the DDS without buying any of the competing deals.

Existing solutions to deal size prediction [8,14] produce global predictors that often neglect complex interactions between deals, such as competition for customer preference. In contrast, we propose to model the attractiveness of a deal relatively to other available deals from the same market. In particular, we propose a topic modeling approach to identify sets of deals that are likely to attract the interest of similar consumers. Furthermore, we propose a context-aware expectation-maximization approach to deal size prediction by considering (i) features associated with the target deal, and (ii) contextual features associated with the other deals in the same market of the target deal. To the best of our knowledge, our proposed approach is the first learning model that takes into account the whole catalog of deals when performing deal size prediction.

To assess the effectiveness of our proposed approach, we perform a systematic evaluation involving real usage data obtained from major DDSs such as Groupon and LivingSocial. In order to evaluate the extent to which our market segmentation strategy is language-dependent, we perform additional experiments with usage data obtained from Peixe Urbano,[3] the largest Brazilian DDS. The results attest the effectiveness of our proposed approach, with precision improvements ranging from 8.18% to 17.67% when compared to existing deal size predictors.

---

[3] http://www.peixeurbano.com.br

2 Related Work

DDSs have recently attracted the attention of researchers in multidisciplinary fields. Regarding the economic aspect of DDSs, Byers et al. [7] presented evidence that Groupon strategically optimizes their deal offerings, giving customers incentives other than price to make a purchase, including deal scheduling and duration, deal featuring, and limited inventory. Groupon's business model was further examined by Arabshahi [2]. An empirical analysis of the experience of merchants that used Groupon was performed by Dholakia [10]. Several studies addressed the propagation effect of daily deals in online social networks. More specifically, they were interested in assessing the impact that a deal had on a merchant's subsequent ratings in social review sites such as Yelp. Different from past research that observed a decrease in the ratings for merchants using Groupon [9], Potamias [17] argued that this effect was overestimated. Finally, Kumar and Rajan [13] analyzed, among other economic aspects, the profitability of social coupons and concluded that they yield profits for merchants.

Another line of research addressed algorithmic problems in DDSs. In particular, these data-driven approaches focus on identifying and understanding the main characteristics of DDSs by analyzing historical purchase data. Two main interrelated problems arise in this scenario: (i) deal ordering, aimed at selecting a set of deals that should be featured in the DDS catalog on a given day, and (ii) deal size prediction, aimed at estimating the number of coupons that are expected to sell for a given deal. As previously discussed, the deal size prediction problem is the focus of this paper. In this line of research, Ye et al. [22] modeled the popularity of group deals as a function of time. Byers et al. [8] modeled deal size prediction as a linear combination of deal features and used ordinary least squares regression to fit their model. Instead of learning a single, global predictor, Lappas and Terzi [14] proposed to learn multiple predictors, one for each market identified using a hierarchical clustering algorithm.

Similarly to the prediction approach of Lappas and Terzi [14], we also seek to identify multiple markets and learn market-targeted deal size predictors. However, in contrast to their approach, we leverage several weighting schemes and structural properties of deals as discriminative features for market identification. Moreover, we introduce a normalization factor learned via expectation-maximization to account for competing interactions between deals from the same market in order to produce context-aware predictions. In our investigations in Section 5, both the global prediction approach of Byers et al. [8] and the segmented prediction approach of Lappas and Terzi [14] are used as baselines.

3 Context-Aware Deal Size Prediction

In this section, we introduce our context-aware deal size prediction approach. Firstly, we present a topic identification strategy to group deals into markets, which determine the context for each available deal. Then, we present a deal size prediction strategy that employs multiple SVR predictors: there are as many predictors as markets, and each predictor is specifically designed to one market. Finally, we present a contextual expectation-maximization strategy that reduces the prediction error by taking into account the competition between deals in the same market and also the representativeness of different markets.

3.1 Identifying Markets

Discovering meaningful markets from textual features associated with the deals is an important step for determining the context of competition among deals. In the following, we detail our approach for representing deals as well as for automatically identifying markets via latent topic modeling.

Representing deals. Our approach represents each deal d as a vector $\boldsymbol{d_s}$ in an n-dimensional space $\{t_1, t_2, \ldots, t_n\}$, where n is the number of unique terms in a given feature space s. In particular, we consider four different features spaces, comprising terms appearing on the *merchant's name*, the *title of the deal*, the *description of the deal*, or *all of these fields* concatenated in a single space.

In order to weigh the relative importance of each term t in a given deal vector $\boldsymbol{d_s}$, we consider four alternative weighting schemes: the *term frequency* TF, denoting the raw frequency of t in d; the *term spread* TS, denoting the spread of t in d, as a measure of the descriptive power of t; and the products TF×IFF and TS×IFF, with the *inverse feature frequency* IFF denoting the rarity of t among all cataloged deals represented in the feature space s.

Identifying latent markets. Our proposed approach defines a market as a set of deals that are likely to attract the interest of a similar group of customers. Under the assumption that customers have a limited budget, our intuition is that deals belonging to the same market are more likely to compete for customer preference. A simple strategy to market identification would be to analyze the customers' purchase history, in order to identify deals that were purchased by a similar set of customers. Unfortunately, as discussed in Section 1, such historical purchase data is very limited due to the scarceness of recurrent or regular customers. As an alternative, we propose a content-based approach to market identification using latent Dirichlet allocation (LDA [5]). LDA is a generative model used to identify latent topics in textual documents, and has been largely used in a variety of tasks, including matrix factorization [1], influential user identification [21], tag recommendation [12], and word sense disambiguation [6]. In our particular case, using LDA to identify latent markets overcomes the lack of historical purchase data by instead leveraging the textual representation of each of the available deals. As an illustrative example, Table 1 shows markets identified from one of the DDS datasets used in our investigations in Section 5.

3.2 Predicting deal size

In order to predict the deal size, we learn multiple predictors, each one targeted to a different market, identified according to the approach described in Section 3.1.

Table 1. Top five terms from each of $k = 5$ markets in the Groupon dataset

"Gym"	"Hair Salon"	"Sports"	"Dentistry"	"Ice Cream"
class	salon	camping	dental	tour
fit	hair	week	teeth	chocolate
body	look	sport	care	cake
train	services	day	value	sweet
workout	cut	academia	whitening	ice

To account for the competition within each market, these initial predictions are further adjusted through an iterative expectation-maximization procedure.

Specialized SVR. Support vector regression (SVR [11]) is an established non-linear regression technique that has been applied successfully to a variety of numeric prediction problems. In order to apply SVR to deal size prediction, we represent deals as follows. Let $\mathcal{D} = \{\mathcal{D}_{m_1}, \mathcal{D}_{m_2}, \ldots, \mathcal{D}_{m_k}\}$ be the collection of all past deals, and each $\mathcal{D}_{m_i} = \{d_1, d_2, \ldots, d_q\}$ be a partition of $\mathcal{D}$ composed of all q deals that belong to market m_i. Further, each deal d_j is represented as a feature vector, using features generally available from DDSs, such as the deal face and discounted values, the day of the week when the deal is launched, etc.

The training set used to build an SVR predictor to market m_i is composed of (deal, size) pairs of the form $\{(d_1, s_1), (d_2, s_2), \ldots, (d_q, s_q)\}$, that is, each pair is composed of a deal $d_j \in \mathcal{D}_{m_i}$ and its corresponding size s_j. For each deal $d_p \in \mathcal{D}_{m_i}$, a specialized SVR predictor takes the form $f(d_p) = \langle w, \Phi(d_p) \rangle + b$, where $w \subset \Re^n$, $b \subset \Re$, and Φ denotes a nonlinear transformation from $\Re^n$ to a high-dimensional space. The SVR objective is to find the minimum value of w and b by solving the following regularized optimization problem:

$$\min_w \frac{1}{2} w^T w + C \sum \xi_\epsilon(w; d_j, s_j), \tag{1}$$

$$\xi_\epsilon(w; d_j, s_j) = \max(|w^T d_j - s_j| - \epsilon, 0)^2, \tag{2}$$

where $C > 0$ is the regularization parameter, and ξ_ϵ is the ϵ-insensitive loss associated with (d_j, s_j), with ϵ given so that the loss is zero when $|w^T d_j - s_j| \geq \epsilon$ [19]. We use a radial basis function (RBF) as the transformation function Φ, and set all other parameters through cross-validation [15].

The CBMP model. The specialized SVR predictor does not take into account complex interactions that may exist between deals in the catalog. For instance, deals in the catalog of the day, denoted S, may compete with each other. As a result, SVR predictions may be overestimated. In order to model competition, we first partition the catalog into k markets, so that $S = \{S_{m_1}, S_{m_2}, \ldots, S_{m_k}\}$. Every deal in S must be assigned to one of these markets, and we say that S_{d_j} is the market to which deal d_j is assigned. Given the SVR prediction $f(q)$ for any deal $q \in S$, the estimated size of deal $d_j \in S$ is given as a combination of two factors: the SVR prediction for d_j, and an average factor that encompasses all deals within the same market. In order to combine these two factors, we introduce the following expectation-maximization (EM) formulation:

$$E: \quad \rho(S) = \frac{\sum_{q \in S} \sigma(q)}{\sum_{d_j \in C} \sigma(d_j)}, \tag{3}$$

$$M: \quad \sigma(d_j) = \alpha \times \frac{\sum_{q \in S_{d_j}} f(q) \times \rho(S_{d_j})}{|S_{d_j}|} + (1 - \alpha) \times f(d_j), \tag{4}$$

where $\alpha = 0.5$ to weigh equally both factors, and $\rho(S_{d_j})$ is an unknown parameter which re-scales the representativeness of market S_{d_j}. This EM procedure uses the training set to find the value for $\rho(S_{d_j})$ that minimizes the loss function:

$$eqn: rmse(y, \hat{y}) = \sqrt{\frac{1}{n} \sum (y_i - \hat{y}_i)^2}, \tag{5}$$

where n is the number of predictions performed, and y_i and $\hat{y}_i$ are the actual and the predicted deal sizes, respectively. Intuitively, this prediction model, named Competitive Business Market Predictor (CBMP), accounts for intra-market competition by re-scaling the predicted size of each deal relatively to the predicted size of other deals from the same market. In the next sections, we assess the effectiveness of the CBMP model compared to both a global as well as market-targeted predictors which do not account for competing relationships.

4   Experimental Setup

This section details the experimental setup that supports the validation of our proposed context-aware deal size prediction approach, including the datasets, the baselines, and the evaluation procedure that we use.

Daily deals datasets. Our evaluation uses data collected from three commercial DDSs: Groupon, LivingSocial, and Peixe Urbano. Groupon and LivingSocial were crawled using as seeds the datasets used by Byers et al. [8]. In particular, Groupon was crawled from Jan 3rd, 2011 to Jul 3rd, 2011, while LivingSocial was crawled from Mar 21st, 2011 to Jul 3rd, 2011. In both cases, the collected data includes English textual features used to group deals into markets, as described in Section 3.1. For Groupon, we collected a total of 16,409 deals comprising 119,525 unique terms. For LivingSocial, we obtained 2,610 deals with 19,102 distinct terms. In addition, to enable our basic prediction model using SVR, we collected a number of non-textual deal features, including (1) the price of the deal, (2) the price after the discount, (3) the tipping point of the deal (i.e., minimum number of coupons that must be sold to enable the deal), (4) the day of the week in which the deal was launched, (5) the category of the deal, (6) the city in which the corresponding merchant is located, and boolean values indicating (7) whether the deal is running for multiple days, (8) whether the deal is featured on the DDS website, and (9) whether the inventory is limited. Finally, for Peixe Urbano, the same features were obtained directly through an API made available to us. Textual features in Portuguese were obtained for 4,309 deals during the entire year of 2012, comprising a total of 31,163 unique terms.

Prediction baselines. We compare the effectiveness of CBMP, our context-aware deal size prediction model, against the following baseline predictors:

- Global Predictor (GLPR) [8], learned using the whole training set, ignoring the existence of markets. The size of an arbitrary deal q is given as:

$$\sigma(q) = 2^{\beta_0 + \sum \beta_i \times f_i},\tag{6}$$

  where features f_i correspond to the ones described earlier in this section and weights β_i are determined using ordinary least squares [15].
- One Predictor per Business Market (OPBM) [14], learned using an SVM-based regression method proposed by Shevade et al. [18]. In particular, they introduced a two-layer clustering algorithm that is based on LDA [5] and a flat clustering algorithm based on the Kullback-Liebler distance to separate deals into markets [16]. Note that, in contrast to our proposed model, OPBM does not exploit competing relationships between deals in a given market.

Evaluation procedure. Our evaluation uses the interleaved test-then-train methodology [4], in which each deal in the catalog on day t is evaluated, and then is included into the historical data that becomes available at day $t+1$. Our intention is to mimic as close as possible the production system of a DDS, which predicts the size of each available deal in the beginning of the day, and incorporates the feedback received on these deals in the end of the day, to be used in the predictions of the next day. We evaluate the effectiveness of all predictors using the root mean squared error (RMSE) [15]. As shown in Equation (5), $\hat{y}_i$ is the predicted value of the i-th sample, while y_i is the corresponding true value. Finally, n is the number of instances in the dataset for which the prediction is performed. The results to be reported correspond to an average over all days. Statistical significance was verified using a paired t-test with $p < 0.05$.

5   Experimental Results

In this section, we empirically validate our proposed context-aware deal size prediction model in contrast to existing deal size predictors from the literature. In particular, we aim to answer the following research questions:

Q1. How effective is CBMP compared to existing deal size predictors?
Q2. How effective is our topic modeling approach to market identification?
Q3. How effective are our various strategies for deal representation?

To answer question Q1 and Q2, we contrast the effectiveness of our CBMP model to GLPR as a global, market-agnostic predictor, as well as to OPBM as a

Table 2. RMSE figures for CBMP and the GLPR and OPBM baselines

	GLPR	OPBM	CBMP$_l$	Δ_{GLPR}	Δ_{OPBM}	CBMP$_c$	Δ_{GLPR}	Δ_{OPBM}
Groupon	1.3864	1.3544	1.1332	17.7%	16.3%	1.2563	8.7%	7.2%
LivingSocial	1.1287	1.1112	1.0203	9.6%	8.2%	1.0931	3.2%	1.6%
Peixe Urbano	1.2956	1.2575	1.1430	11.8%	9.1%	1.1841	8.6%	5.8%

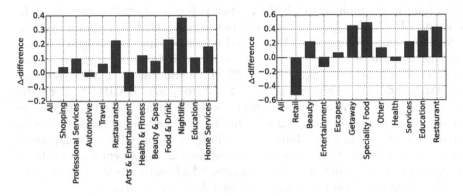

Fig. 2. Market-specific prediction model vs. global model

local, market-aware predictor that does not exploit the competing nature of deals in the context of an individual market. Table 2 shows the prediction performance of these three models across the three considered DDS datasets (Groupon, LivingSocial, and Peixe Urbano) in terms of RMSE. Our CBMP model is deployed using either latent topics (CBMP$_l$) or explicit categories (CBMP$_c$) for market identification. The CBMP$_l$ variant identifies latent markets (50 for Groupon; 30 for LivingSocial and Peixe Urbano) with deals represented by a concatenation of all their terms weighted using TF, which was the best performing setting identified during training.[4] For each of the two variants of CBMP, namely, CBMP$_l$ and CBMP$_c$, the additional columns Δ_{GLPR} and Δ_{OPBM} show the percentage RMSE improvement compared to GLPR and OPBM, respectively.

From Table 2, we first observe that both variants of our CBMP model outperform both deal size prediction baselines. In particular, CBMP$_c$ outperforms the global GLPR predictor with significant improvements in RMSE ranging from 3.2% (LivingSocial) to 8.7% (Groupon). These gains are further illustrated by the breakdown analysis shown in Fig. 2 for deals across multiple categories of Groupon and LivingSocial.[5] In the figure, a positive delta indicates an improvement of CBMP against GLPR, whereas a negative delta indicates otherwise. From the figure, we observe that most of the observed differences are positive,

---

[4] A complete analysis of the impact of different deal representations for a varying number of target markets is presented later in this section.

[5] Results on Peixe Urbano show similar trends and are omitted for brevity.

indicating that it is indeed better to use multiple market-specific predictors than a single, global predictor for all markets. Notable exceptions can be observed for categories such as "Retail" and "Entertainment", when a global predictor performs better. Such categories are arguably less cohesive, which may indicate weaker intra-market competition relationships. Further exploiting such a nuanced view of markets is a direction for future investigation.

A promising alternative to arbitrarily defined markets is the latent topic identification approach employed by $CBMP_l$. Indeed, as shown in Table 2, $CBMP_l$ further improves compared to $CBMP_c$, which is based on explicit categories. In particular, $CBMP_l$ significantly outperforms the global GLPR predictor by up to 17.7% (Groupon). Compared to the OPBM baseline, which also leverages topic modeling to produce market-specific deal size predictions, our approaches are also effective, with gains ranging from 8.2% (LivingSocial) to 16.3% (Groupon) for $CBMP_l$ and 1.6% (LivingSocial) to 7.2% (Groupon) for $CBMP_l$. Recalling question Q1, these results attest the effectiveness of our proposed context-aware deal size prediction model compared to both global as well as market-specific, content-agnostic predictors. Moreover, recalling question Q2, the comparison between the two variants of our model also attest the effectiveness of our topic modeling approach to identify latent markets, which outperforms a hard partition of markets based on the category of each deal in the catalog.

To address question Q3, we further evaluate our topic modeling variant (henceforth referred to as CBMP) in light of different strategies for deal representation. To this end, we break down the performance of CBMP for different feature spaces (merchant's name, deal title, deal description, concatenation of all terms), weighting schemes (TF, TS, TF×IFF, and TS×IFF), and target number of markets. In particular, Figs. 3, 4, and 5 show the results of this breakdown analysis for the Groupon, LivingSocial, and Peixe Urbano datasets, respectively.

Regarding the impact of different feature spaces, we observe a generally superior performance of CBMP in denser spaces, such as those induced with terms obtained from the description or the concatenation of all terms comprised by each deal. In addition, of these two feature spaces, the concatenation of all terms appears to be less sensitive to the variation in the number of chosen markets. Considering the other two spaces, merchant's name and deal title, both yield generally similar prediction performances across the three datasets. Regarding the different weighting schemes considered, TF and TS are consistently the best performers across all datasets. Furthermore, we can also see that the weighting schemes that are combined with IFF are more sensitive to the number of markets. For instance, when considering the deal description on Groupon (Fig. 3), or the deal title on LivingSocial (Fig. 4) and Peixe Urbano (Fig. 5).

Recalling question Q3, the results in Figs. 3, 4, and 5 show that denser feature spaces with pure frequency-based weighting functions are generally preferred for representing deals within our latent market identification approach. The target number of markets, on the other hand, is a key parameter that must be carefully tuned when deploying our proposed context-aware deal size prediction model for different datasets. Finally, it is also worth noting that our approach performs effectively for datasets in different languages, as exemplified by the English and Portuguese DDS datasets considered in our investigations.

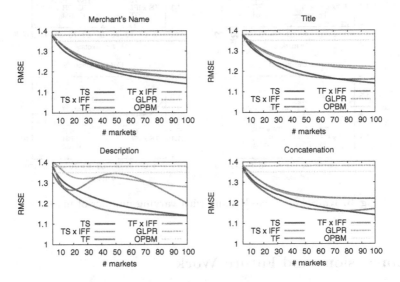

Fig. 3. RMSE on Groupon with varying number of markets

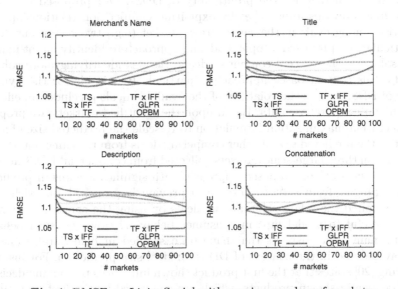

Fig. 4. RMSE on LivingSocial with varying number of markets

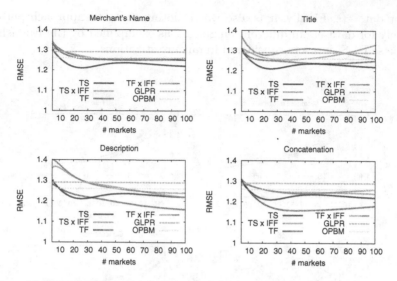

Fig. 5. RMSE on Peixe Urbano with varying number of markets

6   Conclusions and Future Work

We introduced CBMP, a novel context-aware deal size prediction model. Predicting the size of a deal (i.e., the number of coupons that will be sold for the deal) is a crucial task for the profitability of DDSs. Our proposed model improves upon previous approaches by exploiting competition relationships that may arise among deals in the same target market (e.g., two restaurant deals). In particular, we proposed a topic modeling approach to identify latent markets based solely on the textual content of deals. Besides identifying more cohesive markets, this content-based approach is particularly suitable for the dynamic nature of DDSs, where the volatility of the available deals precludes an effective use of historical purchase data. Based upon the identified markets, we proposed an expectation-maximization formulation to re-scale the predicted size of a deal in light of the predicted size of other competing deals from the same market. Experiments on three large-scale datasets collected from commercial DDSs attested the effectiveness of our proposed approach, with significant gains in prediction accuracy ranging from 8.2% to 17.7% over previously proposed approaches.

We exploited the relationship among deals by using the concept of markets, which is useful to model flips in consumer behavior. However, such behavior presents many other aspects that have been studied in behavioral economics and may be used in the context of DDSs for deal size prediction. For instance, anchoring [20] states that the first product shown influences the buying decisions for subsequently shown products, which are compared to the first one. Hence, in the context of DDSs, the ordering of the products presented in the web page may affect deal size, i.e., the top product may also have an anchoring effect.

Acknowledgements. We thank the partial support given by the Brazilian National Institute of Science and Technology for the Web (grant MCT-CNPq 573871/2008-6) and authors' individual grants and scholarships from CNPq and CAPES.

References

1. Agarwal, D., Chen, B.-C.: fLDA: matrix factorization through latent dirichlet allocation. In: ACM WSDM, pp. 91–100 (2010)
2. Arabshahi, A.: Undressing groupon: An analysis of the groupon business model (2011), http://www.ahmadalia.com/blog/2011/01/undressing-groupon.html
3. Basak, D., Pal, S., Patranabis, D.C.: Support vector regression. Neural Information Processing-Letters and Reviews 11(10), 203–224 (2007)
4. Bifet, A., Holmes, G., Kirkby, R., Pfahringer, B.: MOA: Massive online analysis. The Journal of Machine Learning Research 11, 1601–1604 (2010)
5. Blei, D.M., Ng, A.Y., Jordan, M.I.: Latent dirichlet allocation. Journal of Machine Learning Research 3, 993–1022 (2003)
6. Boyd-Graber, J.L., Blei, D.M., Zhu, X.: A topic model for word sense disambiguation. In: ACM ACL, pp. 1138–1147 (2010)
7. Byers, J., Mitzenmacher, M., Potamias, M., Zervas, G.: A month in the life of groupon. CoRR, abs/1105.0903 (2011)
8. Byers, J., Mitzenmacher, M., Zervas, G.: Daily deals: Prediction, social diffusion, and reputational ramifications. In: ACM WSDM, pp. 543–552 (2012)
9. Byers, J.W., Mitzenmacher, M., Zervas, G.: The groupon effect on yelp ratings: A root cause analysis. In: ACM EC, pp. 248–265 (2012)
10. Dholakia, U.M.: How effective are groupon promotions for business (2010), http://www.ruf.rice.edu/~dholakia
11. Drucker, H., Burges, C.J., Kaufman, L., Smola, A., Vapnik, V.: Support vector regression machines. In: NIPS, pp. 155–161 (1997)
12. Krestel, R., Fankhauser, P., Nejdl, W.: Latent dirichlet allocation for tag recommendation. In: ACM RecSys, pp. 61–68 (2009)
13. Kumar, V., Rajan, B.: Social coupons as a marketing strategy: A multifaceted perspective. Journal of the Academy of Marketing Science 40(1), 120–136 (2012)
14. Lappas, T., Terzi, E.: Daily-deal selection for revenue maximization. In: ACM CIKM, pp. 565–574 (2012)
15. Mitchell, T.M.: Machine learning, vol. 45. McGraw Hill, Burr Ridge (1997)
16. Pinto, D., Benedí, J.-M., Rosso, P.: Clustering narrow-domain short texts by using the kullback-leibler distance. In: Gelbukh, A. (ed.) CICLing 2007. LNCS, vol. 4394, pp. 611–622. Springer, Heidelberg (2007)
17. Potamias, M.: The warm-start bias of yelp ratings. CoRR (2012)
18. Shevade, S.K., Keerthi, S.S., Bhattacharyya, C., Murthy, K.R.K.: Improvements to the SMO algorithm for svm regression. IEEE Transactions on Neural Networks 11(5), 1188–1193 (2000)
19. Smola, A.J., Schölkopf, B.: A tutorial on support vector regression. Statistics and Computing 14(3), 199–222 (2004)
20. Tversky, A., Simonson, I.: Context-dependent preferences. Management Science 39(10), 1179–1189 (1993)
21. Weng, J., Lim, E.-P., Jiang, J., He, Q.: Twitterrank: finding topic-sensitive influential twitterers. In: ACM WSDM, pp. 261–270 (2010)
22. Ye, M., Sandholm, T., Wang, C., Aperjis, C., Huberman, B.A.: Collective attention and the dynamics of group deals. In: WWW, pp. 1205–1212 (2012)

Simple and Efficient String Algorithms for Query Suggestion Metrics Computation

Alexander Loptev[1,2], Anna Selugina[1], and Tatiana Starikovskaya[1,*]

[1] National Research University Higher School of Economics (HSE), Russia
aoselyugina@edu.hse.ru, tstarikovskaya@hse.ru
[2] Yandex, Russia
alonger@yandex-team.ru

Abstract. In order to make query suggestion mechanisms more efficient, it is important to have metrics that will estimate query suggestions quality well. Recently, Kharitonov et al. [7] proposed a family of metrics that showed much better alignment with user satisfaction than previously known metrics. However, they did not address the problem of computing the proposed metrics. In this paper we show that the problem can be reduced to one of the two string problems which we call Top-k and Sorted-Top-k. Given an integer k and two sets of pairwise distinct strings (queries) with weights, Q and Q_{test}, the Top-k problem is to find, for each query $q \in Q_{test}$, its shortest prefix $q[1..i]$ such that q belongs to the list of k heaviest queries in Q starting with $q[1..i]$. The Sorted-Top-k problem is to retrieve, for each $q \in Q_{test}$ and $1 \leq i \leq |q|$, a position of q in the sorted list of the k heaviest queries in Q starting with $q[1..i]$. We show several linear-time solutions to these problems and compare them experimentally.

1 Introduction

Almost every big search engine has a build-in query suggestion mechanism — a special feature that helps a user to type less when submitting a query. When a user submits a new letter $q[i]$ of her query q to a search engine, the mechanism forms a list of queries starting with $q[1..i]$. If the user sees q in the list, she selects it and sumbits it to the search engine. Otherwise, she types the next letter of her query.

To improve quality of search suggestion mechanisms it is essential to have metrics that reflect a user satisfaction level well. Recently, Kharitonov et al. [7] introduced a model of interaction between a user and a query suggestion mechanism and proposed a family of associated metrics called *Saved*. (Prior to that, no model existed.) Under this user model, they performed a thorough experimental comparison of their metrics and the MRR [8], $wMRR$ [1], and MKS [3] metrics that had been known prior to their work. The experiments showed that the *Saved* metrics family is aligned with a user satisfaction level much better than the other metrics.

---

* Tatiana Starikovskaya was partly supported by Dynasty Foundation.

E. Moura and M. Crochemore (Eds.): SPIRE 2014, LNCS 8799, pp. 268–278, 2014.
© Springer International Publishing Switzerland 2014

Hence, usage of the *Saved* metrics family for evaluating quality of a query suggestion mechanism is well-justified. However, in their work Kharitonov et al. did not discuss the problem of computing the metrics. Here we address this gap. In doing so, we reduce the problem of computing a metric of the *Saved* family to one of the two string problems which we call *Top-k* and *Sorted-Top-k*. These problems is a core difficulty, and once the answer for them is known, the metrics can be computed by one linear-time pass through the answer.

In both problems we assume that two sets Q, Q_{test} of pairwise distinct strings (queries) are given, and $|Q| \gg |Q_{test}|$. For each query we are given its weight. The Top-k problem is to define, for each query $q \in Q_{test}$, its shortest prefix $q[1..i]$ such that q belongs to the list of k heaviest queries in Q starting with $q[1..i]$. The Sorted-Top-k problem is to retrieve, for each q and $1 \le i \le |q|$, a position of q in the sorted list of k heaviest queries in Q starting with $q[1..i]$.

For each of the problems we show two algorithms. Let n be the total length, and m be the total number of queries in Q. For the Top-k problem we propose algorithms with $\mathcal{O}(n)$ and $\mathcal{O}(n \log \sigma)$ time, where σ is the size of the alphabet. For the Sorted-Top-k problem we give algorithms with $\mathcal{O}(n \log k)$ and $\mathcal{O}(n \log \sigma)$ time. Each of the four algorithms uses $\mathcal{O}(m)$ memory (not counting the memory needed to store the input and the output). The algorithms we give are very simple, which we consider as an advantage given that calculating the metrics is supposed to be a basic tool in the area of research related to query suggestion mechanisms.

The rest of the paper is organized as follows. In Section 2 we remind a reader the definitions of the user model and the metrics of the *Saved* family. In Section 3 we show that the metrics can be calculated by reduction either to the Top-k or to the Sorted-Top-k problem. In Section 4 we show four solutions to these problems. We report the results of experimental analysis of the algorithms in Section 5. Finally, we conclude in Section 6.

2   User Model

We briskly remind the user model introduced in [7]. It is supposed that a query suggestion mechanism stores a set of pairwise distinct queries, Q. The set Q contains all queries previously submitted, and for each query in Q the number of times it has been submitted is known (weight). In the process of submitting a query q to a search engine, a user types q letter by letter. After each letter, the list of suggested queries is updated to contain k heaviest queries in Q starting with the typed prefix of q. The queries in the list are sorted in the decreasing order of their weights.[1] The user inspects the list, and if she sees q in the list, she selects it and that ends the interaction with the query suggestion mechanism. Otherwise, the user types the next letter. It is assumed that the user's behaviour

---

[1] In general, the weights can be arbitrary. For instance, they can also depend on the user's geographical location. The algorithms we propose work for any choice of weights.

satisfies the Markovian assumption meaning that the behaviour of the user at the current step does not depend on her earlier decisions.

Formally, the user model can be described as follows. Let $q_{i,j}$ be the jth suggested query for a prefix $q[1..i]$. We introduce the following Boolean random variables:

- N_i: was $q[i]$ submitted;
- $E_{i,j}$: is $q_{i,j}$ examined by the user;
- $S_{i,j}$: is the user satisfied with $q_{i,j}$ after submitting $q[1..i]$.

Firstly, we define the order in which the letters of q are submitted and what the user sees in response: (a) The first letter is always submitted; (b) Letters are typed sequentially; and (c) If $q[i]$ is not submitted, the user does not see suggestions for $q[1..i]$.

(a) $N_1 = 1$
(b) $N_i = 0 \Rightarrow \forall k \geq i+1: \ N_k = 0$
(c) $N_i = 0 \Rightarrow \forall j: \ E_{i,j} = 0$

Secondly, the user is satisfied with $q_{i,j}$, that is, she selects $q_{i,j}$ from the list and stops the interaction with the query suggestion mechanism if and only if she examined $q_{i,j}$ and $q_{i,j} = q$:

(e) $S_{i,j} = 1 \Leftrightarrow E_{i,j} = 1, \ q_{i,j} = q$

Finally, we need a set of equations defining when the user stops typing: (f) If the user is satisfied with $q_{i,j}$, the interaction stops; (g) If the user is not satisfied with suggestions for $q[1..i]$ and $i < |q|$, she types $q[i+1]$; (h) If the user typed all letters of q, the interaction stops.

(f) $\exists j: S_{i,j} = 1 \Rightarrow N_{i+1} = 0$
(g) $\forall j \ S_{i,j} = 0, i < |q| \Rightarrow N_{i+1} = 1$
(h) $N_{|q|+1} = 0$

The probability of examining the query suggested at the jth position is a function of the user's state and defines the model. Kharitonov et al. [7] considered five different probability functions, which can be classified as follows:

- The function is equal to one for any i, j, i.e. $f_1(i, j) = 1$;
- Dependency on j only ($f_{rr}(i, j) = \frac{1}{j+1}$, $f_{\log}(i, j) = \frac{1}{\log_2 j+2}$, $f_l^i(i, j) = A_j$);
- Dependency on both i and j ($f_l^d(i, j) = B_{i,j}$).

The function f_1 corresponds to the case when the user always examines all suggested queries. The parameters A_j and $B_{i,j}$ are learned by a simple one-time scan of the set Q. It can be assumed that these values are computed once for a fixed language, a search engine, and a query suggestion mechanism.

3   Metrics

Let Q_{test} be a set of distinct queries observed during some fixed period of time, and $w(q)$ be the number of times a query q was submitted during this time (weight). To estimate the quality of query suggestions for this period of time, Kharitonov et al. [7] proposed a family of metrics defined as the expectation of a utility function at a position where the user is satisfied by a query suggestion averaged over all queries in Q_{test}:

$$M = \frac{1}{W(q)} \sum_{q \in Q_{test}} w(q) \sum_{i=1}^{|q|} \sum_{j=1}^{k} U(i,j) P(S_{i,j} = 1) \tag{1}$$

where $W(q) = \sum_{q \in Q_{test}} w(q)$. The first utility function is defined to be equal to 1 if the user is satisfied by a query suggestion, and 0 otherwise. The resulting metric is equal to the probability of using the query suggestion mechanism and is denoted by $pSaved$.

$$pSaved = \frac{1}{W(q)} \sum_{q \in Q_{test}} w(q) \sum_{i=1}^{|q|} \sum_{j=1}^{k} P(S_{i,j} = 1) \tag{2}$$

The second function is defined as $U(i,j) = 1 - \frac{i}{|q|}$ to reflect the effort it takes the user to find q in the suggestions. The resulting metrics is referred to as $eSaved$.

$$eSaved = \frac{1}{W(q)} \sum_{q \in Q_{test}} w(q) \sum_{i=1}^{|q|} \left(1 - \frac{i}{|q|}\right) \sum_{j=1}^{k} P(S_{i,j} = 1) \tag{3}$$

Both proposed metrics, $pSaved$ and $eSaved$ are defined by the examination probability function as $S_{i,j} = 1$ if and only if $E_{i,j} = 1$ and $q_{i,j} = q$. (See Equation (e).)

Kharitonov et al. [7] showed experimentally that the proposed metrics are aligned with the quality of a query suggestion mechanism much better than the previously known MRR [8], $wMRR$ [1], and MKS [3] metrics. However, for this family of metrics to become widely used, it is essential to show that they can be easily computed.

4   Reduction to Stringology

We distinguish between two cases: the examination probability function is defined as f_1 and the examination probability function is defined as f_{rr}, $f_{\log}$, f_l^i, or f_l^d.

To compute the metrics in the first case, it is sufficient to know, for each query, the minimal prefix length when the query first appears among suggestions. Indeed, $P(S_{i,j} = 1)$ for a query q will be equal to one if $q[1..i]$ is the shortest prefix for which q belongs to the list of suggested queries, and to zero otherwise.

More information is required in the second case: for each prefix $q[1..i]$ of each query q we need to know the position p_i of the query among the suggested queries. Then

$$P(S_{i,j} = 1) = \begin{cases} 0 & \text{if } j \neq p_i, \\ P(E_{i,p_i} = 1) \prod_{\substack{r \in [1,i-1]: \\ p_r \text{ is defined}}} (1 - P(E_{r,p_r} = 1)) & \text{otherwise.} \end{cases} \quad (4)$$

If $P(S_{i,j} = 1)$ is known, the metrics $eSaved$ and $pSaved$ can be computed in a straightforward way: For each query in Q_{test} we iterate over its prefixes and calculate the metrics according to Equation 1.

Hence, we restrain our attention to two string problems. In both problems we assume that two sets Q, Q_{test} of pairwise distinct strings (queries) are given, and $|Q| \gg |Q_{test}|$. For each query we are given its weight. The *Top-k* problem is to define, for each query $q \in Q_{test}$, its shortest prefix $q[1..i]$ such that q belongs to the list of the k heaviest queries in Q starting with $q[1..i]$. The *Sorted-Top-k* problem is to retrieve, for each q and $1 \leq i \leq |q|$, a position of q in the sorted list of the k heaviest queries in Q starting with $q[1..i]$.

Algorithm 1. Computation of the metrics of the *Saved* family

---

1. Compute the examination probability function f
2. **if** $f = f_1$ **then**
3.   Solve the Top-k problem for Q and Q_{test}
4.   Write the output into an array *MinPref*
5.   Compute the metric using Equation 1
6. **else**
7.   Solve the Sorted-Top-k problem for Q and Q_{test}
8.   Write the output into an array *Pos*
9.   Compute the metric using Equations 1 and 4
10. **end if**

The basis of the algorithms is a compacted trie T for the set Q. The compacted trie T is a tree such that each of its edges corresponds to a non-empty string. The concatenation of strings corresponding to the edges in a path is called a *label* of this path. For each query $q \in Q$ there must exist a root-to-node path with the label equal to q, and the label of each root-to-leaf path must be equal to one of the queries in Q. We also require the degree of each inner node not corresponding to a query from Q to be at least two and the first letters on edges outgoing from any node to be distinct. Strings corresponding to the edges of T are not stored explicitly — as it follows from the definition, each of the strings is a substring of a query from Q, and hence it is sufficient to store the starting and the ending positions of the substring in the query. Under this assumption, T can be constructed in $\mathcal{O}(n)$ time and $\mathcal{O}(m)$ space, where n is the total length and m is the total number of queries in Q [5]. We assume that for every node of T the length of its label (*string depth*) is known, which can be achieved by a linear-time post-processing.

Consider a prefix $\rho = q[1..i]$ of a query $q \in Q_{test}$. If Q contains queries starting with ρ, then there is a path in T that starts at the root of T and is labelled by ρ, and all queries starting with ρ end in the subtree rooted at the lower end of this path. Therefore, to determine whether q is in the list of suggested queries for ρ, it is sufficient to retrieve the list of the k heaviest queries in the subtree. To determine the position of q in the list of suggested queries for ρ, it is sufficient to retrieve the *sorted* list of the k heaviest queries in the subtree. Note that the lists of the k heaviest queries in a subtree do not change on edges, and we need to examine only those prefixes of q that correspond to nodes of the trie.

The high-level idea of our solution of the Top-k problem is as follows. We maintain an array $MinPref$ of length $|Q_{test}|$ with all values initialized by the maximal length of a query in Q_{test}. For each node of the trie T corresponding to a prefix of a query from Q_{test} we retrieve the k heaviest queries in its subtree. If a retrieved query is in Q_{test}, we compare $MinPref[q]$ and the string depth of the current node. If $MinPref[q]$ is bigger than the string depth, we update it, otherwise we do nothing. For the Sorted-Top-k problem we do practically the same, but retrieve sorted lists of the k queries and then, for each retrieved list, iterate over queries in it and write the positions of the queries that belong to Q_{test} into an array Pos.

We present two different techniques for retrieving the lists. The first one uses binary heaps, while the second one uses sorted lists merge. Time complexities of the proposed algorithms differ by logarithmic factors, and the space complexities differ by a multiplication constant. We compare their behaviour experimentally in Section 5.

4.1 Binary Heaps Algorithms

We start by traversing the trie T pre-order. When we see an end of a path starting at the root of T and labelled by a query from Q, we write out the weight of this query. The resulting array is denoted by W. Note that weights written out while traversing a fixed subtree of T form a continuous fragment of W. For each subtree we compute the endpoints $\ell(v), r(v)$ of such fragment and store them at the root v of the subtree. These two steps take $\mathcal{O}(m)$ time. We also build a range maximum query data structure [6] on top of W using $\mathcal{O}(m)$ time and space. The data structure allows to locate the maximal value in any fragment of W in $\mathcal{O}(1)$ time.

After the preprocessing, we traverse the trie with each query $q \in Q_{test}$ and mark nodes in the path labelled by q. For each marked node v we are to retrieve the list of the k heaviest queries in $W[\ell(v), r(v)]$. As noticed in [2], it can be done in $\mathcal{O}(k)$ time. We lazily build the Cartesian tree [9] on $W[\ell(v)], W[\ell(v) + 1], \ldots, W[r(v)]$ using the range maximum data structure. Since the Cartesian tree is a binary heap, we can use Frederikson's algorithm [4] to select the k maximum values in the Cartesian tree in $\mathcal{O}(k)$ time. Hence, the k heaviest queries for v can be retrieved in $\mathcal{O}(k)$ time (in unsorted order).

Lemma 1. *The Top-k problem can be solved in $\mathcal{O}(n + |Q_{test}| \cdot k)$ time and $\mathcal{O}(m)$ space.*

We now perform a more careful analysis of the algorithm. Note that if a node's subtree contains $k' < k$ queries, then only k' queries will be retrieved and it will take $\mathcal{O}(k')$ time. Hence, the running time of the algorithm can be bounded by the time required for construction of the trie and the range maximum data structure plus the sum of the number of queries below each marked node. Consequently, to estimate the running time of the algorithm it is sufficient to upper-bound the sum of the number of queries below each marked node. We give even a stronger bound:

Lemma 2. *The sum of the number of nodes labelled by queries from Q in all subtrees of T is at most n.*

Proof. Consider all paths starting at the root of T and labelled by queries from Q. Let the total length of the paths be equal to ℓ. The union of the paths contains all nodes of the tree, moreover, nodes that have i queries in their subtrees are counted i times. Hence, ℓ equals the sum of the number of queries below each node of T. At the same time, ℓ does not exceed the total length of all queries in Q, n. The claim follows.   □

Theorem 1. *The Top-k problem can be solved in $\mathcal{O}(n)$ time and $\mathcal{O}(m)$ space.*

We note that if we want to solve the Sorted-Top-k problem instead of the Top-k problem, we can simply sort the retrieved weights at each node. This will result in additional logarithmic factor in the time complexity. Consequently,

Theorem 2. *The Sorted-Top-k problem can be solved in $\mathcal{O}(n + n \log k) = \mathcal{O}(n \log k)$ time and $\mathcal{O}(m)$ space.*

4.2   Sorted Lists Algorithms

We will now show how to retrieve queries in the subtrees using sorted lists. As before, we start by traversing the trie with each query $q \in Q_{test}$ and marking nodes in the path associated with q. The algorithms for the Top-k problem and for the Sorted-Top-k problem retrieve sorted lists for every marked node, and only the information they extract from the lists is different — for the Top-k problem it is only whether a query is in the list of suggestions, and for the Sorted-Top-k problem it is a position of the query in the list.

To retrieve the sorted lists we traverse the tree bottom-up, starting from the leaves of T. For each node we maintain a list of queries sorted according to their weights. The length of the list is the maximum of k and the number of queries below the node. Suppose that the lists for a node's children are already known. If the node is labelled by a query from Q, we create a list containing a single element q, otherwise we create an empty list. Then we merge the lists associated with the children and the node into a single list in the following way.

We maintain a binary search tree on the values stored in the heads of the children's lists. At each step we take the maximum value in the tree and move it from its list to the resulting list. Then we update the tree and proceed. When the size of the resulting list becomes equal to k or the merged lists become empty, we delete the children's lists and the node's list and stop. If σ is the size of the alphabet and k' is the number of queries ending below a node v, the algorithm needs $\mathcal{O}(k' \log \sigma)$ time to compile the list for v. From Lemma 2 it follows that

Theorem 3. *The Top-k and the Sorted-Top-k problems can be solved in $\mathcal{O}(n \log \sigma)$ time and $\mathcal{O}(m)$ space.*

5   Experiments

The metrics computation algorithms we proposed differ only in the way the problems Top-k and Sorted-Top-k are solved. The time and the space needed to compute the values of the examination probability functions, the probabilities $P(S_{i,j} = 1)$, and the metrics are fixed for any user model. Hence, we restrict our attention to the Top-k and Sorted-Top-k problems.

We start with a short description of the dataset we used. Sets Q and Q_{test} were randomly sampled from the query log of a commercial search engine so that $|Q| \approx 30|Q_{test}|$. The log contained 47 M unique queries in two languages submitted by users from a European country in March 2014. Misspelled queries were filtered out in order to reduce noise in the evaluation. The average length of a query in the log was 27.57 letters. The value of k was taken equal to 10. The experiments were performed on a server equipped with 256 GB of RAM and one Intel Xeon E5-2660 processor.

We performed four series of experiments — two for the Top-k problem and two for the Sorted-Top-k problem. In the first series we compared the total time (Fig. 1a) and the total space (Fig. 1b) required by the algorithms we proposed for the Top-k problem. The binary heaps algorithm builds and stores a range maximum query data structure and hence it can be suggested that it will be outperformed by the sorted lists algorithm, which is confirmed by the experiments (see Fig. 1).

In the second series (see Fig. 2) we did not count the time required for building the data structures, i.e. the trie and the range maximum data structure as they can be constructed once and then used for evaluating the metrics for different test sets. The space consumption does not change, but the situation is different for the time — if the construction time is not included, the binary heaps algorithm outperforms the sorted lists algorithm. This can be explained in the following way: after the data structures are constructed, the binary heaps algorithm inspects only the nodes labelled by prefixes of queries in Q_{test}, and the sorted lists algorithm has to consider all nodes of T. Also, the sorted lists algorithm does unnecessary job of sorting the retrieved queries.

For the Sorted-Top-k problem the situation is similar: the total time consumption is higher for the binary heaps algorithm, but if we do not count the

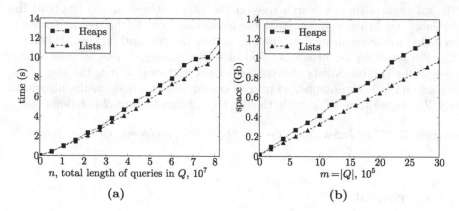

(a)   (b)

Fig. 1. Total time and space consumptions of the solutions to the Top-k problem. The space is measured in terms of $m = |Q|$ similar to the theoretical bound we gave before.

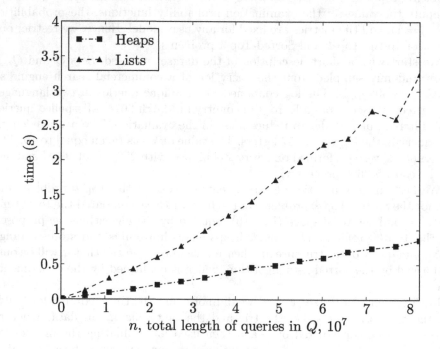

Fig. 2. Time consumptions of the solutions to the Top-k problem, construction time is not included

construction time, the binary heaps algorithm outperforms the sorted lists algorithm (Fig. 3). The space consumption is the same as for the Top-k problem, so we do not give the figure.

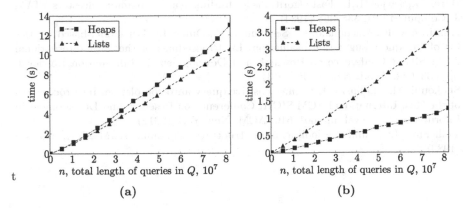

Fig. 3. Time consumption of the solutions to the Sorted-Top-k problem: (a) Total time consumption, (b) Time consumption without construction time

6   Conclusions

In this paper we showed that the problem of computing the metrics of the *Saved* family [7] estimating query suggestion mechanism quality can be reduced to one of the two string problems: Top-k or Sorted-Top-k. For each of these two problems we presented two solutions, one based on binary heaps and one — on sorted lists merge. For each of the solutions we showed theoretical upper bounds on their complexity. We also compared the solutions experimentally. The results of the experimental evaluation showed that the binary heaps algorithm can be used in the case when the database of queries is fixed and we use it to evaluate the quality of the query suggestion mechanism for different periods of time. If this is not the case, the sorted lists algorithm should be the choice.

References

1. Bar-Yossef, Z., Kraus, N.: Context-sensitive query auto-completion. In: Proceedings of the 20th International Conference on World Wide Web, pp. 107–116. ACM, New York (2011)
2. Brodal, G.S., Fagerberg, R., Greve, M., López-Ortiz, A.: Online sorted range reporting. In: Proceedings of the 20th International Symposium on Algorithms and Computation, pp. 173–182 (2009)
3. Duan, H., Hsu, B.-J.P.: Online spelling correction for query completion. In: Proceedings of the 20th International Conference on World Wide Web, pp. 117–126. ACM, New York (2011)
4. Frederickson, G.N.: An optimal algorithm for selection in a min-heap. Inf. Comput. 104(2), 197–214 (1993)
5. Gusfield, D.: Algorithms on Strings, Trees and Sequences: Computer Science and Computational Biology. Cambridge University Press (1997)

6. Harel, D., Tarjan, R.E.: Fast algorithms for finding nearest common ancestors. SIAM J. Comput. 13(2), 338–355 (1984)
7. Kharitonov, E., Macdonald, C., Serdyukov, P., Ounis, I.: User model-based metrics for offline query suggestion evaluation. In: Proceedings of the 36th International ACM SIGIR Conference on Research and Development in Information Retrieval, pp. 633–642. ACM, New York (2013)
8. Shokouhi, M., Radinsky, K.: Time-sensitive query auto-completion. In: Proceedings of the 35th International ACM SIGIR Conference on Research and Development in Information Retrieval, pp. 601–610. ACM, New York (2012)
9. Vuillemin, J.: A unifying look at data structures. Commun. ACM 23(4), 229–239 (1980)

Author Index